World Soils Book Series

Series Editor

Alfred E. Hartemink
Department of Soil Science, FD Hole Soils Laboratory
University of Wisconsin–Madison
Madison, WI
USA

The World Soils Book Series publishes peer-reviewed books on the soils of a particular country. They include sections on soil research history, climate, geology, geomorphology, major soil types, soil maps, soil properties, soil classification, soil fertility, land use and vegetation, soil management, soils and humans, soils and industry, future soil issues. The books summarize what is known about the soils in a particular country in a concise and highly reader-friendly way. The series contains both single and multi-authored books as well as edited volumes. There is additional scope for regional studies within the series, particularly when covering large land masses (for example, The Soils of Texas, The Soils of California), however, these will be assessed on an individual basis.

More information about this series at http://www.springer.com/series/8915

Roshan Babu Ojha • Dinesh Panday
Editors

The Soils of Nepal

Springer

Editors
Roshan Babu Ojha
National Soil Science Research Center
Nepal Agricultural Research Council
Lalitpur, Nepal

Dinesh Panday
Department of Biosystems Engineering
and Soil Science
The University of Tennessee-Knoxville
Knoxville, TN, USA

ISSN 2211-1255 ISSN 2211-1263 (electronic)
World Soils Book Series
ISBN 978-3-030-80998-0 ISBN 978-3-030-80999-7 (eBook)
https://doi.org/10.1007/978-3-030-80999-7

This Springer imprint is published by the registered company Springer Nature Switzerland AG
The registered company address is: Gewerbestrasse 11, 6330 Cham, Switzerland

Foreword

Government of Nepal
NEPAL AGRICULTURAL RESEARCH COUNCIL
(Established by the Government of Nepal under Nepal Agricultural Research Council Act, 2048 B.S.)
(.. Division)

Ref. No.:-
Dispatch No.:-

NARC Building
Singhdurbar Plaza
Kathmandu
NEPAL

Achieving food and nutrition security is always a challenge. Increased pressure to grow quality food under increasing human population, market preferences, and climate change are some of the triggering factors that demand to make soil sustainably productive. Understanding soil and making it more productive has been a prime concern in the agriculture sector. Nepal Agricultural Research Council (NARC) is conducting dynamic research on soil science since its establishment and has developed and generated several technologies for soil management. Attempts to understand the complexities and interactions among different factors in soil systems have resulted in broad research and publication record. I believe findings of soil research from the beginning of soil science research in Nepal have been included in this book with wide coverage in the soil science sector.

Climatic variation in the country has resulted in diverse soil types and systems. This demands the wide knowledge and use of multiple strategies for better management of soil. Major problems noted across the country are soil erosion, soil acidity, and nutrient mining which accelerates land degradation. Knowledge, both basic and applied, in soil science is therefore needed to make the diverse soil systems more dynamic and productive. Nepalese soil scientists must be applauded for their relentless efforts in developing comprehensive information regarding the soil status of Nepal through this book entitled "The Soil of Nepal". I am confident that this book helps national and international audiences about soil types and class of Nepal. I also hope that this resource has met the learning goals of researchers, students, and development professionals to be well acquainted with the foundations of soil genesis, management, utilization, and conservation.

I am very pleased with the efforts put by the Nepalese soil scientists in bringing this excellent publication as an outcome of their long years of experience and knowledge. I thank the editors, authors, and co-authors for their hard work and the strong team spirit they demonstrated in developing and bringing out this publication. I believe that this book will serve as a valuable reference for researchers, development professionals, students, academicians, farmers, and relevant stakeholders not only in Nepal but globally as well.

Deepak Bhandari, PhD
Executive Director **EXECUTIVE DIRECTOR**
Nepal Agricultural Research Council
Singhadurbar Plaza, Kathmandu

Mailing Address: P.O. Box 5459, Kathmandu, NEPAL
Phone: (+977-1) 4262663, 4262585, 4262567, 4262504, 4257805 Fax: (+977-1) 4262500
E-mail: ednarc@ntc.net.np, Website: http://www.narc.gov.np

Preface

Soil formation in Nepal is dynamic due to continuing active weathering that makes Nepalese soils young and fragile. Nepalese farming is largely based on the soil medium and must ensure food security for around 29 million people. Globally, soil solely produces 95% of the food required for human beings, making soil vital for human food and nutrition security along with delivery of numerous eco-system services. The importance of soil is even more pronounced for food production in Nepal due to its landlocked geography. To ensure food and nutritional security, maintaining and enhancing soil fertility is of prime importance. Soil fertility is governed by a multitude of factors, including parent materials, soil types, input management, soil microbial activity, and many others. In this book, we elucidate different aspects of soils to inform national and international readers about the key features of the soils of Nepal. We also aim to make this effort a national soil reference book that will help others understand the young and dynamic soils of Nepal.

There have been numerous soil studies conducted in Nepal but most are produced as gray literature or inaccessible journal articles, and very few open-access articles are available. Because researchers and academics have always struggled to find consolidated information on the soils of Nepal, our contributing authors have together collected more than 1500 references to assemble the key information about Nepalese soils in one place. We have decided to publish this book as part of the "World Soil Book Series" to make this information available to global audiences who are eager to learn about the soils of Nepal. We believe this book will fulfill readers' quests to some extent.

The authors in this book are subject matter specialists for their respective chapters who understand Nepalese soils from both local and national perspectives. Authors from geology, environmental, and soil backgrounds together combined their efforts to make this book informative. The affiliation of authors to different institutions, such as the Nepal Agricultural Research Council, Kathmandu University, Institute of Forestry, Institute of Engineering, the International Fertilizer Development Center, the University of Tennessee-Knoxville, New Mexico State University, University of New England, and Chinese Academy of Sciences, among others, showed a wide range of collaboration. This collaborative work helps to shape the book chapters concisely and comprehensively for the global academic community.

Lalitpur, Nepal — Roshan Babu Ojha
Knoxville, USA — Dinesh Panday

Acknowledgments

This book, *The Soils of Nepal*, would not be possible in the current shape without the support of our valuable authors who agreed to write each of its chapters. Our sincere thanks go to all lead authors and co-authors of this book. We are grateful to Dr. Renuka Shrestha, Chief of Agronomy Research Center, NARC, for her valuable suggestions; Prof. Roshan Man Bajracharya for reviewing chapter "soil issues and future perspective"; and are equally grateful to Dr. Deepak Bhandari, Executive Director of NARC, for providing a graceful foreword. We would also like to thank Mr. Ian Rogers for his English editing service, Er. Bikesh Twanabasu for mapping, Mr. Suresh Rettagunta facilitating during book publication process, and series editor Prof. Alfred E. Hartemink for providing critical review and feedback on the book. Additionally, this book would not come to this shape without the reference work and data collected by several researchers and institutions. We would like to pass our immense gratitude to all those who involved including National Land Use Planning project, Nepal. Finally, we received immense support and best wishes from our senior soil scientists, fellows, and family members, and wish to send them all a great round of applause.

Roshan Babu Ojha
Dinesh Panday

Contents

1 **Introduction** 1
Roshan Babu Ojha and Dinesh Panday

2 **History of Soil Research** 5
Krishna Bahadur Karki and Shree Prasad Vista

3 **Climate** 15
Basanta Paudel, Dinesh Panday, and Kundan Dhakal

4 **Geology and Physiography** 29
Basanta Raj Adhikari and Roshan Babu Ojha

5 **Land Use and Land Cover** 41
Basanta Paudel and Basanta Raj Adhikari

6 **Soil Formation** 53
Roshan Man Bajracharya

7 **Soil Types, Soil Classification, and Mapping** 63
Subhasha Nanda Vaidya, Dil Prasad Sherchan, Krishna Raj Tiwari, Subash Subedi, Krishna Bahadur Karki, Dinesh Panday, and Roshan Babu Ojha

8 **Soil Properties** 91
Shree Prasad Vista, Krishna Bahadur Karki, Yam Kanta Gaihre, Sonisa Sharma, and Bandhu Raj Baral

9 **Soil Fertility and Nutrient Management** 111
Krishna Bahadur Karki, Dil Prasad Sherchan, Dinesh Panday, and Rajan Ghimire

10 **Land Degradation** 127
Krishna Bahadur Karki and Roshan Babu Ojha

11 **Soils and Humans** 143
Dinesh Panday, Shree Prasad Vista, Dikshit Poudel, Arjun Chhetri, and Bharat Sharma Acharya

12 **Soil Issues and Future Perspective** 151
Roshan Babu Ojha, Krishna Bahadur Karki, and Dinesh Panday

Index 159

Editors and Contributors

About the Editors

Roshan Babu Ojha (M.Sc. Ag., 2014; B.Sc. Ag., 2012 Tribhuvan University), a Ph.D. candidate at the University of New England, Australia, is a soil scientist to the National Soil Science Research Centre, Nepal Agricultural Research Council (NARC) since 2015. Before joining NARC, he worked as an Assistant Professor at Purwanchal University and taught several courses of soil science. Soil fertility, soil carbon, and pedology are his field of expertise. As a national focal point from Nepal, he represented the Global Soil Partnership (GSP) program of Food and Agriculture Organization of the United Nations (UNFAO), Rome, Italy from 2016 to 2018. He is an awardee of "Young Scientist" award (2017) presented by Society of Agriculture Innovation and Development, India and "Outstanding Contribution Scientist in Soil Conservation" award (2018) jointly presented by the GSP, UNFAO, Italy and the organizing committee of soil health and sustainable development international symposium, China. Realizing the immense potentiality of soils in our life, he is keen to learn and share soil science language to the scientific and local community. Besides, he is fond of having expeditions and excursions, tasting new foods, triumphing with nature, and understanding people and culture.

Dinesh Panday Dinesh Panday is a Post-doctoral Research Associate at the University of Tennessee-Knoxville, United States since 2020. He completed his undergraduate studies at Tribhuvan University (B.Sc. Ag., 2012) in Nepal and went on to graduate studies in Environmental Sciences at Lincoln University of Missouri (M.S., 2015) and Soil and Water Sciences at University of Nebraska-Lincoln (Ph.D., 2020) in the United States. Soil fertility, soil nitrogen, biogeochemistry, greenhouse gas emissions, and digital soil mapping are his field of expertise. He received Alltech Young Scientist Awards in 2010, 2011, and 2012. He also received 2018 Maize-Asia Youth Innovators Awards, organized by CGIAR Research Program on Maize and Young Professionals for Agricultural Development (YPARD). He serves as Coordinator at Maize Youth Task Force and Communication office at YPARD Asia and Pacific Regions. He also serves as an Editor in Journal of Nepal Agricultural Research Council, Nepal and reviewer in many journals, including Nature Scientific Reports, Journal of Environmental Quality, Agronomy Journal, PLoS ONE, and Remote Sensing. As a researcher, he is interested in deeper understanding of soil biogeochemical controls of greenhouse gas emissions to mitigate carbon footprint of agricultural systems. He is fond of music, travel, and nature.

Contributors

Bharat Sharma Acharya Department of Mines, Oklahoma City, OK, USA

Basanta Raj Adhikari Institute for Disaster Management and Reconstruction, Sichuan University-The Hong-Kong University, Sichuan, China;
Department of Civil Engineering, Pulchowk Campus, Institute of Engineering, Tribhuvan University, Kathmandu, Nepal

Roshan Man Bajracharya Department of Environmental Science and Engineering, Kathmandu University, Dhulikhel, Kavrepalanchowk district, Nepal

Bandhu Raj Baral Nepal Agricultural Research Council (NARC), Kathmandu, Nepal

Arjun Chhetri Department of Biosystems Engineering and Soil Science, The University of Tennessee-Knoxville, TN, Knoxville, USA

Kundan Dhakal Noble Research Institute, LLC, Sam Noble Parkway, Ardmore, OK, USA

Yam Kanta Gaihre International Fertilizer Development Center (IFDC), Kathmandu, Nepal

Rajan Ghimire Agricultural Science Center, New Mexico State University, Las Cruces, NM, USA

Krishna Bahadur Karki Nepal Agricultural Research Council (NARC), Kathmandu, Nepal

Roshan Babu Ojha Nepal Agricultural Research Council, Kathmandu, Nepal

Dinesh Panday Department of Biosystems Engineering and Soil Science, The University of Tennessee-Knoxville, Knoxville, TN, USA

Basanta Paudel Key Laboratory of Land Surface Pattern and Simulation, Institute of Geographic Sciences and Natural Resources Research, Chinese Academy of Sciences, Beijing, China

Dikshit Poudel Department of Agricultural and Applied Economics, University of Georgia, Athens, GA, USA

Dil Prasad Sherchan Nepal Agricultural Research Council, Kathmandu, Nepal

Sonisa Sharma Department of Plant and Soil Sciences, Oklahoma State University, Oklahoma, USA

Subash Subedi Genesis Consultancy Pvt. Ltd./Geoinformation and Earth Observation Services, Lalitpur, Nepal

Krishna Raj Tiwari Institute of Forestry, Tribhuvan University, Pokhara, Nepal

Subhasha Nanda Vaidya Nepal Agricultural Research Council, Kathmandu, Nepal

Shree Prasad Vista Nepal Agricultural Research Council (NARC), Kathmandu, Nepal

List of Figures

Fig. 2.1 An example of early soil survey work done by the survey team. Previous days soil maps were hand-crafted but in the later decades they are digitized in Geographical Information System platform 10
Fig. 2.2 Soil science program development in Nepal. 12
Fig. 3.1 Topography and spatial distribution of meteorological and hydrological stations across Nepal. Meteorological and hydrological station data were obtained from the Department of Hydrology and Meteorology (DHM), Government of Nepal 16
Fig. 3.2 Major climate zones of Nepal. 18
Fig. 3.3 Observed climatic trends (temperature and precipitation) in Nepal between 2000 and 2015. The trend calculations were based on observed daily weather records between 2000 and 2015 (115 stations for temperature and 272 stations for precipitation). 20
Fig. 3.4 Annual total precipitation distribution of Nepal between 1980 and 2000. The dotted symbols show the distribution of meteorological stations within the country 21
Fig. 3.5 Delineated agro-ecozones of Nepal, adopted from Carson (1992) and Gauchan and Yokoyama (1999). 23
Fig. 4.1 Geological longitudinal subdivision of the Himalayas from west to east (Redrawn after Gansser 1964) 30
Fig. 4.2 Geological subdivision of the Nepal Himalayas modified from (Amatya and Jnawali 1994). STDS: South Tibetan Detachment System. MCT: Main Central Thrust. MBT: Main Boundary Thrust. 32
Fig. 4.3 Different rock types of the different geological zones of the Nepal Himalayas. **a** Sand and gravel in the Indo-Gangetic Plain at the Karnali River, Kailali; **b** Contact between gravel and sandstone beds in the upper Siwalik, Puntura River, Dadeldhura; **c** Highly fractured quartzite beds of the Lesser Himalayas, Kaligandaki River, Myagdi; **d** Garnetiferous schist of the Higher Himalayas, Kaligangaki Valley, Myagdi; **e** Fold and fault of Tibetan sedimentary rocks, Kaligandaki River, Mustang. 33
Fig. 4.4 Physiographic map of Nepal. 35
Fig. 4.5 Physiographic landforms of different regions. **a** Flat alluvial land of Tarai, Ratu River, Mahottarai; **b** Fragile landscape of the Siwaliks, Kailali; **c** Panoramic view of Hills in the Nepal Himalayas (looking south from the Dolkha); **d** Steep mountains in the Middle Mountain region, Kalikot; **e** Snowcapped mountains in the High Himalayas region, Pokhara 36
Fig. 5.1 Land use and land cover pattern in Nepal 42
Fig. 5.2 Terrace farmland in Ilam district, Nepal. 43
Fig. 5.3 Forest cover in the Mahabharat range of Dadeldhura district, Nepal 44

Fig. 5.4 Built-up area of Phidim, the district headquarters of Pachthar district, Nepal 44
Fig. 5.5 Snow/glacial cover on **a** Mt. Dhaulagiri in Mustang district and the **b** Dudhkoshi River in the border area of Okhaldhunga and Khotang district, Nepal 45
Fig. 5.6 Grassland on Chirdung Mountain in Dolakha district, Nepal 46
Fig. 5.7 **a** Barren land in Mustang district and **b** sand and boulders in the Karnali River in the Bardiya district of Nepal 46
Fig. 5.8 Land use and land cover change between 1978 and 2010 47
Fig. 7.1 Map of Nepal showing different physiographic zones 65
Fig. 7.2 Soil map of Nepal based on the United States Department of Agriculture classification system 69
Fig. 8.1 Soil moisture index (SMI) variation in different seasons and physiographic regions of Nepal 96
Fig. 8.2 Soil pH status of different regions of Nepal 97
Fig. 8.3 Organic matter content in soils across different agro-ecological zones of Nepal 101
Fig. 8.4 Total nitrogen content (%) of soils across different agro-ecological zones of Nepal 102
Fig. 8.5 Available phosphorus (kg P_2O_5 ha^{-1}) of soils across different agro-ecological zones of Nepal 103
Fig. 8.6 Available potassium (kg K_2O ha^{-1}) of soils across different agro-ecological zones of Nepal 104
Fig. 10.1 **a** Land degradation due to tree cutting, **b** Goat grazing in the public forest as the forest is pushed away, and **c** Sloping land cultivation without any scientific knowledge 131
Fig. 10.2 Farmers grow maize crops slopes such as this one, where pre-monsoon torrential rain washes away all the topsoil and manure applied as fertilizer 132
Fig. 12.1 Framing of soil threats or soil issues within the Nepalese perspective 153

List of Tables

Table 3.1 Agro-ecozones and farming promises in Nepal 23
Table 4.1 Geological subdivision and their key features of Nepal Himalayas 31
Table 4.2 Physiographic zones of Nepal and their key features 37
Table 5.1 The conversion statistics of major land use between 1978 and 2010 . . . 47
Table 6.1 A short list of common rocks found in Nepal and the dominant minerals comprising them 55
Table 6.2 Elevation ranges of the cross-section from South to North of the Himalayan Region 55
Table 6.3 Dominant rocks and landform processes in different physiographic zones of Nepal 56
Table 6.4 Climate types and area coverage across physiographic regions of Nepal 57
Table 6.5 Major soil orders and dominant soil types in different physiographic regions of Nepal 60
Table 7.1 Relationship between elevation, climate zone, and temperature 67
Table 7.2 Relationship between soil temperature regime and elevation 67
Table 7.3 Local soil classification based on color that indicates soil productivity 67
Table 7.4 Local terminology used to classify soils by farmers based on clay content 67
Table 8.1 Soil quality rating in Nepal based on soil pH value 98
Table 8.2 Recommended lime requirements by the Government of Nepal based on soil type and physiographic region 100
Table 8.3 Status of micronutrients in soils across different NARC research stations in Terai, Nepal 104
Table 9.1 Soil texture from surface soil of the 18 different districts of Nepal 113
Table 9.2 Particle size distribution in Rhodic Ustochrept soil profile opened at Kallaritar of Dhading District 114
Table 9.3 Nutrient content in some indigenous green manure plants 115
Table 9.4 Plant nutrient contents of organic manures as extracted by Aqua Regia 115
Table 9.5 Plant nutrient contents in dung, biogas slurry, and biogas compost content 116
Table 9.6 Rice crop response to some green manuring plants 117
Table 9.7 Nutrient uptake by rice and wheat crops compared to organic manure in the Eastern Hills of Nepal 118
Table 9.8 World demand for fertilizer nutrients in 2014–2018 120
Table 9.9 Fertilizer distribution for 3 years in Nepal 121
Table 10.1 Nutrient losses under different land-use categories in Nepal 134

Table 10.2 Soil nutrient status in Nepal. The percentage coverage indicates the percentage of the total cropping area . . . 135
Table 10.3 Percentage of livestock population and grazing land availability in different ecological regions . . . 138
Table 12.1 Major soil threats of Nepal . . . 152

Introduction

1

Roshan Babu Ojha and Dinesh Panday

Abstract

Agriculture, the cultivation of food and goods through farming, began in Nepal thousands of years (yr) ago with bench terraces in the hillslopes, one of the most unique cropping landscapes in the world. Manipulation of soil in these bench terraces exhibits the soil reference group 'Anthrosols' as per world soil reference base, or 'Anthropic epipedon'as per the United States Department of Agriculture soil classifications. However, active weathering is still prevalent in the soils of Nepal and generally referred to as 'young soil'. That is why the soils of Nepal lack a large area of old-aged soils such as 'Oxisols,' 'Histosols,' and 'Vertisols.' Climate and geology affect the formation of soil and different types of soils can be found in different climate and geological settings (Chaps. 3 and 4). Newly developed and fragile soil including 'Entisols'' and 'Inceptisols'' with 'colluvial' or 'alluvial' materials covered more than 50% of the total soil area in Nepal (Chaps. 6 and 7). These soils are abundantly distributed in Tarai (also spelled as Terai), Siwaliks, and the middle-Mountains (Chaps. 4 and 5) regions of Nepal. Due to their fragile nature, they are easily eroded by water, and require a proper land management strategy (Chap. 10). For proper land management, it is necessary to understand the intrinsic soil properties and inherent fertility of the soil (Chaps. 8 and 9). Additionally, we valued soil as it is worthy to human beings (Chapter 11). In this book, we delve into the wide array of young Nepalese soils that are of particular interest to the global academic fraternity. For these reasons, we have organized this book into 12 chapters.

R. B. Ojha (✉)
Nepal Agricultural Research Council, Kathmandu, Nepal

D. Panday
Department of Biosystems Engineering and Soil Science,
The University of Tennessee-Knoxville, Knoxville, TN, USA

Keywords

Book overview • Chapter summary • Nepal • Young soil

1.1 Introduction

In the second chapter the authors provide a brief history of soil science development in Nepal. As stated earlier, cultivation in Nepal began ages ago. The authors present evidence that cultivation began in the region starting in 1,000 BC with reference to the '*Vedic*' (the infamous Eastern mythological document) land classification system. Our ancestors were aware of soil conservation, and thus built bench terraces to control soil erosion. However, our government realised the importance of the soil institution very late. Only after 1957 was a formal soil unit established under the Department of Agriculture. In this chapter, the authors focus on how the soil science institute was established and how soil science research was developed in Nepal. The authors then present the progress of soil science development under different political regimes. They mentioned soil research institute (the soil science division, now known as the National Soil Science Research Centre) was established under the Nepal Agricultural Research Council in 1991 and is the only formal government institution with a soil research mandate. The authors then summarize the research work of different research and academic institution of Nepal. In the similar manner, the authors create an overview of soil science education in Nepal.

In the third chapter, the authors present an overview of the climatic system in Nepal. Five major climatic zones ('Tropical,' 'Subtropical,' 'Temperate,' 'Sub-alpine,' and 'Alpine') exist in Nepal, with a drift in climate within a short gradient toward the north. The authors illustrate this drift in maps according to physiographic regions. This chapter informs readers about the climatic zones, their spatial extent, and major cultivated crops in the respective climatic zones.

R. B. Ojha and D. Panday (eds.), *The Soils of Nepal*, World Soils Book Series,
https://doi.org/10.1007/978-3-030-80999-7_1

The authors explained the climatic seasons, climate, and precipitation trends of Nepal. Information on natural hazards such as floods, droughts, frosts, and cold wave trends of Nepal are supported by the literature. It is interesting to understand the soil development in different agro-ecozones and how climate change impacted the soil-forming process in Nepal, and this chapter informs readers about the climatic zones of Nepal and their relationship with these soil-forming factors.

In the fourth chapter, the authors present the geology of Nepal. Authors provide evidence of the origin of the Nepal Himalayan region due to the collision of the Indian and Eurasian plates, which occurred around 30 million (M) yr ago (late Eocene to Oligocene period) during the period new/modern life began. Since that time, the geology of Nepal has been continuously moving and soil formation still accelerating due to this dynamic geology. This dynamism is well depicted in maps and is described in this chapter. The authors describe the five geological sub-divisions of Nepal ('Tibetan-Tethys zone,' Higher Himalaya,' 'Lesser Himalaya,' 'Siwaliks,' and 'Indo-Gangetic Plain') along with the five physiographic zones ('High Mountain,' 'Middle mountain,' 'Hill,' Siwaliks,' and 'Tarai'). The authors provide field images of the geological zones and physiographic zones that manifest the clarity of this chapter. Furthermore, readers will learn how the geologic divisions of Nepal differ in terms of geologic origin and their temporal and spatial extent. Similarly, the authors explain the differences in altitude, parent materials, and dominant climate types between the physiographic zones.

In the fifth chapter, the authors illustrate the land use and land cover change in Nepal. Land is a vital and scarce resource, and the interaction between human activities and land use affects the ecosystem, landscape dynamics, and overall economy of a nation. These interactions are both positive and negative, and the authors present evidence covering the past 100 yr of land use and land cover change in Nepal. The dynamics of change between agricultural land and forested land is a worldwide issue, and Nepal is no exception. This chapter depicts how deforestation is linked with the expansion of agricultural areas as the authors review land use and land cover patterns of Nepal along with their change over time. Authors categorize land use and land cover into agricultural land, barren land, forested land, built-up area, shrub/grassland, snow/glacier/water bodies, and other types of land. The chapter further discusses the land tenure system and land use policies of Nepal as the authors identify the constraints and strengths of sustainable land use management and provide perspectives for better management of the land. In summary, this chapter informs the land use and land cover status of Nepal over the past 100 yr and provides a way forward for sustainable land use management.

In the sixth chapter, the author explains the formation of the soil and its associated factors in Nepal. Soil formation in Nepal is active and the recent products of weathering can be found there. Hill slopes are continuously eroding, which results in the loss of soil sediment from the source and deposition of the sediment across the river valleys and Tarai plain. Different types of parent materials are found in Nepal in the different physiographic zones. The author lists all the dominant rocks and minerals found in Nepal. In this chapter, it will be interesting for readers to learn about the driving factors of soil formation and resultant soil types specific to the different physiographic regions of Nepal. The author concludes the chapter with the imperatives for sustainable land management in Nepal focusing on the outlook of soil formation. In summary, this chapter reports on key soil-forming factors, dominant parent materials and soil types and future priorities and outlook for sustained land management from the point of soil formation.

In the seventh chapter, the authors classify the soils of Nepal and present the types of soils found in the different physiographic zones. Each of the physiographic regions is subdivided into different land morphological units. Soils in each morphological unit are different and the dominant soils in each morphological unit are well described. In this chapter, soil types are explained in terms of both the United States Department of Agriculture soil classification system and the World Reference Base soil classification system. This is the chapter that carries the essence of this book, and as a result, the authors have gone to great lengths to make it more informative. While a large quantity of information is available on the soil classification of Nepal, the authors only highlight the dominant soils found in each land morphological unit. Furthermore, the authors add examples of the soil profile descriptions typical to each dominant soil type along with location information.

In the eighth chapter, the authors explain the soil properties of Nepal, including the interaction of physical, chemical, and biological properties of soil governed by the distribution of soil particles or size fractions and influenced by soil-forming factors. This interaction regulates most of the soil processes that deliver different ecosystem functions and are a source of food for terrestrial life. The soils of Nepal are attributed with a mostly low to medium clay content. This attribution governs most of the soil characteristic features. In this chapter, the authors gather information on the physical, chemical, and biological properties of the soils of Nepal. Soil acidity is one of the major issues facing Nepalese soils. The authors explain why acidity is a major factor in Nepalese soils and provide ways to manage this problem. They review the status of major and micronutrients throughout Nepal with some interesting research results. The information on soil microbial diversity, which itself is a scanty work in the Nepalese context, is of great importance to understanding the functioning of the soil. The authors collect as much information as possible from the real ground data about the microbial status of Nepal. The chapter provides information about the soil's physical, chemical, and biological properties, particularly these problem soils and their management in the Nepalese context.

In the ninth chapter, the authors highlight the soil fertility status of Nepal. Food production is an absolute function of soil fertility, which is directly linked with food security and the gross economy of the country. Soil fertility management practices influence the sustained food production system. The authors provide evidence of local soil management practices in Nepal and explain the research needs for improving soil fertility management practices for optimum crop production. Readers will be interested to learn how different factors affect soil fertility status and about the specific problems relating to fertility management in crop production specific to Nepalese soils. The authors find different ways to manage the soil fertility problems in this chapter. Fertilizer availability has always been an issue in Nepal, and it is relevant to study the nation's fertilizer demand and figure out the potential sources of fertilizers to fulfil this demand. The strength of this chapter relies on calculating national fertilizer needs and ascertain these new fertilizer sources. Additionally, the chapter provides some policy recommendations for improving soil fertility in Nepal. Throughout this chapter, readers will find information about soil fertility governing factors, constraints to soil fertility, their management, and policy recommendations for better sustainable soil fertility management practices.

In the tenth chapter, the authors describe the status of land degradation in Nepal. Nepal is on the verge of land degradation mostly due to soil erosion. Being geologically active with steep topography and fragile soil, the natural rate of soil erosion is high compared to accelerated erosion. The authors review the literature documenting the different causes of land degradation in Nepal and provide evidence of water erosion, wind erosion, physical degradation, and chemical land degradation. Forest degradation is also an important issue in Nepal, and while this may sound contradictory, the country's forested land is in a state of degradation due to its steep slope and history of unscientific forest management. These factors are explained in this chapter about the state and types of land degradation in Nepal.

In the eleventh chapter, the authors explain the interplay between soils and humans. Around 125 different caste-ethnic groups reside in Nepal, and every ethnic group has cultural values that are linked with soil. In this chapter, the authors collect soil facts related to our culture and how it carries on through the generations. Similarly, the authors elucidate the role of soil in food production, food security, human nutrition, and establish a close connection between soil and human health. Healthy soils are the wealth of a nation, as healthy soil produces food with enriched nutrients that are directly linked with human health. Furthermore, the authors review the land cover changes in Nepal. The authors also present some information on soil issues resulting from heavy metals, organic pollutants, and pathogens, and provided brief recommendations to guide and improve soil health and quality pollution in Nepal. The chapter informs readers about the cultural values of soils, the importance of soil in food security and human nutrition, and land cover change in Nepal.

In the twelfth chapter, the authors summarize and conclude the whole essence of the book. The authors also highlight researchable issues regarding the soils of Nepal, allowing readers to understand the future perspective of soils of Nepal. Soils in Nepal are relatively younger compared to the rest of the world, and the active stage of soil weathering provides an avenue of soil research for understanding this dynamism. The authors emphasize the potential of Nepalese soil as a research hub for global researcher. In total, this chapter covers the research potency of Nepalese soil with concluding remarks to the book.

2 History of Soil Research

Krishna Bahadur Karki and Shree Prasad Vista

Abstract

The knowledge of soil science is as old as civilization and was used even from Vedic times around 700 BC. Kautillyas' Arthashastram (400 BC) mentioned the improvement of soil fertility and growing rice and wheat crops in the fertile Indo-Gangetic valley. Information regarding the systematic study of western agriculture dating from the fourth century is available. Studies by Robert Boyle, Francis Bacon, Arthur Young, Justus von Liebig, and Birkland–Edie and Haber, among others, on the development of soil fertility and fertilizer are remarkable. It was Vasily Dokushaev's work in soil genesis in 1983 that the International Union of Soil Sciences recognized and led to their including soil science in International scientific society. The Nepalese history of soil science dates to 1957 when soil science was established as a unit under the Department of Agriculture to conduct soil sample analysis and soil fertility experiments. Later other units were added and gradually the number of people working in soil science increased but not sufficient. Scientists with soil science as an academic qualification are lacking in the world by 40%, Nepal has the same fate. While the national priority has been on higher food production, little thought has been given to the negative consequences of our actions, including land degradation and desertification. It is imperative that we should work hard and convince society and policy-maker that proper attention is given to restore soil fertility and soil health before it is too late, therefore, soil will continue to provide the nation with goods and sustainable services. In this chapter, we present a brief history of soil science established in Nepal and discuss how it develops with time.

Keywords

Agriculture • Nepalese farmers • Panchayat system • Rana regime • Soil Research • Soil Science education

2.1 Introduction

It is believed that the science and art of agriculture began with the evolution of human beings. Agriculture was believed to be the main component of economics since the third century. Farmers in those days classified land into four types based on their productivity, and in general, farmers could identify these different land classes: best, intermediate, moderate, and marginal. The system of green manuring in soil was also practiced, and locally available potential green manuring crops were often incorporated during tillage operation. Farmyard manure (FYM), sheep and goat manure, poultry manure, kalimati (black clay), etc. were the main sources of nutrients, and these methods have been practiced since ancient times. To keep the soil more fertile, crop rotation or the integration of legumes in the cropping system was practiced. Terracing was also practiced in the land that was sloped. Most often, Nepalese farmers worked in groups for preparing land, facilitating irrigation channels, transplanting rice, weeding, harvesting, threshing, winnowing, and even in safeguarding farms from wild animals.

When we consider the history of agriculture in Nepal, systematic agriculture dates to the Rana reign where several seeds, saplings, and animal breeds were imported from the United Kingdom during 1850 by Janga Bahadur Rana. Nepal had its first agriculture office or Agriculture Council established vis-a-vis an experimental area for agricultural research in 1922. There is some information about the irrigation projects developed even during the Ranas' autocratic regime. An irrigation channel (*Chandra Nahar*) at Saptari was established for agricultural purposes between 1926 and 1927. In 1937, an agriculture technical school at Chhauni

K. B. Karki (✉) · S. P. Vista
Nepal Agricultural Research Council, Kathmandu, Nepal

R. B. Ojha and D. Panday (eds.), *The Soils of Nepal*, World Soils Book Series,
https://doi.org/10.1007/978-3-030-80999-7_2

was established to impart agricultural knowledge, and in the same year, an agricultural research area was also established at Baishdhara, Balaju. In 1947, one agriculture farm at Parwanipur of Bara District, and another small farm at Kakani were established in 1951. Parwanipur is still an important Agriculture Research Station, though Kakani was handed over to other government agencies after the 1990s movement (in 2029 BS - Nepalese date). The substantial irrigation development during the early period of Nepal is well documented by Gautam (2012). Apart from irrigation development, no other achievements are mentioned in the history of agriculture development and research activities. The low population during that time suggests that the rice crops harvested on that land were sufficient for the population. Population statistics show that there were only 8,256,625 people in Nepal and production was sufficient (Maharjan and Khatri-Chhetri 2006).

We get food, fodder, and renewable energy from soil, including buffering, filtering, and the transformation of clean groundwater. However, to meet the demands of a growing population the natural way of meeting humankind's needs has been exploited. M K Gandhi said, "Earth provides enough to satisfy every man's need but not every man's greed." Presently humans have overused the amount of natural resources and paved the way for a global tragedy (Singh 2011). Thus, civilization's challenge is to reconcile the demands of human developmental intolerances with nature.

It has been established that settlement began in the fertile alluvial soils by riverbanks and gained development momentum through river valleys in Asiatic regions. Human settlements seem to have developed at almost the same time in both western and eastern societies. Indian literature makes clear that cultivation of the subcontinent began sometime between 1000 to 500 BC. In Hindu mythology the Ramayan illustrates that King Janak (ca 500 BC) plowed the drought-stricken field and unearthed Sita, indicating that cultivation was already present. Even in Veda (1000 BC to 800 BC), the land was classified based on its level of fertility, such as '*urvara*' for fertile land and '*unurvara*' for infertile land (Abrol and Nambiar 1997; Blume 2014). Soils that are suited for rice, sesame, and other cereals differed. Another piece of Vedic mythology, Kautillya's Arthahsastra (Chanakya c. 350 to 283 BC) notes the need to increase agricultural productivity and farmers' income. Similarly, rice, wheat, and barley were first grown in the Indus and Ganges river valleys. Evidence shows that soil was plowed several times before seeds were sown and bullocks were used to plow the field. While animal husbandry was practiced and cow dung was used as manure during the same period, there is no record of any one individual pioneering these activities.

2.2 Soil Fertility in the Western World

Agricultural development and soil fertility maintenance in the Western world are well documented, in Aristotle's Human Theory (350 BC) among other sources. Winiwarter (2006) covers the ancient Greek and Roman scientists who documented agricultural development at that time. Among them were Cato (200 BC), Varro, and Columella during the early period of the Roman and Greek Empire, who covered agricultural practices, soil husbandry, and the relationship between humans and the soil. Later, van Helmont (1574 to 1664) experimented with willow plants growing in soil for 5 yr while adding only water, concluding that water was the source of plant nutrients as he found that only 200 g of soil from the original 5 kg was lost. Robert Boyle (1661) conducted a similar experiment and published his results on the importance of humus, which came from the decay of plants and animals. In the 1800s Humboldt recommended the use of guano in agriculture. Later, Justus von Liebig made an important discovery on the use of synthetic fertilizer in agriculture (1830). Birkland–Edie (1903) processed nitrogen (N) synthesis, and Fritz Haber (1918) manufactured ammonia fertilizer. Despite the many scientific publications on soil and related fields, soil science was not considered a science until Dokushaev (1883) published a report on soil genesis.

The soil in most countries has been always related to agricultural production. It is accepted that increased food demand can be met in three ways: increasing the amount of land for cultivation; creating higher crop yields per unit of land area; and reducing postharvest losses. Although the substantial but marginal amount of land area can be increased, scientists and planners believe that a greater part of the increase must come from a higher yield. The United Nations Food and Agriculture Organization (UNFAO/ FAO) predicted that by the year 2050 world food production of cereal must increase by 940 Million tons (M t) to reach a total of 3 billion (B) t, meat production must increase by 196 M t to reach a total of 455 Mt, and oilseeds must increase by 133 M t to reach a total of 282 M t (Alexandratos and Bruinsma 2012). Since the ability to expand the land area in most countries is limited, efforts must be made to increase crop yield per unit area. The FAO has estimated that more suitable land could also be brought under cultivation, though the exact location of this land has yet to be identified (Young 1999). The FAO estimates that to meet the targeted production of food by the year 2050, prime agricultural land of 132 M hectares (ha) can be brought under cultivation in sub-Saharan Africa and Latin America, leaving 1.3 B ha as free land (Alexandratos and Bruinsma 2012).

2.3 Agriculture Research Development in Nepal in the Post-Rana-Regime (1951–1961)

Prior to the 1951 democratization of Nepal, the Ranas in 1935 initiated some development activities, though these efforts were not enough to reach the people. They created the Agriculture Board, the Bureau of Mines, the Forest Office, and the Industrial Promotion Office, but these entities never materialized. Similarly, in 1949 the National Planning Committee also was formed, and a 15-yr plan was formulated but never executed (Dhital 1970). Due to country-wide dissatisfaction and the people's movement, the Ranas' autocratic rule ended in 1951. When they left, they had built up practically no development infrastructure during their 104-year rule.

After Nepal's democratization, it was up to the country's democratic leaders to transfer the ancient society to a modern one, though the king was still the head of state. The country's administrative infrastructure was well established, whereas developmental activities needed to start from scratch. Technical manpower was needed in every sector, ranging from housing, road, education, irrigation, forest, and agriculture. Nepal and the United States signed the Four Point Agreement on Technical Cooperation just a few months before the collapse of the Rana Regime in 1950, under which many Nepalese scholars were sent for training and higher education to different countries (Skerry et al. 1991). At this time, the Department of Agriculture was established in 1952 with very little technical manpower. At the same time, Nepal joined the Technical Cooperation Scheme of the Colombo Plan in 1952, received scholarship funding, and sent many Nepalese to study in foreign countries (Dhital 1970; Pant 1956). Similarly, in 1956 another agreement was signed with the US International Cooperation Administration to train 1,700 people for technical manpower building through education and training. It was the beginning of technical manpower building in Nepal. With the initiation of USOM, a soil laboratory was created in Kathmandu (Min Bhavan) in 1953 (2010 BS), which was later shifted to Harihar Bhavan and in 1971 to Khumaltar, where the Soil Science Division has operated since then.

Following the visit of Mrs. Eleanor Roosevelt, the former first lady of America (31 December to April 1952) during the monsoon season of 1952, there was a massive flood and landslides. In 1954, there was another landslides and flood that killed more than 1,000 people and left 100,000 homeless. The people in western hills suffered the most, and were in desperate need of food grain (Annonymos 1959). In the first year of disaster relief was provided in the form of food aid by the U.S. Operations Mission (USOM), and the following year Indian Aid provided fertilizer that coincided with Malaria Eradication Program in the Tarai and river valleys. As mentioned by Skerry et al. (1991), an American agriculture scientist and five former American Extension Agents were engaged in different agriculture improvement programs. He invited an American Soil Scientist (Surveyor), who formed a soil survey team with two Indian soil surveyors and surveyed the valley floor to understand soil fertility and resettle the flood and landslide victims from the hills. During that time, they established a field-level model soil laboratory in 1953. Some 5,000 families resettled on 12,000 ha of land, although there was 100,000 ha of land available to reclaim in the Rapti Valley. The Chitwan Valley was one such area where these flood victims were resettled, and modern agriculture was practiced there using improved crop varieties and fertilizers.

The fertilizer donated by the USOM and the Indian Aid Mission was mostly ammonium sulfate (NH_4SO_4), single superphosphate (SSP) and muriate of potash (MoP), and was distributed through Tribhuvan Gram Vikash Samiti (TGVS), an integrated rural development program that ran from 1956 to1962 (FAO 2010; Gurung 2011). The Gram Sevak (extension agents from TGVS) demonstrated the use of the fertilizer and distributed it to the farmers in some of the selected districts. The TGVS program was later run by the Youth Program of the Department of Agriculture in 1962 (Dhital 1970). After the program of TGVS, it was handed over to the Youth Program, the District Agriculture Development Office (DADO) included farmers' field fertilizer demonstrations in their overall program.

When the Department of Agriculture was first created in 1951 there was limited technical manpower available. The USOM established soil laboratories in the field in 1953 and at Kathmandu in 1956, while technicians trained in chemistry were recruited for soil sample analysis. In addition to Parwanipur, the Rampur model farm that was established in 1953, agriculture research stations in Bhairahawa, Nepalgunj, Janakpur, and Tarahara were instituted in 1956 and 1957. After 1957, qualified agriculture graduates began returning from abroad, soil fertility experiments and pocket soil surveys of the area's agricultural potential were carried out. A cartography unit was also developed to prepare a soil map. Soil fertility experiments were extended to other farms as well. More graduates returned from abroad after receiving their bachelor's degrees and joined the Department of Agriculture, and some were recruited into the Soil Science Section. All the soil scientists went to the field for soil surveys and after their fieldwork conducted laboratory analysis of soil samples. The soil survey team began writing reports and began other soil fertility experiments, though they were very limited. In 1960, King Mahendra sized control and suspended elected parliament and multi-party politics, and in 1962 promulgate a new constitution and one-party Panchayat system was implemented.

2.4 Soil Science Under the Panchayat System (1962–1970) to Present

There has been a gradual increase in the number of soil fertility experiments in Nepal since they began. Variety cum fertilizer trials were the major experiments in different crops in different research stations. The Kakani, Khumaltar, and Rampur farms mostly concentrated around maize and soybean experiments. Other farms and stations experimented mainly on wheat and rice crops. This was the period when the Agriculture Association was registered and the Nepal Agriculture Journal began publication once a year thereafter. Several of the experiments conducted in different research farms and stations were published. Fields such as crop research, horticulture, livestock (including veterinary and fisheries), and agriculture extension were all included under the single Department of Agriculture but were headed by section chiefs who were academically qualified and had experience in their respective fields. This was followed by a major change in the agricultural structure, in which the single department was divided into five departments, viz. Crop Science, Horticulture, Livestock and Veterinary, Fisheries, and Agriculture Extension. Soil Science was included under the Crop Science section. Until 1970, graduates in agriculture from India and some from the USSR joined the Soil Science Division, after which scientists were assigned to a different unit in the section. This led to the strengthening of the soil physics laboratory and soil microbiology unit.

Although work in soil science began as early as 1953, actual recognition was made in 1957 when the Soil Science Section was set up as a unit under the Department of Agriculture, then His Majesty's Government (HMG) of Nepal headed by Bidur Kumar Thapa. This unit conducted a soil survey, soil sample analysis, and soil fertility experiments from various resettlements programs such as the Rapti Valley Development Program and the Banke and Bardia resettlements program. Later, units such as soil survey, soil physics, soil microbiology, plant nutrients, and recently Geographic Information System (GIS) and remote sensing were added. Soil scientists were posted to the Khumaltar and Parwanipur stations only. More soil scientists were added to Tarahara Agriculture Farm, Janakpur Agriculture Farm, Rampur Agricultural Station, Bhairahawa Agricultural Station, and Khajura Agricultural Station in 1972.

Farms established for the purpose of agricultural development felt the lack of soil scientists and recruited additional soil scientists in their command area. These farms and stations were GADP Khairenitar, JADP Nakatajhij, Agriculture Center Lumle and Pakhribas, and Hill Agriculture Station Kavre. Other farms and stations where soil scientists were recruited were Agricultural Station Doti, Jumla Agricultural Station, National Citrus Research and Development Centre Paripatle, Kirtipur Horticulture Station, Malepatan Horticulture Farm, and the Tobacco Development Program. The farms and stations were assigned to conduct soil fertility experiments mostly variety cum fertilizer trials and to collect soil samples for analysis to send to Khumaltar. The soil laboratory at Khumaltar was overloaded and additional soil labs were set up in some farms and stations, including Tarahara, Parwanipur, Rampur, Bhairawa, Khajura, and Khairenitar and later Lumle Agriculture Centre and Pakhribas Agriculture Centre.

After the partition of the Department of Agriculture (DoA) and Nepal Agricultural Research Council (NARC) in 1991, additional soil laboratories were set up. Nepal Agricultural Research Council concentrated on soil related to research and DoA in development activities such as providing soil analytical services. The soil laboratories that DoA presently owns are at Surunga (Jhapa), Jhumka (Sunsari), Hetauda (Makawanpur), Harihar Bhawan (DoA, Lalitpur), Pokhara (Kaski), Nepalgunj (Banke), and Sundarpur (Kanchanpur), with soil scientists and required staff. Service provided by these government laboratories can sufficiently cater to the needs of farmers if all are run properly. Because the governmental laboratories could not fulfill the need for soil sample analysis, private soil laboratories were set up in Kathmandu Valley and outside as well.

Nepalese farmers applied organic manure mostly Farmyard Manure (FYM) to upland crops where maize and millet were grown. Rice crops depended on the organic matter-rich flood sediments that entered the field through irrigation. One source reported that 60 t ha^{-1} of FYM had been applied by the farmers (Karki et al. 2007). A higher amount of FYM/compost is also applied in the Lumle Agriculture Research Command areas (Subedi and Gurung 1991). Use of green leaves of ashuro (*Adhatoda vasica*), tite pati (*Artemesia vulgaris*), masyang (*Vigna umbellate*), dhaincha (*Sesbania cannabina*), siris (*Albizia amluki*), and other succulent wild plants and leaves are still used to fertilize the rice nursery in the hills and mountain. In situ manuring by halting flocks of animals on different fallow land and shifting animal sheds during winter was practiced in early Nepalese agriculture.

2.4.1 Human Resources in Soil Science

The public profile of soil science and soil scientists in Nepal including other countries is on a level of the soil profile that is underground and largely invisible. Globally, some soil scientists have held the highest positions in international organizations such as the World Bank, the International Union of Soil Science, and International Agricultural

Research Institutes such as ICRISAT, IRRI, etc. These soil scientists also have received prestigious awards such as the Nobel Prize, as mentioned by White (1997), Australian Prizes, and the Wolf Prize for their work in soil science. Nationally, higher positions that soil scientists have held include Director-General of the DoA and Deputy Director-General (DoA), Joint Secretary at the agriculture ministry, Directors at NARC, the Dean of the Institute of Agriculture and Animal Science, and Dean of the Institute of Forestry. However, very little work has been done to persuade the public as well as the government to create new posts of soil scientists, whereas the need for soil scientists exists in every district and region and should be remedied if we are to save the soil and maintain sustainable food production in Nepal. It is not effective for the Nepalese government to recruit fertilizer inspectors with no knowledge of fertilizers and soil fertility from other faculties. Another example stemming from our lack of knowledge is that while major soil is lost from cultivated land and soil conservation activity is needed in upland cultivated fields, the Department of Soil Conservation is established under the Ministry of Forest. In short, professionals from other backgrounds are running soil programs without proper training and technical know-how.

With the limited number of soil scientists, we have implemented soil and fertilizer-related programs and are training technicians and farmers to balance the use of fertilizers for higher crop yield and soil management. However, farmers tend to forget soil fertility management when the program training and follow-up are terminated. There have been ample opportunities to create posts for soil scientists and extend our activities to each district, but this has not been possible. The number of scientists working in soil science has been decreasing every year, and this may be due to the government and society ignoring the importance of soil in Nepal. The same situation has been occurring throughout the world, and this has been a concern of the International Union of Soil Science (IUSS) as well.

In the developed world, soil science as a discipline has been slowly merged into other disciplines such as crop science, geology, ecology, and environmental sciences. Now, soil-related studies have been diverted through agronomic, environment, and their relation to human health. Though the number of soil scientists in the world has been lowered by 40% (Baveye et al. 2006), the number of research publications in soil science is increasing (Hartemink and McBratney 2008), indicating the importance of soil science. In Nepal also the number of soil scientists and students in soil science is diminishing even though we have hardly completed any work in saving our soils. This could be due to less attention from the Government of Nepal in the care and management of soil and land use. Recently the Ministry of Land Reform and Development is using soil pedology and soil fertility evaluation as a major component in its land-use zoning program country-wide. There is a high demand for soil science activities to save the most fragile mountain ranges in the world, which are increasingly threatened by large-scale human activities. Extensive deforestation and intensive farming on steep slopes, along with heavy population pressure on natural resources have resulted in overall environmental degradation (Shengji and Sharma 1998).

2.4.2 Soil Science Works Before the Formation of NARC

From the beginning, soil science as a unit began working in soil analysis and soil sample collection. Later, soil fertility experiments in cereals, particularly rice, wheat, and maize were conducted and presented in workshops organized within the Department of Agriculture. Around 1965 an expert in soil science from UNDP/FAO supported and guided Nepalese soil scientists in soil survey and mapping, including soil and water analysis. This strengthened soil analytical service, as all the members of soil survey, included soil fertility experiments and carried out soil sample analysis themselves. This assured the quality of the analytical results. Soil surveys of most of the districts in Tarai and Siwaliks have been completed, including some of the important hills and mountain districts such as the Andhikhola valley of Syangja, the Pokhara valley, and later, the Gandaki and Dhaulagiri Zones. However, very few reports have been published (Fig. 2.1).

In 1963, Dr. N. Borlaug came to India to implement his Green Revolution technologies. This revolution revolutionized fertilizer, seeds, plant protection chemicals, agricultural machinery, and water use. Intensive use of High Yielding Crop Varieties and NPK fertilizers including maximum use of water in irrigation no doubt increased crop production, but the project paid little attention to soil biodiversity, micronutrient exhaustion, and water pollution, including environments that were massively deteriorated due to soil mining and the effect of pesticides on downstream ecology. Activities of the Green Revolution were introduced in Nepal as well, and some improved varieties of wheat such as Lerma Roho 52 had already been introduced. As influenced by Green Revolution, DoA launched a UNDP special project, "Increase Use of High Yielding Crop Varieties and Fertilizer" (NEP-12, 1970–75). This project concentrated on improved varieties of rice, wheat, and maize where NPK fertilizer was applied in conjunction with irrigation. Several trials and demonstrations were conducted in the districts of Narayani and Bagmati Zones. This project NEP-12 impacted positively in the Tarai districts but had little impact in the hills. The remarkable impact of this project was the report of a reconnaissance soil survey of the Bagmati and Narayani

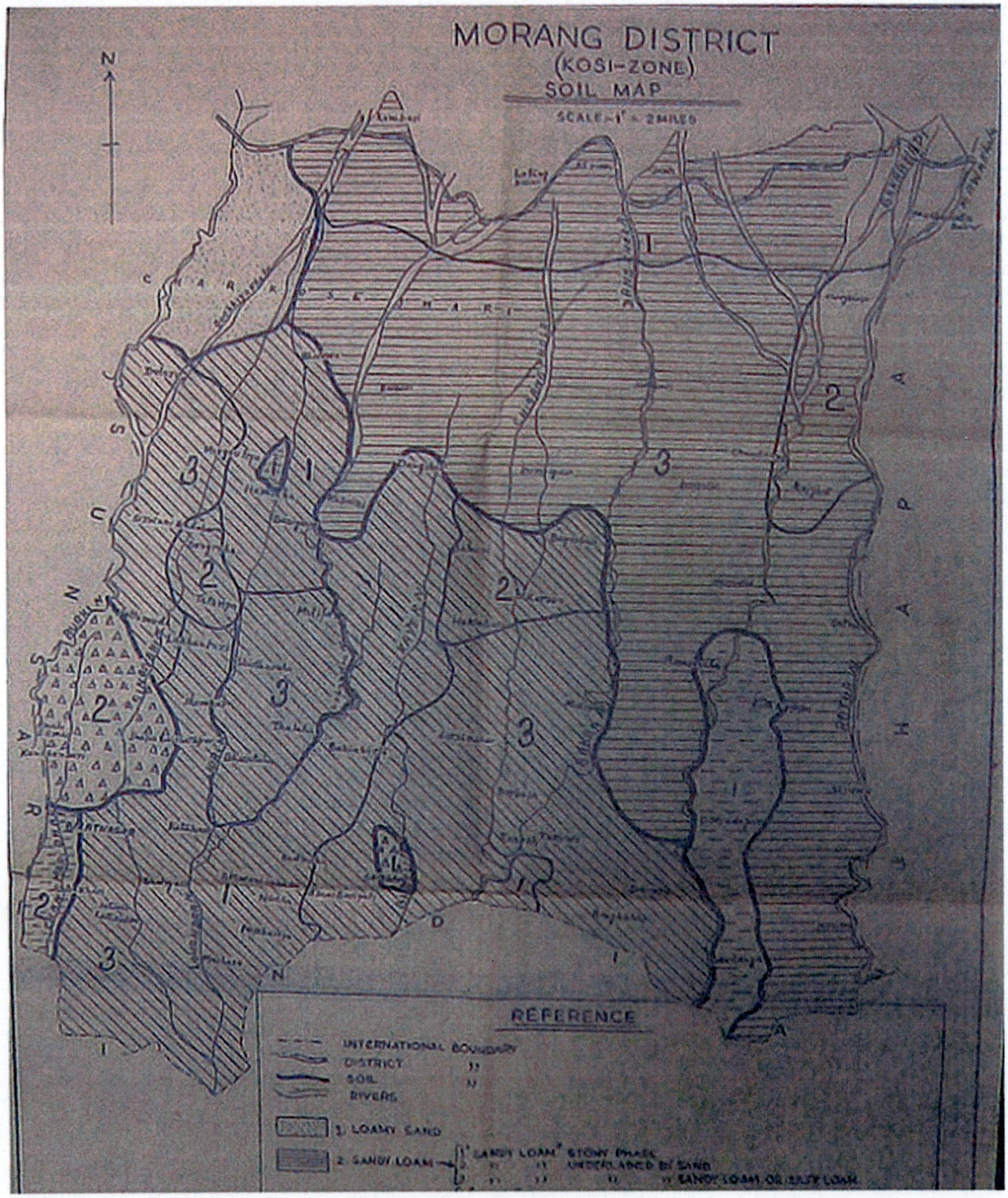

Fig. 2.1 An example of early soil survey work done by the survey team. Previous days soil maps were hand-crafted but in the later decades they are digitized in Geographical Information System platform

zone of the project area. This was the only reconnaissance-level soil survey report and is still a good reference for the study of soil science in this area. For the hills, another FAO-supported project named FRIP was launched in 22 hill districts of the Western, Central, and Eastern Regions from 1982 to 1992. Similar activities of NEP-12 were replicated. This project came up with site-specific recommendations of NPK fertilizers for rice, wheat, maize, and potato.

Fertilizer did not respond well in the absence of irrigation and hence international organizations such as FAO/UNDP, ADB, the World Bank, and other multilateral and bilateral projects supported Nepal in its water use and constructed large irrigation plans. Many of these projects included feasibility and pre-feasibility studies and prepared soil maps of the command area for their use. Government soil scientists had very little knowledge of them. In addition, foreign universities have conducted academic as well as non-academic research on soil science and related fields in a scattered way. Most of the irrigation projects kept one demonstration farm within each irrigation project command area, which was later handed over to the Department of Agriculture to carry out demonstrations to farmers and produce quality seeds. Jhumka is one such farm in the command area of the Morang-Sunsari Irrigation Project. Later, the Department of Irrigation (DoI) implemented feasibility studies of major and minor irrigation projects where a soil survey was obligatory. However, most of the irrigation engineers did soil surveys and made soil maps as they liked.

2.4.3 Nepalese Soil Science at Present

Presently, soil survey in Nepal has been digitalized and computer-aided programs are used to prepare maps with the help of a global positioning system (GPS), which is much easier than in the days of hard-copy toposheets and aerial photographs. Laboratory procedures, including atomic absorption spectrometry (AAS), make use of highly expensive equipment, to detect individual elements. In just a short time, a soil survey report with a fully digitized map can be prepared.

Fertilizer experiments are a major task of the Soil Science Division (National soil science research center) in NARC. Some initial work in fertilizer use efficiency has also begun. Chemical and biological testing of some fertilizer products are included in these activities, along with digitizing previous soil survey work and participatory fertilizer experiments. Soil science in the Department of Agriculture mainly consists of concentrating soil fertility evaluation and mapping at the district level and providing soil analytical services to farmers. The Soil Service Directorate is also involved in helping the agriculture ministry to formulate fertilizer policy (Fig. 2.2).

2.4.4 Soil Research Carried Out by Education Institutes

The oldest agricultural institute (Institute of Agriculture and Animal Science) in Nepal is at Tribhuvan University (TU) where the master's in soil science program was started late and students conducted master's-level thesis research mostly through agronomic studies. Some academic studies are also related to soil erosion and conservation, including micronutrients, but all are agronomic.

The Institute of Forestry at TU also offers a master's degree in forestry where graduate students carry out research. They concentrate mostly on soil conservation, whereas students at Kathmandu University research environmental science. Several foreign universities also have conducted thesis research and/or collaborative projects with counterparts in Nepal, but the authors have access to very few records. Carbon (C) mapping studies, Likhu Khola Watershed studies, and many others are some examples.

2.5 Future Soil Science in Nepal

Soil science can be viewed through the perspective of a society's health and hygiene. Disease pathogens such as helianthus organisms and other worms can be housed in the soils and easily contaminate humans. The presence of radon-causing cancer is related to poorly drained soils and increases infant mortality rates (Oliver 1997). We do not know the health risk of poorly drained soils, despite most towns in the lower plains of Tarai being flooded every year. An excess of micronutrients and heavy metals such as aluminium (Al), arsenic (As), cadmium (Cd), copper (Cu), fluorine (Fl), iodine (I), lead (Pb), selenium (Se), thallium (Tl), and zinc (Zn) can be toxic when they contaminate food. In addition, soil deficiency in micronutrients results in low plant uptake, which ultimately creates micronutrient deficiency in humans. One of our results showed that most micronutrients (Boron (B), Zinc (Zn), and molybdenum (Mo)) are deficient in Nepalese soils (Karki et al. 2005). These deficiencies might have consequences on the health of our people.

Though we cannot take the right turn and begin these activities right away, some thought needs to be given on how we can increase food nutrients through biofortification. Biofortification of micronutrients could help improve the health of infants, children, and pregnant women (Bouis and

Fig. 2.2 Soil science program development in Nepal

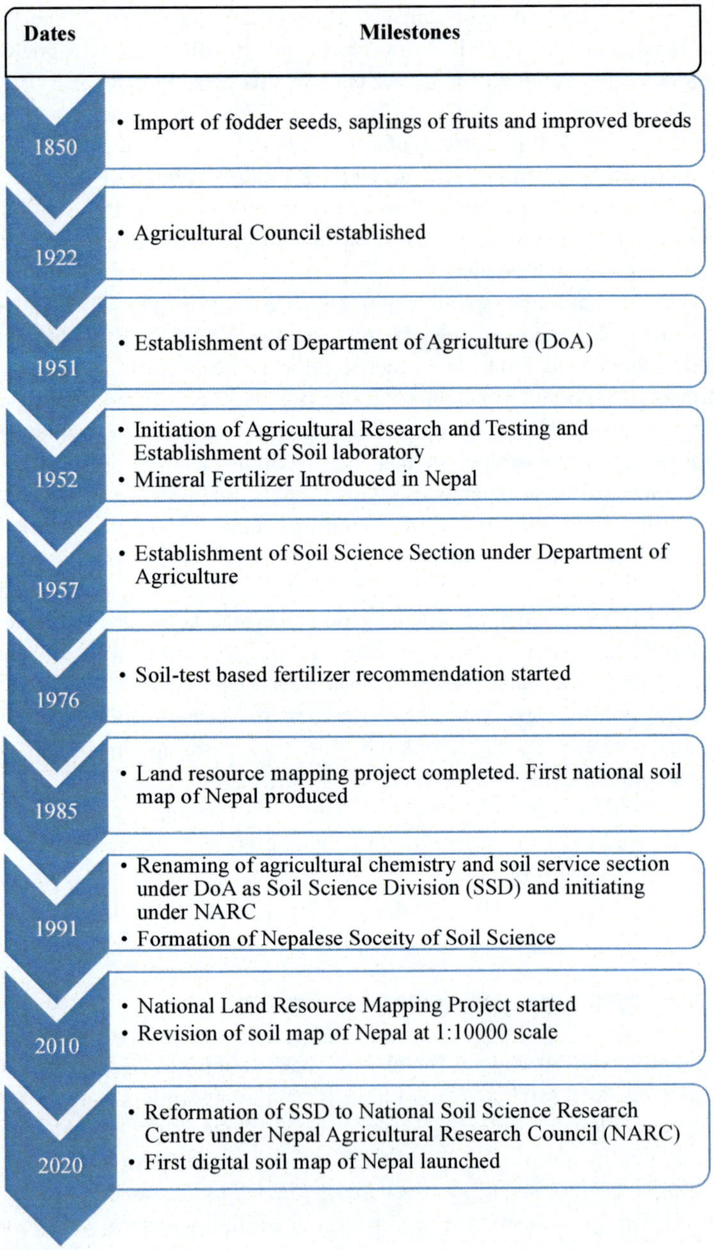

Welch 2010). Similarly, other avenues for increasing food nutrients need to be explored.

2.6 Conclusion

The soil history of Nepal dates to 5,000 years ago during the Vedic period, though it received formal address in the government sector only when trained soil scientists joined the Department of Agriculture in 1957. The first phase of soil science development was quite slow and mostly spent developing in-house facilities. Despite limited manpower, some remarkable work in the field of soil fertility has been conducted, including site-specific fertilizer recommendations for cereals, soil surveys to produce soil fertility maps, and soil survey reports of all the Tarai districts including some potential production pockets of the hills. Later fertilizer recommendations were included for potatoes, legumes, and some vegetables. As time passed, soil survey work was digitized, soil microbiology work was initiated, and soil micronutrient studies were reported. Still, soil research has been more focused on food security. We need to begin studying soil differently from a production perspective that includes the human medicinal value of soil, nutrient-enriched biofortification, and soil biodiversity. To explore different aspects of the soils of Nepal, studies in soil science need to be collaborated and documented by national and international agencies.

References

Abrol I, Nambiar K (1997) Fertility management of Indian soils-A historical perspective. Adv Geoecology 29:293–310

Alexandratos N, Bruinsma J (2012) World agriculture towards 2030/2050: the 2012 revision. FAO, Rome

Annonymos (1959) The Sub-continent of South Asia: Afganistan, Cylon, Indian, Nepal, Pakistan. In: Affairs BoP (ed) US Department of State Publication No 6839, Near and Middle East Series 41, vol 41. Uinited State Department of State Bulletin-Nepal, Washington DC, pp 48–62

Baveye P, Jacobson AR, Allaire SE, Tandarich JP, Bryant RB (2006) Whither goes Soil Science in the United States and Canada? Soil Sci 171:501–5188

Blume HP (2014) Die Wurzeln der Bodenkunde. In: Handbuch der Bodenkunde, pp 1–32

Bouis HE, Welch RM (2010) Biofortification—a sustainable agricultural strategy for reducing micronutrient malnutrition in the global south. Crop Sci 50:S-20-S-32

Dhital BP (1970) Role of agriculture XE "Agriculture" in economic development of Nepal. Aowa State University, Ames, Iowa

FAO (2010) Agricultural extension services delivery system in Nepal. Food and agriculture XE agriculture. Organization (FAO), Kathmandu

Gardner RAM, Gerrard AJ (2003) Runoff and soil erosion on cultivated rainfed terraces in the Middle Hills XE "Hills" of Nepal. Appl Geogr 23:23–45

Gautam U (2012) Nepal: food security, a localized institutional irrigation perspective on public irrigation systems. Hydro Nepal: J Water Energ Environ 95–99

Gurung N (2011) Local democracy in the political transition of Nepal. Soith Asian institute and alliance for social dialogue.

Hartemink AE, McBratney A (2008) A soil science renaissance. Geoderma 148:123–129

Karki KB, Joshi BK et al (2004) Status of agrobiodiversity in upper arun river Valley, Shankhuwashabha. In: Agricultural research for enhancing livelihood of Nepalese peoplel. vol 30, pp 265. NARC, Kathmandu

Karki KB, Sherchan DP, Bhandari RC (2007) Residual effect of organic and inorganic fertilising materials especially NO3 (nitrate) in different crops under different agro climatic condition. In: National Seminar on organic agriculture. Department of Agriculture, GoN, Kirtipur, Kathmandu, pp 7

Karki KB, Tuladhar JK, Upreti R, Maskey SL (2005) Distribution of micronutrients XE "Micronutrients" available to plants in different ecological regions of Nepal. In: Andersen PT, Karki KB, Maskey SL (eds) Micronutrients in South and South East Asia. ICIMOD/NARC/Bergen University, Norway, Kathmandu, Nepal, pp 17–29

Kunwar RM, Duwadee N, Shrestha K, Bhandari G, Gupta V, Vetaas O, Chaudhary R (2008) Plant diversity and soil characteristics of Shorea robusta and Castanopsis hystrix forests and slash and burn habitats of Arun Valley, Eastern Nepal. Pleione 2:87–97

Maharjan KL, Khatri-Chhetri A (2006) Household food security in rural areas of Nepal: Relationship between socio-economic characteristics and food security status. In: Poster paper prepared for presentation at the international association of agricultural economics conference, Gold Coast, Australia

Maskey R, Joshy D (1991) Soil and nutrition loses under different soil management practices in the middle mountains of central Nepal. In: The Workshop Proceedings of soil fertility and erosion issues in the middle mountains of Nepal held in Kathmandu, Sponsored by IDRC, Canada (Shrien HASP ed)

Maskey R, Thakur N, Shrestha A, Rai S (2001) Management of soil erosion consortium (MSEC): an innovative approach to sustainable land management in Nepal. Journal Article. agris.fao.org

Maskey S, Bhattarai S, Karki K (2004) Long-term effect of different sources of organic manures on wheat soybean rotation. SAARC J Agric 2:243–256

Newnham RE (1994) Essentiality of Boron for healthy bones and joints. Environ Health Perpectives 102(Suppl 7):83–85

Nielsen FH, Meacham SL (2011) Growing evidence for human health benefits of Boron. J Evid Based Complement Altern Med 16:169–180

Oliver MA (1997) Soil and human health: a review. Eur J Soil Sci 48:573–592

Pant YP (1956) Nepal has a five-year plan. Econ Wkly 8:113–117

Shengji P, Sharma UR (1998) Transboundary biodiversity conservation in the Himalayas. In: Ecoregional co-operation for biodiversity conservation in the Himalaya. UNDP/WWF/ICIMOD, Kathmandu

Sherchan DP, Gurung GB (1995) Soil and nutrient losses under different crop husbandry practices in the hills of East Nepal. In: Soil fertility and erosion issues in the middle mountains Nepal, ISS Kathmandu, Kathmandu

Shrestha DP (1997) Assessment of soil erosion in the Nepalese Himalaya: a case study in Likhu Khola Valley, Middle Mountain XE "Middle Mountain" Region. Land Husbandry 2:59–80

Singh K (2011) Tragedy of the global commons: causes, impacts and mitigations. India Natural Resource Economics and Management (INREM) Foundation Anand, Gujarat, India

Skerry C, Moran K, Calavan K (1991) Four decade of development: the history of US assistance to Nepal 1951–1991. United States Agency for International Development (USAID) in Kathmandu, Kathmandu Nepal

Subedi K, Gurung G (1991) Soil fertility thrusts towards sustainable agriculture: experiences of Lumle agriculture research center. In: Soil fertility and Erosion Issues in the Middle Mountain of Nepal, Jhiku Khola Watershed. IDRC Kathmandu

White RE (1997) Soil Science: raising the profile. J Soil Res 35:961–977

Winiwarter V (2006) Soil scientists in ancient Rome. In: Footprints in the Soil; people and ideas in soil history. (Wenkentin BP ed) pp 3–16. Elsevier Amsterdam/Oxford

Young A (1999) Is there Really Spare Land? A critique of estimates of available cultivable land in developing countries. Environ Dev Sustain 1:3–18

Climate

Basanta Paudel, Dinesh Panday, and Kundan Dhakal

3

Abstract

Nepal is a relatively small, mountainous country in the Central Himalayas with a diverse climate within a short aerial distance due to its unique topography and altitudinal variation. The country is characterized mainly by six climatic zones, ranging from tropical in the southern plains to tundra/nival in the northern part with perpetual snow cover. In this chapter, a brief explanation of the climate zones; a description of the climate seasons; and overall climatic trends including temperature, rainfall, floods, drought, and frosts and cold waves of the country is provided. Further, the chapter covers the status of agro-ecozones, soil climate, and climate change and its impact. In recent decades, Nepal's climate shows trends of increasing temperature and decreasing rainfall. The country's agro-ecozones are divided into five major zones, including Tarai, River Basins, Lower Hills, High Hills, and High Mountains, ranging between 60–4800 meters (m) above mean sea level (asl). Climate is known to be one of the five soil-forming factors and has a significant influence on soil properties. For example, the properties of soils found in the High Hills and High Mountains are different than in soils found in other agro-ecozones of Nepal. Finally, we discuss the climate change impacts in Nepal and the significant risks they pose, especially on agriculture, food security, and people's livelihoods. A broad-brush description of various climatic scenarios of Nepal Himalayas is presented.

Keywords

Climate of Nepal • Climate zones • Agro-ecozones • Soil climate • Climate change

3.1 Introduction

Geographically Nepal is located in the central Himalayan region of South Asia (Khanal et al. 2020), with climatic and topographic variations that are unique in the world (Shrestha and Aryal 2011). Mountainous terrain covers most of the country, including the world's highest peak Sagarmatha (Mount Everest), reaching 8848.86 m asl (Nepal et al. 2020). The climatic variation within a short aerial distance is remarkably high due to vast altitudinal differences, meaning ranging from a tropical climate in the south to a tundra/nival/trans-Himalayan climate in the north (Karki et al. 2017; Paudel et al. 2020). The vast altitudinal variation from 60 m asl in the southern Tarai plains to 8848.86 m asl in the northern Himalayan region directly affects overall weather, climate, and their variation. Further, the climatological and topographical factors (i.e., air temperature, solar radiation regime, slope, and aspect) are directly associated with the overall variability of the country's climate (Collier and Immerzeel 2015; Karki et al. 2020; Talchabhadel et al. 2018).

The country is divided into five physiographic regions, including Tarai, Siwalik, Middle Mountain, High Mountain, and High Himalaya (LRMP 1986). This classification shows that most of the country's territory is covered by mountainous areas, which acts as a barrier. This natural barrier largely impedes humid air mass circulation, resulting in precipitation on the southern slope or sunward side of the mountain and an area of rain shadow on the leeward side of the

B. Paudel (✉)
Key Laboratory of Land Surface Pattern and Simulation, Institute of Geographic Sciences and Natural Resources Research, Chinese Academy of Sciences, Beijing, China
e-mail: paudelb@igsnrr.ac.cn

D. Panday
Department of Biosystems Engineering and Soil Science, The University of Tennessee-Knoxville , Knoxville, TN, USA

K. Dhakal
Noble Research Institute, LLC, Ardmore, OK, USA

R. B. Ojha and D. Panday (eds.), *The Soils of Nepal*, World Soils Book Series,
https://doi.org/10.1007/978-3-030-80999-7_3

trans-Himalayan region (Kansakar et al. 2004). For example, the Manang, Mustang, and Dolpa districts of western Nepal are on the Himalaya Mountains' leeward side and receive less precipitation (Karki et al. 2017). This results in the formation of arid and semi-arid environments and directly impacts agricultural activities. However, the sunward side of the mountains receives more precipitation, which is largely beneficial for farming activities and people's livelihoods (Shrestha et al. 2000). Further, it has been reported that every 1000 m increment in altitude reduces temperature by 6.5 °C, which is widely known as the standard (average) lapse rate (Chapman et al. 2002; Tabony 1985)—the primary driver of temperature variability within the country. Thus, the presence of the high mountain range largely impacts the country's overall climate. Topographical features and the distribution of meteorological/hydrological stations of Nepal are presented in Fig. 3.1.

There are six major climatic zones in Nepal: tropical, subtropical, temperate, subalpine, alpine, and tundra/nival/trans-Himalayan, ranging from the southern Tarai plain to the northern high and trans-Himalaya regions, respectively

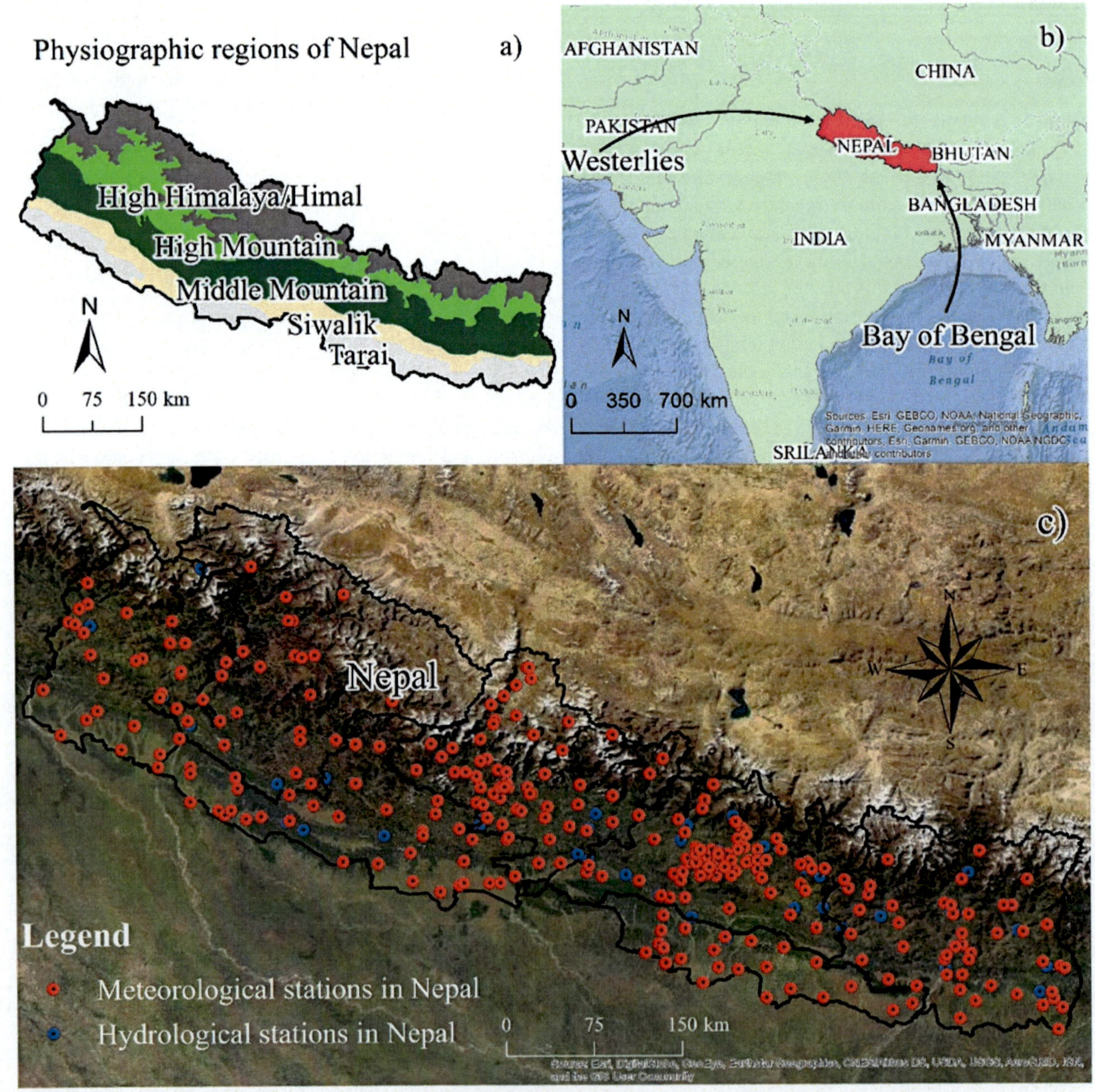

Fig. 3.1 Topography and spatial distribution of meteorological and hydrological stations across Nepal. Meteorological and hydrological station data were obtained from the Department of Hydrology and Meteorology (DHM), Government of Nepal

(Nayava 1975). The districts located in the southern Tarai plain regions experience the tropical climate, and northern districts such as Mustang and Manang in the high Himalaya regions usually experience the trans-Himalaya climate. Further, there are four main climate seasons in Nepal: pre-monsoon, monsoon, post-monsoon, and winter (Karki et al. 2017). The pre-monsoon season occurs in Nepal between March and May, and the monsoon season occurs between June to September. The post-monsoon season falls between October and November, and the winter season extends from December to February (Karki et al. 2020).

Over the last few decades, several studies have revealed a significant rising trend in temperature (Poudel et al. 2020; Shrestha et al. 1999), coupled with a declining precipitation rate (Duncan et al. 2013; Ichiyanagi et al. 2007; Shrestha et al. 2000; Talchabhadel et al. 2018). However, short-term erratic heavy rainfall occurrence has been more frequent in the last several years, creating several flood events in Nepal (Devkota 2014). During the monsoon season, these flood events devastate the rural economy and the livelihood of the farming community, especially in the southern Tarai region (Malla 2008; Pangali Sharma et al. 2019). The overall decreasing trend of annual precipitation and increasing frequency of drought events (Dahal et al. 2016) directly impact the country's agricultural systems (Wu et al. 2019). Moreover, during Nepal's winter season, especially in the Tarai regions, frosts and cold waves have been commonly noted in recent decades, causing human fatalities through exposure to cold waves and impacting the region's winter crops (Pradhan et al. 2019). These climatic characteristics, including the increasing trends of temperature, droughts, frequent seasonal floods, extended frosts, and cold waves, as well as decreasing trends of precipitation clearly show that the climate in Nepal is rapidly deviating from the long-term normal. This suggests the robust negative impact of climate change, which impacts people's livelihoods and the country's overall agricultural systems.

Based on Nepal's local landscape and dominant farming system, several suitable agro-ecozones were developed and categorized, based mainly on elevation, climatic condition, and soil/land type (Gauchan and Yokoyama 1999). These agro-ecozones help researchers formulate suitable strategies and recommend appropriate crops for farmers in their locality. Further, the soil climate is generally characterized by air, temperature, and moisture associated with soil. Warm conditions promote chemical and biological reactions that more quickly convert parent materials into the soil. As the soil is dried out, plant growth is reduced due to poor contact with soil water, which reduces the surface layer's stability (Sindelar 2015). In this context, this chapter mainly discusses the overall climate scenarios of Nepal, including climate zones, seasons, trends, agro-ecozones, soli climate, and climate change impact.

3.2 Climate Zones in Nepal

The mountainous country of Nepal contains vast topographical differences; thus, the climate also varies significantly within different regions. It is mainly categorized into six climate zones ranging from southern tropical to tundra/trans-Himalayan perpetual snow.

3.2.1 Tropical

In Nepal, the tropical climate is observed below 1000 m asl (Nayava 1975). This climate zone is located mainly in the southern part of the country (Fig. 3.2). In the tropical zones, summers are hot and winters are cold (Pradhan et al. 2013). The tropical climate zone encompasses most of the districts located in the Tarai region of the country. This zone is also known as the lower tropical zone, with an average elevation mostly below 300 m asl. The east–west highway of the country primarily runs through this lower tropical climate zone. The area above 300 m asl to 1000 m asl within this climate zone is also recognized as the upper tropical climate zone in Nepal and includes most of the Siwalik Hills. In addition, this climate zone can be observed in the lower elevation river valley areas of the hill and mountain regions. The lower tropical climate zone is a geographically plain area, and most of the soil in this zone is highly fertile. Thus, the lower part of this climate zone is more suitable for agricultural activities and the climate is more ideal for paddy, wheat, and maize crops (Rimal et al. 2018; Sharma 2001). Moreover, due to the fertile soil and tropical climate, the production of cash crops (sugarcane and jute), tropical fruits (mango, papaya, litchi, and banana), and vegetables is higher in this climate zone compared to the other climate zones in Nepal (Amatya 1976; Ghimire and Thakur 2013; Neupane et al. 2017). This climate zone is inhabited by over 50% of the country's population (CBS, 2012), and as such, many people's livelihoods are affected by the hot summer waves and cold winter waves (Pradhan et al. 2019; Pradhan et al. 2013).

3.2.2 Subtropical

The subtropical climate region is found in elevation ranging between 1000 and 2000 m asl. This climate zone is located between the upper part of the tropical climate zone and the lower part of the temperate climate zone (Fig. 3.2) and mostly covers the area within the country's middle mountain physiographic region. The capital of Nepal, Kathmandu, is located within this climate zone, and it is the second most populous climate zone after the tropical climate zone (CBS 2012). In this subtropical climatic characteristic

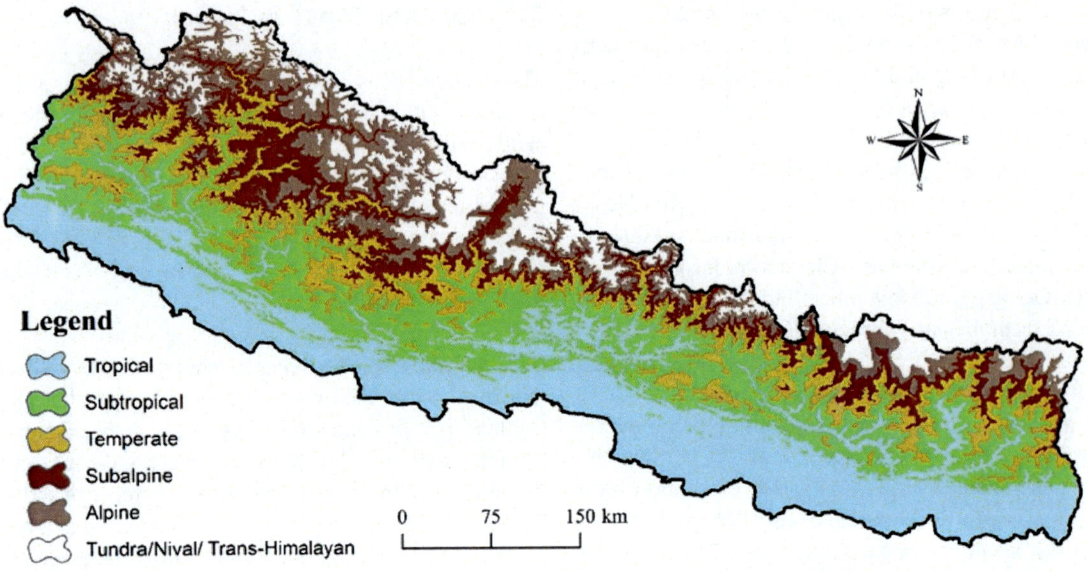

Fig. 3.2 Major climate zones of Nepal

zone, the major crops are maize, millet, rice, wheat, and potato (Pokharel 2019). Farmers in the flat and lower elevation of this climate zone have access to irrigation facilities. Farmers mainly cultivate rice and wheat in fields with very gentle slopes and produce maize, millet, potato crops, and citrus fruits and vegetables in gentle slopes (Devkota 1999; Sharma 2001).

3.2.3 Temperate

The elevation ranges between 2000 and 3000 m asl in Nepal are referred to as the temperate climate zone (Fig. 3.2). It shares boundaries with the majority of the subtropical climate zone area in the southern part of the country and the subalpine climate zone in the northern part (Devkota 1999). This climate zone is mainly situated from the upper part of the Middle Mountain physiographic region to the lower part of the High Mountain region. This zone is known for its mild climate, and the major cultivated crops within this climate zone are maize, barley, wheat, potato, rice, and buckwheat (Sharma 2001). Further, this zone is suitable for apple cultivation (Manandhar et al. 2014).

3.2.4 Subalpine

The subalpine climate zone ranges between 3000 and 4000 m asl in Nepal (Fig. 3.2). The majority of this region is situated within the High Mountain physiographic regions of the country, and some areas are extended into the High Himalayas/Himal region (Devkota 1999). The climatic condition of the subalpine climate zone consists mostly of cold weather, and as there are many areas of pastureland/grassland, inhabitants practice transhumance adaptation in different climatic seasons. During the summer season, people shift/migrate to the higher elevation from the lower elevation with their livestock for grazing, and move back to the lowland area in the winter seasons (Aryal et al. 2014; Gentle and Thwaites 2016). People in this climate zone mainly domesticate yak, sheep, and goats, and due to the cold climate, mainly cultivate buckwheat, potato, and barley (Gauchan and Yokoyama 1999; Pokharel 2019). Further, some within this climate zone in Nepal also practice fruit (apple) and vegetable farming.

3.2.5 Alpine

The elevation range from 4000 to 5000 m asl in Nepal is known as the alpine climate zone (Fig. 3.2). Due to the cold climate, few people inhabit this region. As some areas are covered seasonally by snow, farming practices are less common (Devkota 1999; Nayava 1975). However, most land in this area is used for pasture, with the potential for medicinal herbs and livestock farming (Aryal et al. 2014; Bhattarai et al. 2010; Pokharel 2019). Usually in the summer season, herdsmen from lower elevations travel with their livestock for grazing and return in the winter season (Gentle and Thwaites 2016).

3.2.6 Tundra/Nival/Trans-Himalayan

The tundra/nival/trans-Himalayan climate exists in the area located above 5000 m asl in Nepal (Fig. 3.2). Due to the extreme cold climate, the area above 5000 m is not habitable. The nival is also called the tundra in Nepal Himalaya (Nayava 1975), where the land is mostly covered by snow all year round, resulting in minimal vegetation. Within this climate zone, trans-Himalayan climates are observed on the leeward side of the Himalaya mountain range. As discussed earlier, this climate zone receives less precipitation; thus, it is also called the rain shadow area (Shrestha et al. 2000). Moreover, due to the low air temperature and decreased humidity, these areas are less suitable for agriculture (Chapagain 2016), resulting in low human population density. The Manang, Mustang, and Dolpa districts are located within this climate zone, representing regions that receive less precipitation and contain arid and semi-arid landscapes.

Additionally, a revised Global level Köppen–Geiger classification study (Kottek et al. 2006) shows there are mainly eight climate types in Nepal: polar frost climate (EF), polar tundra climate (ET), temperate climate with dry winter and warm summer (Cwb), temperate climate with dry winter and hot summer (Cwa), cold climate with dry winter and warm summer (Dwb), cold climate with dry winter and cold summer (Dwc), arid steppe cold climate (BSk), and tropical savanna (Aw) (Karki et al. 2016).

3.3 Climate Seasons in Nepal

There are mainly four climate seasons in Nepal (Talchabhadel et al. 2018). Pre-monsoon occurs between March and May, followed by monsoon season from June to September, post-monsoon season from October to November, and winter between December and February (Karki et al. 2017).

A wet and warm climate is observed mainly between June and September. During this period, a low-pressure zone is created, which attracts moist air from the Indian Ocean (Shrestha et al. 2000). Further, the period between October and May is marked by a dry season with cold temperatures. These months create high air pressure that produces dry air and moves to other regions (Karki et al. 2017). April and May are noted as the hottest months, and for some part of the country, primarily the southern Tarai tropical region, the temperature peaks around 38 to 40 °C (Karki et al. 2020). The months of June and July are known for their high rainfall intensity and thunderstorms that drive most rain-fed cultivation; however, this also creates landslides in the Hill and Mountain regions, along with flooding. These detrimental events cause severe property damage and losses in the Tarai region of the country and negatively impact people's lives (Karki et al. 2017; Pangali Sharma et al. 2019; Shrestha et al. 2000). The details of each climate season in Nepal are described herein.

3.3.1 Pre-Monsoon (March–May)

The pre-monsoon climate season in Nepal occurs from March to May. The pre-monsoon climatic season is mainly active for three months, with moisture-laden air and winds typically originating from the Bay of Bengal and creating short-term light rainfall in some areas (Nayava 1975; Shrestha et al. 2000). This depends on local weather conditions and the effect of cumulonimbus clouds (Talchabhadel et al. 2018). During this season, thunderstorm and hailstorm events are seen comparatively more often due to the impact of cumulonimbus clouds (Aryal 2018). Usually, the days are sunny, with rising temperatures and scattered thundershowers, especially during the afternoons and evenings. Outside of this season, April–May are the hottest months. The temperature of these months ranges up to 40 °C in the Tarai regions (Karki et al. 2020). Further, around 12.5% of rainfall occurs during Nepal's pre-monsoon season (Karki et al. 2017).

3.3.2 Monsoon (June–September)

The term monsoon is also known as the rainy season (between June and September) in Nepal (Brewin et al. 2000). The monsoon season refers to heavy periodic winds/breezes blowing predominantly from the Bay of Bengal (Nayava 1975), which carry heavy showers and humidity throughout most parts of Nepal. Similarly, some rainfall in the country is due to western Arabian sea winds (Shrestha et al. 2000). The monsoon season begins in the middle of June, peaks in July/August, and ends in September of each year (Karki et al. 2017; Nayava 1975). This season records the highest rainfall in Nepal; however, different regions vary in rainfall intensity (Barros and Lang 2003). Moreover, Nepal's distinct landscape features and associated environmental factors influence this season (Nayava 1974).

3.3.3 Post-Monsoon (October–November)

The post-monsoon season in Nepal lasts between October and November. Out of the four seasons, the post-monsoon season is the shortest. Noticeable characteristics of this season include windy days and the likelihood of cyclones forming in some parts of the country (Shrestha et al. 1999). In this season, a decreasing order temperature is observed due to changing seasonal weather and sometimes several days of continuous rainfall (Shrestha et al. 2000). The prevailing climatic conditions across Nepal during this season are between hot and cold; thus, this season is also known as a mild climatic season. The ending month of this season (November) is quite dry, mainly due to the dry and windy atmosphere (Merz et al. 2006).

3.3.4 Winter (December-February)

Winter is the coldest season in Nepal and occurs between December and February. The winter season happens after the post-monsoon season and before the pre-monsoon season each year. During this season, the majority of the High Mountain and trans-Himalayan regions are covered by snow with freezing temperatures (Nayava 1975). Further, the country's Middle Mountain regions and Hill regions also feel much colder compared to other seasons. Similarly, the southern part of the country (Tarai region) is highly affected by cold waves, and many people in the region lose their lives due to extreme cold events (Pradhan et al. 2019). The cold waves and frost in this region also impact farmers' agricultural activities and livelihoods (Malla 2008). The days become shorter, and the nights are longer than other seasons.

3.4 Climatic Trend in Nepal

3.4.1 Temperature

Historically, the temperature trend in Nepal showed a sharp increasing order by 0.06 °C yr^{-1} between 1977 and 1994 (Shrestha et al. 1999). The increasing trend was also reported until 2000 (Shrestha and Aryal 2011). A recent study from Nepal Himalaya based on data from 115 temperature stations from the Government of Nepal's Department of Hydrology and Meteorology (DHM) noted the temperature trend was rising noticeably by 0.05 °C yr^{-1} between 2000 and 2015 (Paudel et al. 2020) (Fig. 3.3). Thus, the overall long-term temperature trend in Nepal was observed to be rising significantly. The rising trend varies across different physiographic regions of the country. Many studies have reported that the noticeable increasing trend of temperature has impacted the overall livelihood of people and agricultural activities in Nepal (Gentle and Maraseni 2012; Malla 2008; Manandhar et al. 2011).

3.4.2 Precipitation

Long-term-climatic-data-based historical studies about climate change in Nepal reported that Nepal's precipitation trend was decreasing (DHM 2017; Shrestha et al. 2000). A recent

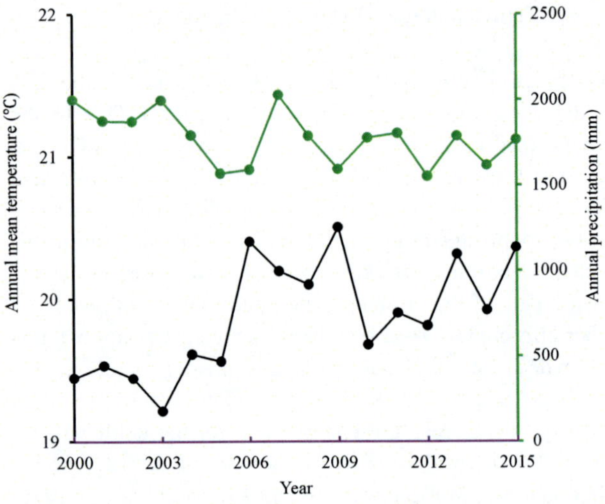

Fig. 3.3 Observed climatic trends (temperature and precipitation) in Nepal between 2000 and 2015. The trend calculations were based on observed daily weather records between 2000 and 2015 (115 stations for temperature and 272 stations for precipitation). *Data source* DHM, Nepal, and modified after Paudel et al. (2020)

study based on DHM Nepal data from 272 rainfall stations revealed that annual total precipitation declined by -16.09 mm yr^{-1} (Paudel et al. 2020) (Fig. 3.3). The long-term precipitation trend by different climatic seasons in Nepal shows that around 80% of precipitation occurs during monsoon season, while the remaining 20% occurs during the other three climatic seasons; with 12.5% during pre-monsoon, 4% during post-monsoon, and 3.5% during the winter season (Karki et al. 2017). Further, the decreasing trend of precipitation directly impacted the overall agricultural activities in Nepal (Malla 2008). The spatial pattern between 1980 and 2000 shows that the surrounding area of Pokhara has higher precipitation records; however, this was less in the upper Mustang and Dolpa regions of the country (Fig. 3.4).

3.4.3 Floods

Flooding events in Nepal are normally seen more frequently during the monsoon season compared to the other seasons because the monsoon season is known as the rainy season, and there has been highest rainfall across the country within this season (Nayava 1974). The studies show that overall flood events mainly impacted the Tarai regions (Pangali Sharma et al. 2019) because the landscape of this region consists of flat plains, allowing the floods to easily spread to more areas compared to the country's Hill and Mountain regions. A recent synthesized study about flood disasters in Nepal reported that there are fourteen districts in the Tarai regions highly affected by floods (Pangali Sharma et al.

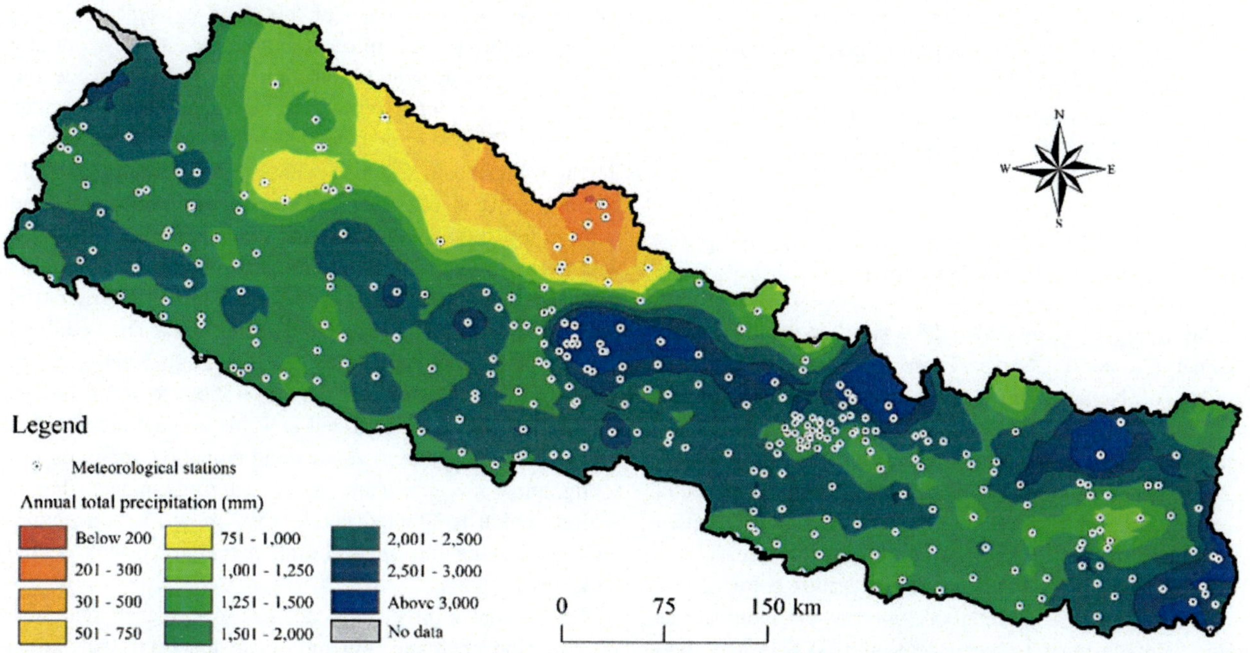

Fig. 3.4 Annual total precipitation distribution of Nepal between 1980 and 2000. The dotted symbols show the distribution of meteorological stations within the country. *Data source* Humanitarian Data Exchange (https://data.humdata.org/)

2019). The studies show that rainfall in Nepal has been decreasing during the past few decades (Paudel et al. 2020; Shrestha et al. 2000) but the flood disaster and event results show an increasing number of events and amount of property loss, as well as an increasing number of casualties (Pangali Sharma et al. 2019). This occurs due to erratic short-term heavy rainfall, which has occurred more frequently in Nepal in recent years (Devkota et al. 2017; Manandhar et al. 2011). During the three-day period from 22 to 25 September 2020, there was short-term heavy rainfall in Nepal, and some areas of south-central Nepal received more than 300 mm of rainfall (http://www.dhm.gov.np/). The tip of western Nepal also experiences very high rainfall and has been the site of several flood and landslides events. These kinds of short-term heavy rainfall, flood, and landslide events have largely impacted people's livelihoods, and many have lost their lives and property.

Further, because the Tarai region of Nepal is known as the floodplain area, the soils of this area are more fertile than those in the Hill and Mountain regions (Bajracharya and Sherchan 2009). Due to the erratic short-term heavy monsoon rainfall, the flood events occur more frequently during the monsoon season, and as a result the Siwalik, Middle Hill and High Mountain regions of Nepal experience high rates of soil erosion (Chalise et al. 2019). The eroded soil is washed away by floods and can be deposited in the Tarai region due to plain topography and lower velocity of the river, increasing the fertility of the topsoil in the Tarai regions compared to the other regions of the country.

3.4.4 Drought

Drought is one of the most complex, harmful, and least understood extreme climatic events. It is dependent on several variables, including precipitation, temperature, land surface parameters, altitude, and wind, among others. Furthermore, our understanding of drought varies depending upon different factors such as intensity, location, and sector (Khatiwada and Pandey 2019). It is well known that the major factor influencing drought evolution in Nepal is the lack of precipitation linked with summer monsoon (southeasterly) and wintertime (westerly) circulations. The amount of annual precipitation generally decreases from east to west (Shrestha 2000), and winter monthly precipitation amounts in western Nepal are on the order of 50 mm or less. This is brought about mainly by synoptic weather disturbances that are dynamically different from the monsoon season (Barlow et al. 2005).

Drought has a devastating impact on rural livelihoods that depend mainly on rain-fed subsistence agriculture. A small reduction in precipitation can have a serious impact on activities such as crop production and may disturb social harmony by creating water-use conflicts (Ghimire et al.

2010). Since the 1990s, droughts have resulted in food deficits, especially since 40 out of 75 districts experienced a food deficit in 2008/2009, leading to serious nutritional crises (MoAC et al. 2009). It is crucial to examine and understand the climate drivers that lie behind the development of drought in Nepal.

3.4.5 Frosts and Cold Waves

Besides drought, frosts and cold waves are another major seasonal disaster in Nepal. Frosts dominate in the high altitudes of Nepal whereas cold waves dominate in low altitudes. The country was ranked as one of the world's most vulnerable countries for a natural disaster by the World Bank in 2011 (MoHA and DPNet-Nepal 2015). Nepal is expected to experience an increase in temperature, more frequent heatwaves, and shorter frost durations in the future.

A dry winter and cool summer (temperature below 22 °C) dominate in the 10% of Nepal occupied by the mountainous region. The elevation between 4000 and 6000 m asl (about 19% of Nepal's area) experiences the polar type of climate, while frost and snow cover the remaining area (9%) above 6000 m asl creating permanent frost and cold desert conditions (Poudel et al. 2020; Sharma 2016).

During the winter, the low-lying districts become colder than the High Mountain region, which experiences lower temperatures for at least half the year. Cold waves that bring unexpected freezes and frosts can kill plants during the early and most vulnerable stages of growth, resulting in crop failure that directly impacts agricultural productivity (Malla 2008; Panday 2012). Such cold waves create vulnerabilities for people's livelihoods and cause many to lose their lives, especially in the southern low land (Tarai) region (Pradhan et al. 2019).

3.5 Agro-Ecozones and Soil Climate

The agro-ecozones of Nepal promote systematic farming and agricultural development. Agro-ecozones are conceptualized by grouping neighboring areas that share similar topography, altitude, climatic conditions, and soil types (Carson 1992; Gauchan and Yokoyama 1999). The concept adopted from Carson (1992), and from Gauchan and Yokoyama (1999) shows five agro-ecozones in Nepal (Table 3.1 and Fig. 3.5). Out of these five, the Tarai agro-ecozone is situated in the southern part of the country, and most of its landscape is flat with fertile soil. This agro-ecozone is conceptualized below 600 m asl and covers some areas of the Siwalik region of the country. The Tarai agro-ecozones are normally subcategorized as lower wetland, mid-wetland, upland/dry land, forest mixed land, and flood-prone land (Gauchan and Yokoyama 1999). The overall climate of this zone is tropical. Paddy, wheat, potato, legume/oilseed, maize, vegetable crops are cultivated in the irrigated lowland area; similarly, maize and mustard are the main crops cultivated in the upland area within the Tarai agro-ecozone (Rimal et al. 2018).

The river basin's agro-ecozones in Nepal are mainly situated between 600 and 1200 m asl with subtropical and with valley floor land types. Most of this zone is located within the Siwalik region and partly within the Middle Mountain region (Fig. 3.5). This zone's primary agricultural potential is subtropical fruits and food grains. The details of the Lower Hills (1200 to 2200 m asl), High Hills (2200 to 3500 m asl), and High Mountain (3500 to 4800 m asl) agro-ecozones are summarized in Table 3.1 and presented in Fig. 3.5. Further, due to its extreme climatic conditions and topography, the landscape of the High Himalayas/Himal region, which is situated above 4800 m asl, is conceptualized as having no agricultural potential (Gauchan and Yokoyama 1999).

Climate has a direct influence on soils. Warm conditions promote the chemical and biological reactions that parent materials into the soil. Similarly, the development of the soil profile and the soil's physical and chemical composition are greatly affected by weather and have a role in crop performance. Furthermore, differences in soil fertility are an essential driver of species turnover in productive environments (Paoli et al. 2006). Soil develops its ability to support the current ecosystems under changing climate—this will lead to changes in the communities of plants growing in different parts of the country.

Not only does climate influence soil, but also soil can influence climate; for example, soils that are wetter or denser hold heat and stabilize their surroundings during temperature changes more than looser, drier soils. In Nepal, local rice varieties are valued, especially in remote mountain areas adapted to diverse ecosystems, including cold stress, water scarcity, flood conditions, and poor soils (Sthapit et al. 2008). Many studies have used estimated realized climate niches to predict potential biogeographical changes in species distribution. This may be due to low temperatures in the temperate zone being directly linked to vital eco-physiological processes. In contrast, warm temperatures, especially in the Tarai region, are more indirectly related to moisture limitations, resistance to pathogens, and species competition (Vetaas 2002).

3.6 Climate Change and Its Impact

Nepal has observed a decreasing trend of precipitation and an increasing trend in temperature (Fig. 3.3) in recent decades (DHM 2017; Paudel et al. 2020), along with a rising trend of drought and flood events (Dahal et al. 2016; Pangali

Table 3.1 Agro-ecozones and farming promises in Nepal

Agro-ecozone	Elevation range, m	Climatic condition	Landform	Farming promise
Tarai	60 to 600	Tropical climate	Wetland, upper and mixed wetland, dry land fans, foot slope fans, and valley floor	Paddy, wheat, potato, legumes/oilseeds, maize, vegetables, and tropical fruits
River basins	600 to 1200	Sub-tropical climate	Valley floor	Foodgrains and subtropical fruits
Lower hills	1200 to 2200	Warm temperate climate	Hill side terraces and slopes	Citrus, temperate fruits, off-season vegetables
High hills	2200 to 3500	Cool temperate climate	Steep slopes and terraces	Temperate fruits and vegetables, seed potato
High mountain	3500 to 4800	Alpine climate	Steep/flat area	Livestock production

Sources Carson (1992) andGauchan and Yokoyama (1999)

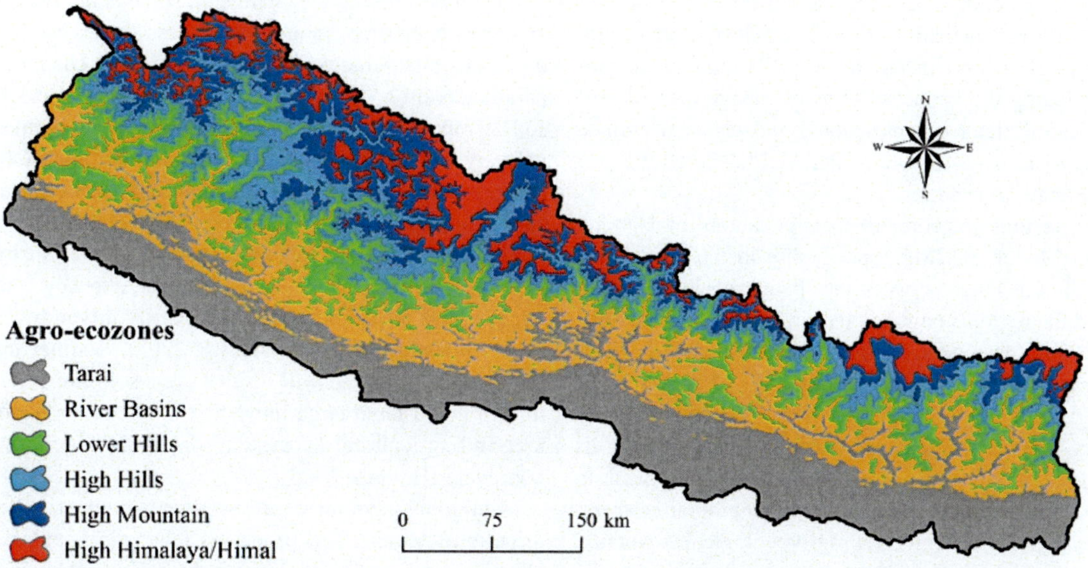

Fig. 3.5 Delineated agro-ecozones of Nepal, adopted from Carson (1992) and Gauchan and Yokoyama (1999)

Sharma et al. 2019). These indicate ongoing climate changes within the region.

Climate change has impacted several sectors in Nepal; however, the impact is most noticeably reflected in the agricultural system (Malla 2008). This impact is more significant on the vulnerable, marginalized communities residing at different ecological zones of the country, such as the communities in the lowland area (Tarai region) that have been the victims of frequent flash-flood events. Likewise, extended cold waves have also largely impacted the Tarai region (Pradhan et al. 2019). However, the Glacial Lake Outburst Flood (GLOF), avalanches, and landslides have more significantly impacted the Mountain and Hill regions of the country. Due to different scales of landslides in different years, especially during the monsoon season, a large number of people have lost their property and their lives (Sudmeier-Rieux et al. 2012). These are all associated with climate change impact driven primarily by anthropogenic activities. To mitigate these climate change impacts in Nepal, suitable adaptation strategies need to be developed based on local geography and conditions. Further, there is an urgent need to develop drought-resilient crops and improved irrigation systems to minimize the impacts of climate change on agriculture and people's livelihood (Paudel et al. 2020).

Climate is one of five factors with a strong influence on soil formation, primarily through the interplay of temperature, moisture, and various patterns of weathering and leaching. Lately, anthropogenic causes have established themselves as above natural variability in the climate system, quickening the rate of climate change over the past century (Crowley 2000). Given accelerated population growth, specifically in urban areas, excessive and indiscriminate use of natural resources has imposed undue pressure on the environment. The assessment report AR5 of the Intergovernmental Panel on Climate Change (IPCC) stated that long-term simulations for the twenty-first century show more frequent adverse weather conditions attributed to climate change (IPCC, 2013). Compared to the late twentieth century, the projected changes for the twenty-first century show a high probability of a decrease in the frequency and magnitude of cold days and nights, along with an increase in the frequency and magnitude of unusually warm days, heat waves, and precipitation (Seneviratne et al. 2012). However, the magnitude and extent of climate change's impact on soils are poorly understood, as soils themselves can be a source or sink of C, at times switching from C sink to C source with antagonistic effects on climate change (Bliss and Maursetter 2010). Predicting the impact of climate change on soil processes is also difficult because global and regional climate models are based on assumptions that yield uncertainties for different climate trajectories.

Another serious impediment in applications of General Circulation Models (GCMs), specifically in the Himalayan region, is the mismatch between the spatial resolution of the GCMs and the region's complex terrain, illustrating the need for high-resolution regional climate models that more realistically represent the region's topography (Panday et al. 2015). Traditionally, Nepal has been a predominantly subsistence-oriented agrarian economy. To a large extent, its wide variety of climatic zones, ranging from tropical to alpine/tundra is due to its large and steep altitudinal gradient, resulting in a high degree of agro-ecological diversity. Orographic forcing is the dominant factor affecting precipitation in the region (Lang and Barros 2004). Various studies using GCMS and regional climate models (RCMs) further indicate that the temperature in Nepal will increase by 2.5 to 5 °C by 2100 (Immerzeel et al. 2012; Nepal 2016).

In addition, more frequent and high-intensity rainfall events are likely to be more common in the future, leading to an increase in rainfall erosivity levels (Mondal et al. 2015). In the Himalayan region, the effects of climate change are very likely to cause a dramatic increase in the number of warm nights, a reduced number of frost-free days, an increase in the number of consecutive dry days, and frequent extreme precipitation (Panday et al. 2015). In addition, climate change is likely to alter the region's hydrological cycle. The cascading effects of altered hydrological phenomena include glacial retreat, inconsistent snow cover change, the bursting of glacial lakes, and flash floods (Kang et al. 2010). The local downstream impact from such climate-change-induced phenomena on people's livelihoods is profound yet largely unknown, mostly due to the lack of pertinent ground truth measurements in this region (Eriksson et al. 2009).

Soil erosion is a natural process of soil transport mainly due to various erosive forces (wind, water, ice, gravity, etc.). Natural soil erosion occurs everywhere—albeit at a slow rate in cases where there is little to no human intervention. Lately, anthropogenic pressure such as improper and unsustainable land use and deforestation have triggered an accelerated rate of soil erosion. This is especially critical to Nepal, as the terrain is characterized by a diversity of fertile but fragile physiographic landscapes. Past research in Nepal has shown different magnitudes of annual soil loss due to erosion in the landscape. Disturbance of soils and the removal of vegetation from landscapes with steep slopes combined with frequent and high-intensity rainfall will only further increase the severity of landslides and soil erosion.

The effects of climate change on soil can be immediate or extended over time, ranging from less than a year to hundreds of years. Change in climate can cause changes in soil parameters on a varying timescale. Soil parameters such as soil temperature, moisture, bulk density, permeability, porosity, and nitrate content fluctuate more quickly (< 1 yr) compared to pH, hydraulic conductivity, cation exchange capacity (CEC), organic matter content (10 to 100 yr) and mineral composition, texture, and particle-size distribution (> 100 yr) (Varallyay 1990). However, the key impact of climate change on soils will be primarily driven by changes to soil-moisture regimes (Bullock 2005). Warmer temperatures and less rainfall will result in less soil moisture, which in turn could have large unwanted implications for the natural and agricultural ecosystem, disrupting the normal carbon cycle (Bullock 2005).

Additionally, an increase in soil temperature will expedite the rate of various soil processes (e.g., nitrification, mineralization, and weathering) given the soil-moisture levels are ample. There is a lack of unanimity among the scientific community regarding the fate of soil organic matter (SOM). One study conducted in the mid-Himalayan region to predict short-, medium-, and long-term trends of soil organic carbon (SOC) dynamics using the CENTURY model (used for simulating carbon and nutrient dynamics for various ecosystems (e.g., grassland, shrubland, pasture, forest) (Parton et al. 1993) found that SOC for A2 and B2 scenarios could drop by up to 19.2% by 2099 (Gupta and Kumar 2017).

3.7 Conclusion

The climatic variation in Nepal within its various agro-ecozones is mostly due to vast topographic variation within a short aerial distance. The altitudinal gradient from the southern Tarai plain to the northern high Himalaya contributes to the six major climatic belts in Nepal, forming a tropical climate in the south to a tundra/nival/trans-Himalayan climate in the north. Nepal has four major climatic seasons: pre-monsoon, monsoon, post-monsoon, and winter. The monsoon season is marked by heavy rainfall, preceded by the high-temperature pre-monsoon season. The increasing trend of temperature, the decreasing trend of precipitation, and the increasing episodes of drought can be observed in the climate archives of the country. The changing climate has primarily impacted the agricultural system and people's livelihood. The magnitude of impact, however, varies by region. Further, the country has five major agro-ecozones with varying premises of farming and these zones are more suitable for formulating farming strategies and policy for the country's overall agricultural development. Moreover, the development of the soil profile and its physical and chemical composition are greatly affected by climate and play a critical role in crop performance.

References

Amatya SL (1976) Cash crop farming in Nepal. Geography instruction committee, institute of humanities and social sciences.

Aryal D (2018) Pre-Monsoon XE "Pre-Monsoon" thunderstorms in Nepal. Int J Rural Dev Environ Health Res 2(3):38–45

Aryal S, Maraseni TN, Cockfield G (2014) Sustainability of transhumance grazing systems under socio-economic threats in Langtang. Nepal. J Mt Sci 11(4):1023–1034

Bajracharya RM, Sherchan DP (2009) Fertility status and dynamics of soils in the Nepal Himalaya: a review and analysis. Soil Fertil 111–135

Barlow M, Wheeler M, Lyon B, Cullen H (2005) Modulation of daily precipitation XE "Precipitation" over Southwest Asia by the Madden–Julian oscillation. Mon Weather Rev 133(12):3579–3594

Barros AP, Lang TJ (2003) Monitoring the monsoon XE "Monsoon" in the Himalayas: observations in Central Nepal. Mon Weather Rev 131(7):1408–1427

Bhattarai S, Chaudhary RP, Quave CL, Taylor RSL (2010) The use of medicinal plants in the trans-himalayan arid zone of Mustang district. Nepal. J Ethnobiol Ethnomed 6(1):14

Bliss NB, Maursetter J (2010) Soil organic carbon XE "Carbon" stocks in Alaska estimated with spatial and pedon data. Soil Sci Soc Am J 74(2):565–579

Brewin PA, Buckton ST, Ormerod SJ (2000) The seasonal dynamics and persistence of stream macroinvertebrates in Nepal: do monsoon floods XE "Floods" represent disturbance? Freshw Biol 44(4):581–594

Bullock P (2005) Climate XE "Climate" change impacts. In: Hillel D (ed) Encyclopedia of Soils in the Environment. Elsevier, Oxford, pp 254–262

Carson B (1992) An agroecological zonation approach to agricultural planning in mountain environments. In: Jodha N, Banskota M, Partap T (ed) Sustainable Mountain Agriculture. Oxford & IBH Publishing Co. Ltd., New Delhi, India

CBS (2012) National population and housing census 2011 (National Report), National Planning Commission Secretariat, Central Bureau of Statistics, Government of Nepal

Chalise D, Kumar L, Kristiansen P (2019) Land degradation by soil erosion in Nepal: a review. Soil Syst 3(1):12

Chapagain PS (2016) Land and livelihood changes in upper Manang valley of Nepal Himalayas. Third Pole 14:1–17

Chapman L, Thornes JE, Bradley AV (2002) Modelling of road surface temperature from a geographical parameter database. Part 2: Numer Meteorol Appl 8(4):421–436

Collier E, Immerzeel WW (2015) High-resolution modeling of atmospheric dynamics in the Nepalese Himalaya. J Geophys Res Atmos 120(19):9882–9896

Crowley TJ (2000) Causes of climate XE "Climate" change over the past 1000 years. Sci 289(5477):270–277

Dahal P et al (2016) Drought XE "Drought" risk assessment in central Nepal: temporal and spatial analysis. Nat Hazards 80(3):1913–1932

Devkota LN (1999) Deciduous fruit production in Nepal. Deciduous fruit production in Asia and the Pacific. Bangkok, Thailand: FAO regional office for Asia and the Pacific

Devkota RP (2014) Climate XE "Climate" change: trends and people's perception in Nepal. J Environ Prot 5(4):255–265

Devkota RP et al (2017) Climate XE "Climate" change and adaptation strategies in Budhi Gandaki River Basin, Nepal: a perception-based analysis. Clim Change 140(2):195–208

DHM (2017) Observed climate trend analysis of Nepal (1971–2014). Department of Hydrology and Meteorology, Government of Nepal

Duncan JMA, Biggs EM, Dash J, Atkinson PM (2013) Spatio-temporal trends in precipitation XE "Precipitation" and their implications for water resources management in climate-sensitive Nepal. Appl Geogr 43:138–146

Eriksson M et al (2009) The changing Himalayas: impact of climate change XE "Climate change" on water resources and livelihoods in the greater Himalayas. International Centre for Integrated Mountain Development (ICIMOD), Kathmandu, p 24

Gauchan D, Yokoyama S (1999) Farming systems research in Nepal: current status and future agenda, 24. National Research Institute of Agricultural Economics, Ministry of Agriculture, Forestry, and Fisheries, Tokyo, Japan

Gentle P, Maraseni TN (2012) Climate XE "Climate" change, poverty and livelihoods: adaptation practices by rural mountain communities in Nepal. Environ Sci Policy 21:24–34

Gentle P, Thwaites R (2016) Transhumant Pastoralism in the context of socioeconomic and climate change in the mountains of Nepal. Mt Res Dev 36(2):173–182

Ghimire T, Thakur N (2013) Constraint and opportunity of raw jute production: a case study of eastern Tarai XE "Tarai" Nepal. Agron J Nepal 3:117–122

Ghimire YN, Shivakoti GP, Perret SR (2010) Household-level vulnerability to drought in hill agriculture of Nepal: implications for adaptation planning. Int J Sust Dev World 17(3):225–230

Gupta S, Kumar S (2017) Simulating climate change XE "Climate change" impact on soil carbon sequestration in agro-ecosystem of mid-Himalayan landscape using CENTURY model. Environ Earth Sci 76(11):394

Ichiyanagi K, Yamanaka MD, Muraji Y, Vaidya BK (2007) Precipitation XE "Precipitation" in Nepal between 1987 and 1996. Int J Climatol 27(13):1753–1762

Immerzeel WW, van Beek LPH, Konz M, Shrestha AB, Bierkens MFP (2012) Hydrological response to climate change XE "Climate change" in a glacierized catchment in the Himalayas. Clim Change 110(3):721–736

IPCC (2013) Summary for Policymakers. In: Stocker TF, Qin D, Plattner GK, Tignor M, Allen SK, Boschung J, Nauels A, Xia Y, Bex V, Midgley PM (eds.) Climate change 2013: the physical science basis. Contribution of Working Group I to the Fifth Assessment Report of the Intergovernmental Panel on Climate Change

Kang S et al (2010) Review of climate and cryospheric change in the Tibetan Plateau. Environ Res Lett 5(1):015101
Kansakar SR, Hannah DM, Gerrard J, Rees G (2004) Spatial pattern in the precipitation XE "Precipitation" regime of Nepal. Int J Climatol 24(13):1645–1659
Karki R, Hasson SU, Schickhoff U, Scholten T, Böhner J (2017) Rising precipitation extremes across Nepal. Climate 5(1):4
Karki R, Talchabhadel R, Aalto J, Baidya SK (2016) New climatic XE "climatic" classification of Nepal. Theoret Appl Climatol 125 (3):799–808
Karki R et al (2020) Rising mean and extreme near-surface air temperature across Nepal. Int J Climatol 40(4):2445–2463
Khanal NR et al (2020) Policy provisions for agricultural development in Nepal: a review. J Cleaner Prod 261:121241
Khatiwada KR, Pandey VP (2019) Characterization of hydro-meteorological drought in Nepal Himalaya: a case of Karnali River Basin. Weather Clim Extremes 26:100239
Kottek M, Grieser J, Beck C, Rudolf B, Rubel F (2006) World map of the Köppen-Geiger climate classification updated. Meteorol Z 15 (3):259–263
Lang TJ, Barros AP (2004) Winter storms in the Central Himalayas. J Meteorol Soc Jpn. Ser II, 82(3):829–844
LRMP (1986) Land utilization report, Land resource mapping project, Kenting earth science Canada and department of Topography, Government of Nepal, Kathmandu, Nepal
Malla G (2008) Climate change and its impact on Nepalese agriculture. J Agric Environ 9:62–71
Manandhar S, Pandey VP, Kazama F (2014) Assessing suitability of apple cultivation XE "cultivation" under climate change XE "Climate change" in mountainous regions of western Nepal. Reg Environ Change 14(2):743–756
Manandhar S, Vogt DS, Perret SR, Kazama F (2011) Adapting cropping systems to climate change XE "Climate change" in Nepal: a cross-regional study of farmers' perception and practices. Reg Environ Change 11(2):335–348
Merz J et al (2006) Rainfall-runoff events in a middle mountain catchment of Nepal. J Hydrol 331(3):446–458
MoAC WFP, FAO, (2009) 2008/09 Winter XE "Winter" Drought XE "Drought" in Nepal Crop and Food Security XE "Food security" Assessment, Ministry of Agriculture XE "Agriculture" and Cooperative. Nepal, Kathmandu
MoHA and DPNet-Nepal (2015) Nepal disaster report 2015, Government of Nepal, Ministry of Home Affairs (MoHA) and Disaster Preparedness Network-Nepal (DPNet-Nepal), Kathmandu, Nepal
Mondal A et al (2015) Impact of climate XE "Climate" change on future soil erosion in different slope, land use, and soil-type conditions in a part of the Narmada River basin. India. J Hydrol Eng 20(6):C5014003
Nayava J (1974) Heavy monsoon rainfall in Nepal. Weather 29 (12):443–450
Nayava JL (1975) Climates of Nepal. Himalayan Rev 7:14–20
Nepal P, Khanal NR, Zhang Y, Paudel B, Liu L (2020) Land use policies in Nepal: an overview. Land Degrad Dev
Nepal S (2016) Impacts of climate change XE "Climate change" on the hydrological regime of the Koshi river basin in the Himalayan region. J Hydro-Environment Res 10:76–89
Neupane PR, Maraseni TN, Köhl M (2017) The sugarcane industry in Nepal: opportunities and challenges. Environ Dev 24:86–98
Panday D (2012) Adapting climate change in agriculture: the sustainable way in Nepalese context. Hydro Nepal: J Water Energ Environ 91–94
Panday PK, Thibeault J, Frey KE (2015) Changing temperature and precipitation XE "Precipitation" extremes in the Hindu Kush-Himalayan region: an analysis of CMIP3 and CMIP5 simulations and projections. Int J Climatol 35(10):3058–3077
Pangali Sharma TP et al (2019) Review of flood disaster studies in Nepal: a remote sensing perspective. Int J Disaster Risk Reduction 34:18–27
Paoli GD, Curran LM, Zak DR (2006) Soil nutrients and beta diversity in the Bornean Dipterocarpaceae: evidence for niche partitioning by tropical rain forest trees. J Ecol 94(1):157–170
Parton WJ et al (1993) Observations and modeling of biomass and soil organic matter dynamics for the grassland XE "Grassland" biome worldwide. Global Biogeochem Cycles 7(4):785–809
Paudel B et al (2020) Farmers' understanding of climate change XE "Climate change" in Nepal Himalayas: important determinants and implications for developing adaptation strategies. Clim Change 158 (3):485–502
Pokharel C (2019) Agricultural diversification in Nepal. In: Thapa G, Kumar A, Joshi PK (eds) Agricultural transformation in Nepal: trends, prospects, and policy options. Springer Singapore, Singapore, pp 323–384
Poudel A, Cuo L, Ding J, Gyawali AR (2020) Spatio-temporal variability of the annual and monthly extreme temperature indices in Nepal. Int J Climatol 1–22
Pradhan B, Sharma P, Pradhan PK (2019) Impact of cold wave on vulnerable people of Tarai region, Nepal, climate change and global warming. IntechOpen
Pradhan B et al (2013) Assessing climate XE "Climate" change and heat stress responses in the Tarai XE "Tarai" region of Nepal. Ind Health 51(1):101–112
Rimal B, Zhang L, Rijal S (2018) Crop cycles and crop land classification in Nepal using MODIS NDVI. Remote Sens Earth Syst Sci 1(1):14–28
Seneviratne SI et al (2012) Changes in climate XE "Climate" extremes and their impacts on the natural physical environment. In: Field CB, Dahe Q, Stocker TF, Barros V (eds) Managing the risks of extreme events and disasters to advance climate change adaptation: special report of the intergovernmental panel on climate change. Cambridge University Press, Cambridge, pp 109–230
Sharma K (2001) Crop diversification in Nepal. Crop Diversification Asia Pac Reg 81
Sharma KP (2016) Climatic extremities in Nepal. Environ Stat 41
Shrestha AB, Aryal R (2011) Climate XE "Climate" change in Nepal and its impact on Himalayan glaciers. Reg Environ Change 11 (1):65–77
Shrestha AB, Wake CP, Dibb JE, Mayewski PA (2000) Precipitation XE "Precipitation" fluctuations in the Nepal Himalaya XE "Nepal Himalaya" and its vicinity and relationship with some large scale climatological parameters. Int J Climatol 20(3):317–327
Shrestha AB, Wake CP, Mayewski PA, Dibb JE (1999) Maximum temperature trends in the Himalaya and its vicinity: an analysis based on temperature records from Nepal for the period 1971–94. J Clim 12(9):2775–2786
Shrestha ML (2000) Interannual variation of summer monsoon rainfall over Nepal and its relation to Southern Oscillation Index. Meteorol Atmos Phys 75(1):21–28
Sindelar M (2015) Soils and climate XE "Climate". Soil science society of America, USA, pp 1–2
Sthapit B, Rana R, Eyzaguirre P, Jarvis D (2008) The value of plant genetic diversity to resource-poor farmers in Nepal and Vietnam. Int J Agric Sustain 6(2):148–166
Sudmeier-Rieux K, Gaillard J-C, Sharma S, Dubois J, Jaboyedoff M (2012) Chapter 7 floods XE "Floods", landslides, and adapting to climate XE "Climate" change in Nepal: what role for climate change models? In: Armando L, Ilan K (eds) Climate change modeling for local adaptation in the Hindu Kush-Himalayan region. Emerald Group Publishing Limited, Community, Environment and Disaster Risk Management, pp 119–140

Tabony R (1985) The variation of surface temperature with altitude. Meteorol Mag 114(1351):37–48

Talchabhadel R, Karki R, Thapa BR, Maharjan M, Parajuli B (2018) Spatio-temporal variability of extreme precipitation XE "Precipitation" in Nepal. Int J Climatol 38(11):4296–4313

Varallyay GY (1990) Chapter 4 influence of climatic change on soil moisture regime, texture, structure and erosion. In: Scharpenseel HW, Schomaker M, Ayoub A (eds) Developments in soil science. Elsevier, pp 39–49

Vetaas OR (2002) Realized and potential climate niches: a comparison of four Rhododendron tree species. J Biogeogr 29(4):545–554

Wu H et al (2019) Spatio-temporal analysis of drought XE "Drought" variability using CWSI in the Koshi River Basin (KRB). Int J Environ Res Public Health 16(17):3100

Geology and Physiography

4

Basanta Raj Adhikari and Roshan Babu Ojha

Abstract

The Nepal Himalayas lies in the central part of the Himalayan arc. The range is a product of the collision between Indian and Eurasian plates in the Miocene period (50 M yr) and is bounded by Indus Tsangpo Suture Zone (ITSZ) to the north and the Ganga basin to the south. The coupling effect of earthquake reoccurrences and strong Asian monsoon has resulted in a steep slope, rugged mountains, deep valleys, intermontane basins, and flat land. Geologically it is divided into five tectonic zones separated by major tectonic discontinuities running parallel to the Himalayan chain. These tectonic zones from south to north are the Indo-Gangetic Plain, the Siwaliks, the Lesser Himalayas, the Higher Himalayas, and the Tibetan-Tethys zone. These zones are formed with different types of rocks including sedimentary, igneous, and metamorphic distributed throughout the Nepal Himalayas, which explains the depositional environmental and evolutionary history of the whole Himalayas range. In addition, the Nepal Himalayas is divided into five physiographic regions based on climatic and geomorphic conditions (Tarai, Siwaliks, Hills, Middle Mountain, and High Mountain). These physiographic regions are attributed to different types of soil, climate, and land use and are mostly characterized based on elevation. The variation on these physiographic zones depends on the micro-climate controlled by peaks and valleys. This chapter discusses the geologic and physiographic classification of the Nepal Himalayas in brief.

Keywords

Climate • Indo-Gangatic Plain • Nepal Himalaya • Physiography • Tectonics

4.1 Evolution of the Nepal Himalayas

The 2400 km Himalayan arc extends from Nanga Parbat (8138 m) in the west to Namche Barwa (7756 m) in the east with a 230 to 350 km width (LeFort 1996). The Himalayan orogenic belt is a product of the collision between the Indian and the Eurasian plates during the late Eocene to Oligocene period (Valdiya 1984). The Indian plate is subducting with different velocities from $\sim$ 0.2 to 28 mm yr^{-1} beneath the Eurasian plate and mountains still being formed (Jade et al. 2017). The collision resulted in numerous tectonic faults and highly deformed rocks that are responsible for triggering numerous earthquakes of varying scales, making the Himalayas a seismically active zone (Bilham et al. 2001; Kobayashi et al. 2015; Kubo et al. 2016; Mencin et al. 2016; Sapkota et al. 2013). Generally, two main geodynamic processes have controlled the evolution of the Himalayas: 1) accretion and subduction in the Trans-Himalayan and Karakoram Mountains along the Shyok Suture Zone (SSZ) and Indus Tsangpo Zone (ITSZ), and 2) continental collision in the Himalayas (Jain et al. 2012). However, Valdiya (1984) elaborated on this by dividing the evolution process into four stages:

(1) Karakoram phase: Convergence takes place between two continental plates between 145 to 55 M (Cretaceous to Paleocene).
(2) Malla Johar phase: Collision and subduction of the Indian plate beneath the Eurasian plate occur between 55 and 23.8 M yr (Late Eocene to Oligocene).

B. R. Adhikari (✉)
Institute for Disaster Management and Reconstruction, Sichuan University-The Hong-Kong University, Sichuan, China
e-mail: bradhikari@ioe.edu.np

B. R. Adhikari
Department of Civil Engineering, Pulchowk Campus, Institute of Engineering, Tribhuvan University, Kathmandu, Nepal

R. B. Ojha
Nepal Agricultural Research Council, Kathmandu, Nepal

R. B. Ojha and D. Panday (eds.), *The Soils of Nepal*, World Soils Book Series,
https://doi.org/10.1007/978-3-030-80999-7_4

(3) Sirmurian phase: Himalayan upheaval with the development of main tectonic features occurs between 23.8 to 5.3 M yr (Miocene to Pontian).
(4) Siwalik phase: The formation of Siwalik in the frontal part of the Himalayas occurs between 5.3 M to 0.01 M yr (late Pliocene to middle Pleistocene).

This process led to the development of different tectonic slices along the principle thrusts, which are stacked one on top of the other, propagating southward and building the architecture of the Himalayas (Molnar 1984). Politically, this tectonic mountain chain consists of several sections of Pakistan, India, Nepal, Bhutan, and China, whereas the region is longitudinally divided into five sections from west to east (Gansser 1964) (Fig. 4.1):

(a) Punjab Himalayas: This section consists of a 550 km long section of the Himalayan chain bounded by the Indus River to the west and the Sutlej River to the east. It includes the Kashmir and Spiti regions.
(b) Kumaon Himalayas: This 320 km long section lies between the Sutlej River to the west and the Mahakali River to the east. It includes the Garhwal Himalayas and parts of southern Tibet.
(c) Nepal Himalayas: This is the central part of the Himalayas consisting of 800 km. It is bounded by the Mahakali River to the west and the Mechi River to the east, covering the entire length of Nepal. This section includes the world's tallest peaks, such as Everest, Kanchanjanga, Lotse, and Annapurna.
(d) Sikkim-Bhutan Himalayas: This section is bounded by the Mechi River to the west and Bhutan to the east and covers Sikkim and Bhutan (400 km).
(e) North-East Frontier Agency (NEFA) Himalayas: This is the easternmost part of the Himalayas, with a length of 400 km extending from the eastern boundary of Bhutan to the Tsangpo-Brahmaputra cross gorges.

The Himalayas is bounded by the Indus Tsangpo suture zone (ITSZ) to the north and the Ganga foreland basin to the south. The northern boundary of the ITSZ is generally exposed by a topographic depression along the Indus and Tsangpo rivers, which flow in opposite directions (Heim and Gansser 1939). Geologically, the Himalayan chain is broadly

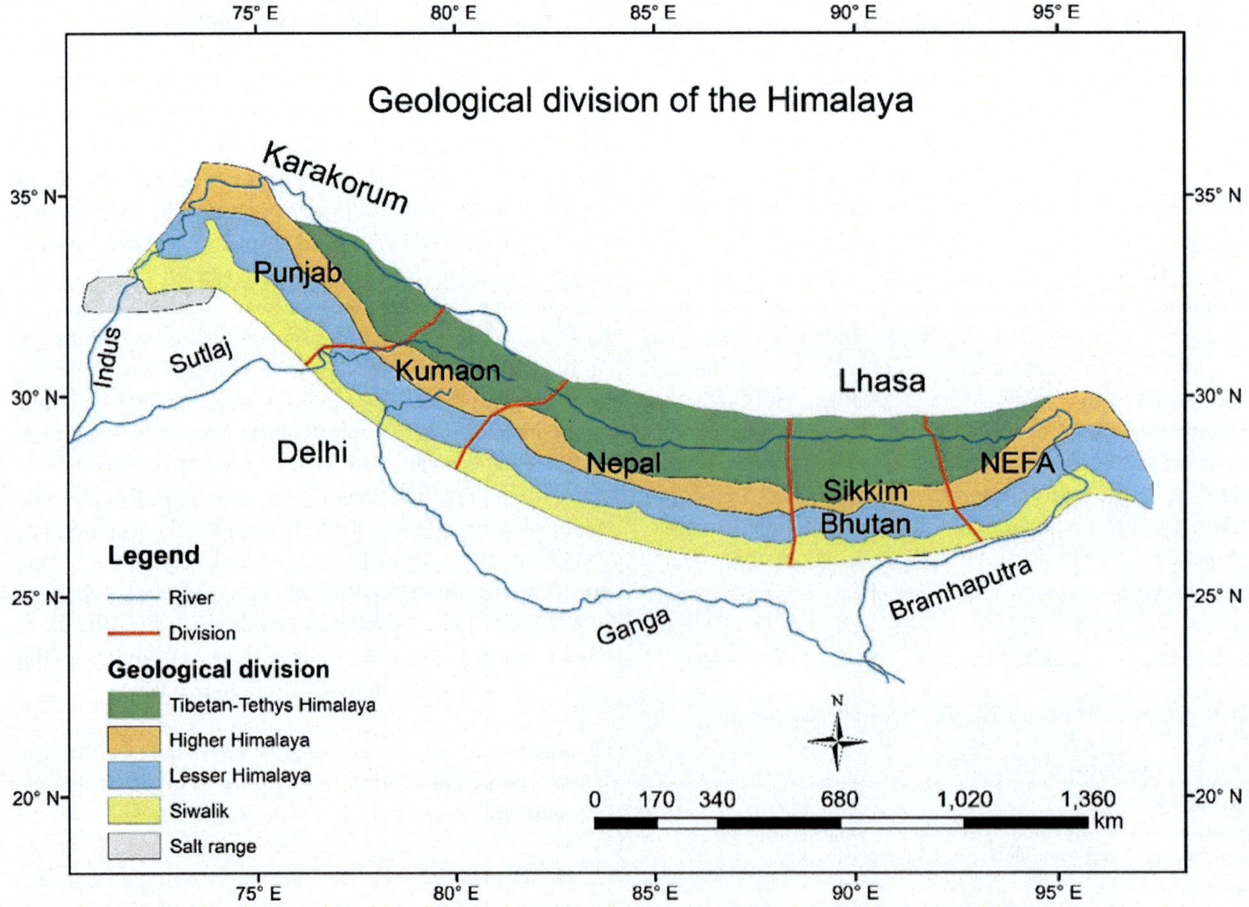

Fig. 4.1 Geological longitudinal subdivision of the Himalayas from west to east (Redrawn after Gansser 1964)

divided into five tectonic units, namely the Indo-Gangetic plain, the Siwalik, the Lesser Himalayas, the Higher Himalayas, and the Tibetan-Tethys Himalayas (Fig. 4.1).

4.2 Geology of the Nepal Himalayas

The Nepal Himalayas lies in the central part of the Himalayan arc covering one-third of the entire Himalayas. The Nepal Himalayas, bounded by latitudes 26°2' and 30°27' N and longitudes 80°11' E and 88°27' E, can be divided into five tectonic zones from south to north based on geological evolution (Table 4.1 and Fig. 4.2). These tectonic joints are separated by faults and thrusts and characterized by their tectonics, structure, and evolutionary history, which are briefly described below.

4.2.1 Indo-Gangetic Plain

The Indo-Gangetic plain lies in the frontal part of the Himalayas with an elevation ranging from 100 to 200 m asl deposited by the tributaries of the Ganga River. This tectonic zone was created beginning in the Pleistocene and continues through recent alluvial deposits (Fig. 4.3a) of approximately 1500 m thickness that already show a significant proportion of stress accumulation and form thrust and thrust-propagated folds beneath the sediments (Bashyal 1998; Mugnier et al. 1999; Upreti 1999). This tectonic zone is separated by the Main Frontal Thrust (MFT) from the Siwaliks to the north. Sedimentation began with the shallow marine environment before changing to estuarine-deltaic and finally to fluvial (Pant and Sharma 1993).

4.2.2 Siwaliks

The Siwaliks is also referred to as the sub-Himalaya and is comprised of the southernmost tectonic zone of the Nepal Himalayas bounded by the MFT to the south and the Main Boundary Thrust (MBT) to the north (Fig. 4.2). Many researchers (Auden 1935; Dhital 1995, 2015; Nakayama and Ulak 1999; Sah et al. 1994; Sharma 1977; Tokuoka et al. 1988) have established and presented different classifications for the Siwaliks sequence. Due to the lack of fossil records in the Nepal Himalayas, these classifications can be grouped and correlated into three basic units based on lithostratigraphy. These units are Lower, Middle, and Upper Siwaliks, and age ranges from the Middle Miocene to the Pleistocene (Gautam and Rösler 1999). In general, the entire Siwalik represents a coarsening upward sequence whereas individual units have fining upward sequence. The thickness of the Siwaliks changes significantly from western to eastern Nepal, with thick sequences in the Karnali section and decreasing as one moves east. The Lower Siwalik is composed of the alternation of fine-grained variegated mudstone and siltstone with some layers of fine sandstone. The domination of sandstone increases in the Middle Siwalik with increasing grain size represented by a wide concentration of black-colored biotite and light-colored quartz and feldspar minerals. The Upper Siwalik is well exposed at the foot of the Lesser Himalayas and consists of coarse-grained sediments with few siltstone and

Table 4.1 Geological subdivision and their key features of Nepal Himalayas

SN	Geological sub-division	Major rocks and minerals	Key geologic features	Age
1	Indo-Gangetic plain	Alluvial deposits consisting of gravel, sand, silt, and clay originating from the Himalayas	Alluvial deposits with thrust and thrust-propagated folds beneath the sediment. Shallow marine environment changed to estuarine-deltaic and now to fluvial earth materials	Pleistocene to recent
2	Siwaliks	Mudstone, Siltstone, Sandstone, Conglomerate	Anticline and syncline folds, fault/thrust	Middle Miocene to Pleistocene
3	Lesser Himalayas	quartzite, limestone, slate, siltstone, schist, gneiss, marble, amphibolite, Nepheline syenite, granites	Nappe/Clippe, Tectonic window, Mabharat Synclinorium, Fold/Thrust	Precambrian to Eocene
4	Higher Himalayas	Schist, Gneiss, Marble, Leucogranite	Folds/Thrust, High-grade metamorphism	Proterozoic to Miocene
5	Tibetan-Tethys zone	Sandstone, Limetone, Quartzite, shale	Folds/Thrust, Fossiliferous layers	Cambrian to Eocene

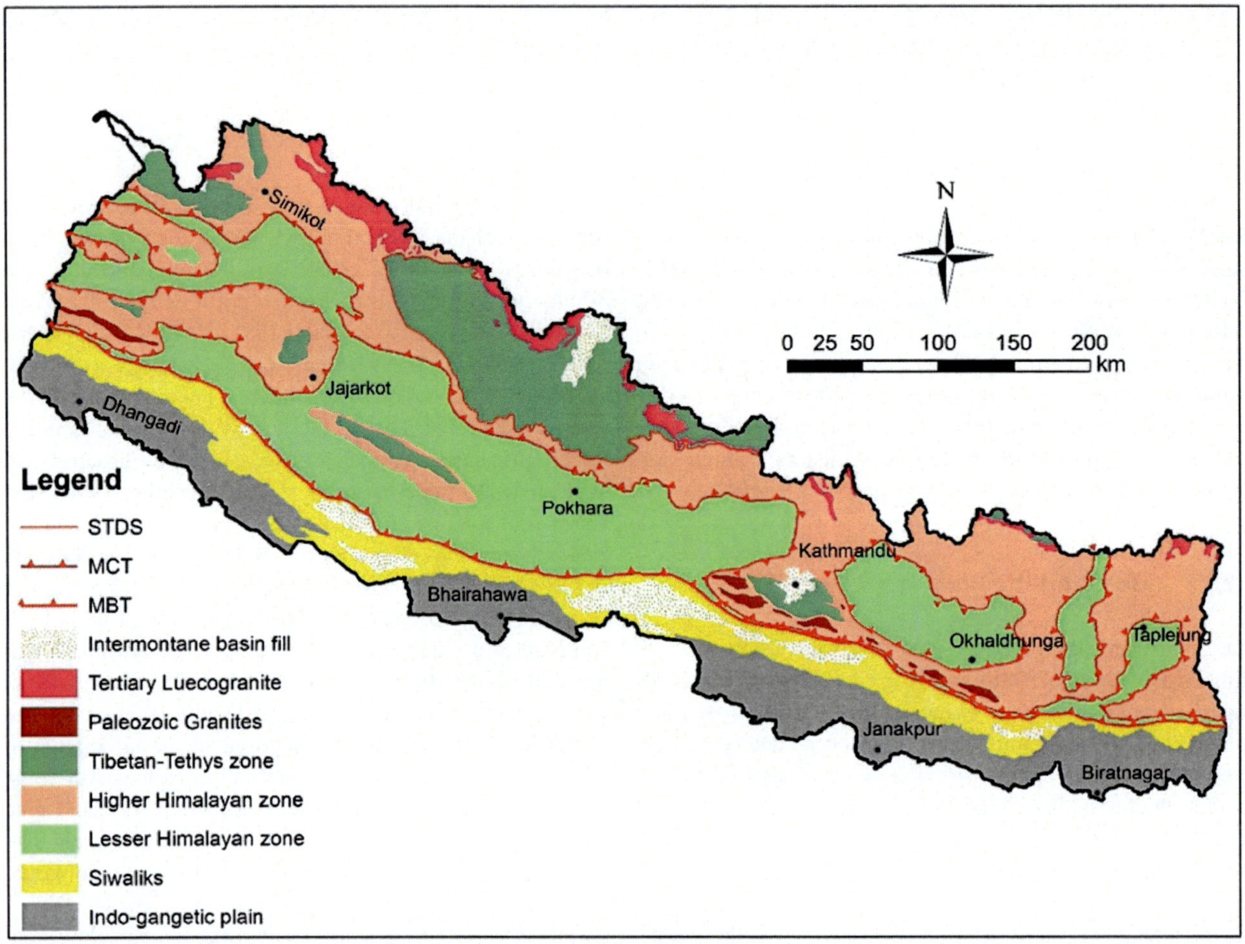

Fig. 4.2 Geological subdivision of the Nepal Himalayas modified from (Amatya and Jnawali 1994). STDS: South Tibetan Detachment System. MCT: Main Central Thrust. MBT: Main Boundary Thrust

mudstone layers (Fig. 4.3b). These sediments were deposited in different depositional environments including meandering, braided, and debris flow (Nakayama and Ulak 1999). The rocks of the Siwaliks are folded and faulted in different directions, resulting in different geological structures.

4.2.3 Lesser Himalayas

The Lesser Himalayas is bounded by the MBT to the south and the Main Central Thrust (MCT) to the north. This sequence consists of sedimentary rocks, low-grade metamorphic rocks, and granitic intrusion distributed throughout the Nepal Himalayas (Fig. 4.3c). This zone is composed of quartzite, limestone, slate, siltstone, schist, gneiss, marble, amphibolite, Nepheline syenite, and granite, ranging in age from the Precambrian to Eocene (Bordet et al. 1961; Hagen 1969; Kohn et al. 2010; Parrish and Hodges 1996), including the oldest Lesser Himalayan Augen gneiss (1.8 B) in the Kuncha Formation (DeCelles et al. 2000; Le Fort and Rai 1999). There is a clear depositional gap between the Proterozoic rocks and the overlying Paleogene beds, indicting a prolonged history of denudation (Dhital 2015). This sequence has a complex pattern of nappes, klippen, and windows due to thrusting, erosion, and folding (Fig. 4.3). The Okhaldhunga, Arun, and Taplejung windows are well exposed in eastern Nepal, where they are surrounded by the Higher Himalayan thrust sheet (Brunel and adnrieux 1977; Dhital 2015). While it is difficult to classify more than a 15 km pile of the Lesser Himalayas, Valdiya (1964) classified araneceous, calcareous and pelitic units that were later applied throughout the Nepal Himalayas (Sharma 1973).

4.2.4 Higher Himalayas

About 10 to 12 km thick crystalline sequences of the Higher Himalayas consist of high-grade metamorphic rocks. This

Fig. 4.3 Different rock types of the different geological zones of the Nepal Himalayas. **a** Sand and gravel in the Indo-Gangetic Plain at the Karnali River, Kailali; **b** Contact between gravel and sandstone beds in the upper Siwalik, Puntura River, Dadeldhura; **c** Highly fractured quartzite beds of the Lesser Himalayas, Kaligandaki River, Myagdi; **d** Garnetiferous schist of the Higher Himalayas, Kaligangaki Valley, Myagdi; **e** Fold and fault of Tibetan sedimentary rocks, Kaligandaki River, Mustang. *Photos credit* Basant Raj Adhikari

zone is confined by the South Tibetan Detachment System (STDS) with the Tibetan-Tethys sedimentary sequence and the MCT to the south marking the border with the Lesser Himalayas. The first appearance of the MCT is marked by the garnet minerals in the schists, which record the temperature and pressure of the thrust movement during the evolution of the Himalayas (Fig. 4.3d). It consists of Tertiary Leucogranites in the upper portions, whereas high-grade crystalline rocks including various kinds of gneisses, schists, and migmatites extend continuously along the entire length of the Nepal Himalayas (Fig. 4.2). Le Fort (1981) classified these leucogranites into two types of granites: older granites from 500 to 1800 Ma and younger leucogranites from 15 to 20 Ma. Detailed classification of the Higher Himalayan crystalline is extremely difficult because it is located in high altitudes in remote locations. Le Fort (1975) studied the Kali-Gangaki section in detail and classified the Higher Himalayas into three formations: Formation I, II, and III (from bottom to top). Formation I is the lower unit consisting of kyanite-garnet-two mica banded gneiss of politic to arenaceous origin. Quartzite beds are at the base of the Formation II, followed by the alternation of pyroxene and amphibole-bearing calc-gneisses and marbles. Formation III is composed of coarse-grained augen gneiss.

4.2.5 Tibetan-Tethys Zone

The Tibetan-Tethys zone is the northernmost tectonic zone of the Nepal Himalayas, extending to the Eurasian plate and the Indus Tsangpo Suture Zone (ITSZ). It consists of sandstone, limestone, quartzite, and shale with fossiliferous layers ranging in age from the Cambrian to Eocene (Dhital 2015). These rocks are strongly folded and faulted in different places (Fig. 4.3e). A pioneering expedition lead by Tony Hagen (Hagen 1959, 1968) first described the Tethys succession in the Thakkhola region followed by Gerhard Fuchs (Fuchs 1964, 1977) in the Thakkhola and Dolpo sections. Bodenhausen et al. (1964) established the basic stratigraphy of the Tethys zone in the Kali-Gandaki section in their expedition in 1962. This succession is also comparable with the sedimentary units exposed to the west in Kumaon, Spiti, and northwestern Himalayas (Gaetani et al. 1986; Gaetani and Garzanti 1991). This zone is highly fossiliferous, consisting of bicalcarenites, pelletal mudstone/wackestone, oolitic grainstone, bivalves, gastropods, brachiopods, crinoids, belemnites, and rare fish fragments (Bassoullet et al. 1986; Bodenhausen et al. 1964) in the Kali-Gandaki section.

4.3 Physiographic Subdivision of the Nepal Himalayas

The Nepal Himalayas extends 800 km from east to west. Nepal occupies 147, 516 km^2 and consists of more than 80% mountains and hills. The altitude varies from 64 m asl (Kechanakal, Jhapa) to 8888.86 m asl (the top of the world, Mount Everest) within an aerial distance of about 150 km (Dhital 2015). The Nepal Himalayas has a steep slope, rugged mountains, deep valleys, intermontane basins, and flat land. It is divided into five major physiographic regions: Tarai (below 200 m asl), Siwalik (100 to 2000 m asl), Hill (200 to 3500 m asl), Middle Mountain (700 to 4100 m asl), and High Mountain (1800 to 8888 m asl) on the basis of elevation and climate (Fig. 4.4) (LRMP 1986).

4.3.1 Tarai

This is the southernmost region of the Nepal Himalayas and the northern edge of the Indo-Gangetic plain lying at an altitude of 64 to 200 m. It consists of Pleistocene to Holocene sediments generated from the Himalayas (Fig. 4.5a). These sediments are deposited by major river systems in the Nepal Himalayas, with the sediment size varying from north to south, with large boulders present in the northern side of the Tarai and the grain size decreasing gradually down to clay on the southern side.

The Tarai region is generally divided into three zones: the Bhabar zone, the Middle Tarai and the Lower Tarai. The Bhabar zone lies in the foothills of the Himalayas (Siwaliks) and consists of poorly sorted boulders, cobbles, sand, and silt derived from the Siwaliks and older rocks. The Middle Tarai has a gentle slope (1 to 5%) and consists of mostly sandy and silty soils formed by braided alluvial deposits. The Lower Tarai is the southernmost part of the Tarai, which consists of pebbles, fine silt, and clays. The climate is subtropical and rainfall occurs mostly in June and July (monsoon season). The winter temperature is not severe and the minimum recorded temperature is around freezing. Erosion is generally low, though gulley erosion can be seen in some areas.

Some elongated valleys lie between the Mahabharat and Siwalik ranges representing Inner Tarai. These valleys, namely Udayapur, Chitwan, Dang, Deukhuri, and Surkhet, lie from east to west, and were formed due to continuous erosion from rivers originating from the Mahabharat Lekh and Siwalik.

4.3.2 Siwaliks

The Siwaliks lies in the frontal part of the Himalayas and is locally known as the Churia Hills in Nepal. It extends from east to west parallel to the Himalayan arc, and many dun valleys lie between the Siwaliks and the Hills. The altitude varies from 100 to 2000 m. This region consists of sedimentary rocks and generally has a fragile landscape (Fig. 4.5b). Rainwater significantly affects the soft and loose soil so that debris flow hazards are common in this region. The region is about 20 km wide near the Mahakali River and less than 1 km wide near the Mechi River.

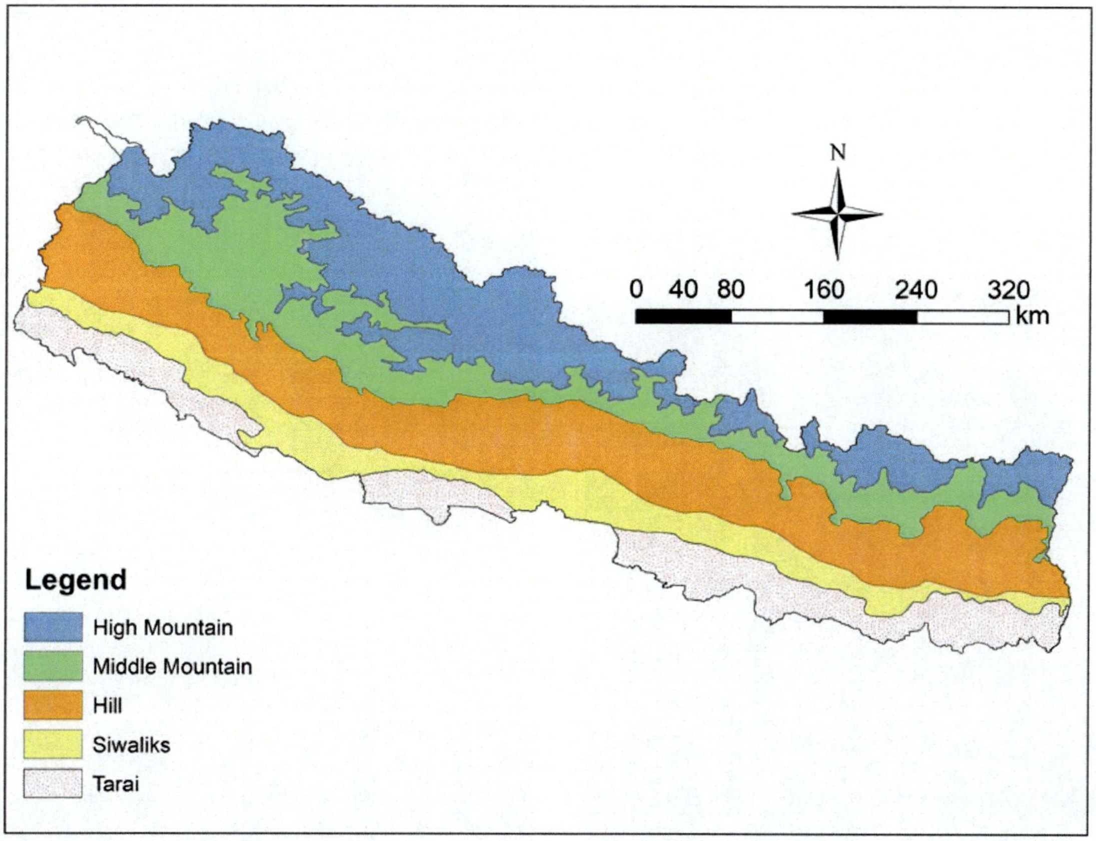

Fig. 4.4 Physiographic map of Nepal. Modified after LRMP (1986)

4.3.3 Hills

The Hills region is located south of the Middle Mountain region and north of the Siwaliks; however, there is no clear boundary between the other two regions. The altitude varies from 200 to 3500 m and extends east to west parallel to the Himalayas. This region consists mostly of agricultural land because of high weathering and erosion rates coupled with rainfall. The extensive erosion in some places has exposed mostly underlying rocks of metamorphic origin. The Hill region consists of Lesser and Higher Himalayan rocks, though some hills are covered by forest, bushes, and grasslands, with some landslides (Fig. 4.5c). This region is dissected by different major river systems in the Nepal Himalayas, resulting in different intermontane valleys (i.e., Pokhara, Kathmandu, and Dhankuta).

4.3.4 Middle Mountain

This region consists of a steep and fragile landscape ranging in altitude from 700 to 4100 m asl (Fig. 4.5d). This region is dissected by two major antecedent rivers (Koshi and Karnali) and is comprised of mostly Higher Himalayan rocks covered by snow in the winter. People have been practicing agriculture on the gentle slopes because the micro-climate is suitable for wheat, maize, millet, and other cash crops. Generally, the climate is warm-temperate, though the micro-climate varies from east to west due to different geomorphology and land use.

4.3.5 High Mountain

The High Mountain region lies in the northern part of the Nepal Himalayas. The altitude ranges from 1800 to 8848.86 m asl and consists of steep slopes with narrow valleys. These high peaks are covered by snow year-round, though some steep slopes at low altitudes are forested (Fig. 4.5e) (Table 4.2). Most of .the valleys at high altitudes consist of glaciers from which all the perennial rivers originate. This region is made up mainly of metamorphic rocks and some sedimentary rocks of the Tibetan-Tethys zone. This region acts as a barrier for Indian monsoons and is responsible for high amounts of rainfall in the southern slope but prevents air moisture from entering the Tibetan Plateau.

Fig. 4.5 Physiographic landforms of different regions. **a** Flat alluvial land of Tarai, Ratu River, Mahottarai; **b** Fragile landscape of the Siwaliks, Kailali; **c** Panoramic view of Hills in the Nepal Himalayas (looking south from the Dolkha); **d** Steep mountains in the Middle Mountain region, Kalikot; **e** Snowcapped mountains in the High Himalayas region, Pokhara. *Photos credit* Basant Raj Adhikari

The climate is cool temperate below zero for more than six months at a time. Agricultural production is mostly concentrated in the valley bottom. The Trans-Himalayan valleys such as the Lo-Manthang, Manang, and Dolpo lie in this region. Many large glaciers and glacial lakes exist in this region and some are potentially in danger of bursting (Veh et al. 2018).

4.4 Conclusion

The evolution of the Nepal Himalayas began around 55 Ma ago and plays an important role in the development of different tectonic discontinuities, which separated different tectonic subdivisions of the Nepal Himalayas. These tectonic

Table 4.2 Physiographic zones of Nepal and their key features (LRMP, 1986)

SN	Physiographic zones	Elevation (m asl)	Area/coverage km^2	Geological/climate/notable feature
1	Tarai	64 to 200	20,220 (14%)	Indo-Gangetic plain, alluvial deposits, subtropical, fertile land
2	Siwaliks	100 to 2000	18,782 (13%)	Siwalik, sedimentary rocks, dissected by rivers, fragile landscape, concentration of settlements
3	Hills	200 to 3500	430,792 (29%)	Lesser and Higher Himalayas, sedimentary, igneous and metamorphic rocks, large-scale folds/thrusts, subtropical
4	Middle Mountains	700 to 4100	29,804 (20%)	Higher Himalayas, steep and fragile landscape, mostly metamorphic rocks, subtropical
5	High Himalayas	1800 to 8848	35,283 (24%)	Higher Himalayas and Tibetan-Tethys Himalayas, metamorphic and sedimentary rocks, steep valleys

zones have different rock types and evolutionary histories. The grade of metamorphism increases from frontal land to hinterland. These tectonic zones coupled with climate and altitude have strong significance on the physiographic division of the Nepal Himalayas. These physiographic divisions have different micro-climates, which control local land use patterns and people's livelihoods.

4.5 Glossary

Ganga Basin: A foreland basin of the Himalayas formed by the Ganga River. It extends from the border between the Ganga and Indus basins and the Aravali mountains to the west, the Brahmaputra basin to the east, and the Vindhyas and Chota Nagpur plateau to the south.

Himalayan arc: The entirety of the Himalayas is 2400 km long and is arc shaped. It extends from the Naga-Parbat (west) syntaxis to the Namcha-Barwa (east) syntaxis.

Indian plate: The Indian plate is one of the major tectonic plates in the earth's crust bordered by the Australian plate, the Arabian plate, and the Eurasian plate. The Indian plate includes most of South Asia and some parts of the Indian Ocean.

Eurasian plate: The Eurasain plate consists of most of Europe and Asia. It shares a boundary with the Indian Plate, the Arabian Plate, and the American Plate.

Lithostratigraphy: An element of stratigraphy that deals with the description and nomenclature of rock based on its lithology and stratigraphic relations. The lithographic classification is based on lithological properties and relations.

Indus Tsango Suture zone (ITSZ): This suture zone is the northern margin of the Himalayas that separates the Tibetan-Tethys zone from the Eurasian Plate. It is the result of a collision between the Indian Plate and the Eurasian Plate. This zone runs parallel with the Indus and Tsangpo rivers. Ophiollite rock is widespread here, showing the remnants of the back-arc basin.

South Tibetan Deatachment System (STDS): A major fault in the Nepal Himalayas, which separates the Higher Himalayas and Tibetan-Tethys zones. This is a low-angle normal fault sequence based on the assumption applied in wedge extrusion and channel flow models. The STDS is well exposed in the Kaligandaki, Manage, and Upper Dolpo sections of the Nepal Himalayas.

Main Central Thrust (MCT): A prominent thrust zone in the Nepal Himalayas, which separates the Lesser Himalayas and Higher Himalayas. The MCT may have been active between 25 and 15 Ma and active until 6 to 8 Ma. The thickness of the MCT is higher in western Nepal compared to eastern Nepal.

Main Boundary Thrust: This thrust separates the Lesser Himalayas and the Siwaliks. It is well exposed throughout the Nepal Himalayas and marked by a pressure ridge. Global Positioning System measurements show that it is still active, creating numerous landslides.

Main Frontal Thrust: This thrust separates the Indo-Gangetic plain from the Siwaliks and extends throughout the Himalayas. It can be identified in the field by non-foliated breccia, fault gouge, and brittle deformation microstructure within the host rock. However, it is very difficult to trace this fault on the surface. It uplifts the Himalayan topographic front.

Intermontaine basin: This basin developed within the mountain landscape during the formation or growing of the mountains. It is generally filled with alluvial deposits transported from the surrounding mountain slopes. Some examples in the Nepal Himalayas include Dang, Deukhuri, Pokhara, Kathmandu, Hetauda, Gaighat, Chitwan, Marin Khola, Puntura Khola and Sindhuli.

References

Amatya K, Jnawali B (1994) Geological map of Nepal, scale: 1: 1,000,000. Department of mines and geology, International centre for integrated mountain development, Carl Duisberg Gesellschaft e. V., and United Nations environment programme

Auden J (1935) Traverses in the Himalaya. Records Geol Surv India 69:123–167

Bashyal R (1998) Petroleum exploration in Nepal. J Nepal Geol Soc 18:19–24

Bassoullet J, Enay R, Mouterde R (1986) La marge nord-himalayenne au Jurassique. *Nancy: Sciences de la Terre*. Mémoires 47:43–60

Bilham R, Gaur VK, Molnar P (2001) Himalayan seismic hazard. Science 293:1442–1444

Bodenhausen J, De Booy T, Egeler C, Nijhuis H (1964) On the geology of central west Nepal—A preliminary note. In: Report on 22nd International Geological Congress Delhi. vol 11, pp 101–122

Bordet P, Bordet P, Bordet P, Géologue F, Bordet P, Geologist F (1961) Recherches géologiques dans l'Himalaya du Népal, région du Makalu, Ed. du Centre national de la recherche scientifique

Brunel M, Adnrieux J (1977) Déformations superposées et mécanismes associes au chevauchement central Himalayan «M. C. T.»: Népal Oriental. *Colloques internationaux du C. N. R. S. No. 268—Écologie et Géologie de l'Himalaya, Paris*, 69–83

DeCelles PG, Gehrels GE, Quade J, LaReau B, Spurlin M (2000) Tectonic implications of U-Pb zircon ages of the Himalayan orogenic belt in Nepal. Science 288:497–499

Dhital M (1995) Geology XE "Geology" and structure of the Siwaliks XE "Siwaliks" and Lesser Himalaya XE "Lesser Himalaya" in the Surai Khola-Bardanda area, mid Western Nepal. Bull Dept Geol Tribhuvan Univ 4:1–70

Dhital MR (2015) Geology XE "Geology" of the Nepal Himalaya XE "Nepal Himalaya": regional perspective of the classic collided orogen. Springer

Fuchs G (1964) Note on the geology XE "geology" of the Palaeozoics and Mesozoics of the Tibetan zone of the Dolpo region (Nepal-Himalaya). Sonderabdruck Aus Den Verhandlungen Der Geologischen Bundesanstalt 1:6–9

Fuchs G (1977) On the geology XE "geology" of the Karnali and Dolpo regions, west Nepal. Mitteilungen Der Geologischen Gesellschaft in Wien 66:22–32

Gaetani M, Casnedi R, Fois E, Garzanti E, Jadoul F, Nicora A, Tintori A (1986) Stratigraphy of the Tethys XE "Tethys" Himalaya in Zanskar, Ladakh. Riv Ital Paleontol Stratigr 91:443–478

Gaetani M, Garzanti E (1991) Multicyclic history of the Northern India continental margin (Northwestern Himalaya)(1). AAPG Bull 75:1427–1446

Gansser A (1964) Geology of the Himalayas

Gautam P, Rösler W (1999) Depositional chronology and fabric of Siwalik Group sediments in central Nepal from magnetostratigraphy and magnetic anisotropy. J Asian Earth Sci 17:659–682

Hagen T (1959) Geologie des Thakkhola (Nepal). Eclogae Geologicae Helveticae 52:709–720

Hagen T (1968) Report on the geological survey of Nepal, vol 2, Geology of the Thakkhola. *Memoircs de la Société Helvetique Sciences Naturellcs* 86

Hagen T (1969) Report on the geological survey of Nepal Preliminary reconnaissance

Heim A, Gansser A (1939) Central Himalaya, geological observations of the Swiss expeditions 1936: Mémoire Société Helvetique Science

Jade S, Shrungeshwara T, Kumar K, Choudhury P, Dumka RK, Bhu H (2017) India plate angular velocity and contemporary deformation rates from continuous GPS measurements from 1996 to 2015. Sci Rep 7:1–16

Jain A, Ahmad T, Singh S, Ghosh S, Patel R, Kumar R, Agarwal K, Perumal J, Islam R, Bhargava O (2012) Evolution of the Himalaya. Proc Indian Natl Sci Acad 78:259–275

Kobayashi T, Morishita Y, Yarai H (2015) Detailed crustal deformation and fault rupture of the 2015 Gorkha earthquake, Nepal, revealed from ScanSAR-based interferograms of ALOS-2. Earth, Planets and Space 67:1–13

Kohn MJ, Paul SK, Corrie SL (2010) The lower lesser Himalayan sequence: a Paleoproterozoic arc on the northern margin of the Indian plate. Bulletin 122:323–335

Kubo H, Dhakal YP, Suzuki W, Kunugi T, Aoi S, Fujiwara H (2016) Estimation of the source process of the 2015 Gorkha, Nepal, earthquake and simulation of long-period ground motions in the Kathmandu basin using a one-dimensional basin structure model. Earth, Planets and Space 68:16

Le Fort P (1975) Himalayas: the collided range. Present knowledge of the continental arc. Am J Sci 275:1–44

Le Fort P (1981) Manaslu leucogranite: a collision signature of the Himalaya: a model for its genesis and emplacement. J Geophys Res: Solid Earth 86:10545–10568

Le Fort P, Rai SM (1999) Pre-Tertiary felsic magmatism of the Nepal Himalaya XE "Nepal Himalaya": recycling of continental crust. J Asian Earth Sci 17:607–628

LeFort P (1996) Evolution of the Himalaya. In: Yin A, Harrison TM (eds) The tectonic evolution of Asia. Cambridge, Cambridge University Press

LRMP (1986) Land resources mapping project. Survey Department, Government of Nepal and Kenting Earth Sciences

Mencin D, Bendick R, Upreti BN, Adhikari DP, Gajurel AP, Bhattarai RR, Shrestha HR, Bhattarai TN, Manandhar N, Galetzka J (2016) Himalayan strain reservoir inferred from limited afterslip following the Gorkha earthquake. Nat Geosci 9:533–537

Molnar P (1984) Structure and tectonics of the Himalaya: constraints and implications of geophysical data. Annu Rev Earth Planet Sci 12:489–516

Mugnier J, Leturmy P, Huyghe P, Chalaron E (1999) The Siwaliks XE "Siwaliks" of Western Nepal: II. Mechanics of the thrust wedge. J Asian Earth Sci 17:643–657

Nakayama K, Ulak PD (1999) Evolution of fluvial style in the Siwalik Group in the foothills of the Nepal Himalaya. Sed Geol 125:205–224.

Pant CC, Sharma A (1993) Quaternary sedimentation in the Indo-Gangetic basin: a review. Curr Sci 855–862

Parrish RR, Hodges V (1996) Isotopic constraints on the age and provenance of the Lesser and Greater Himalayan sequences, Nepalese Himalaya. Geol Soc Am Bull 108:904–911

Sah R, Ulak P, Gajurel A, Rimal L (1994) Lithostratigraphy of Siwalik sediments of Amlekhganj-Hetauda area, sub-Himalaya of Nepal

Sapkota S, Bollinger L, Klinger Y, Tapponnier P, Gaudemer Y, Tiwari D (2013) Primary surface ruptures of the great Himalayan earthquakes in 1934 and 1255. Nat Geosci 6:71–76

Sharma C (1973) Geology XE "Geology" of Nepal. Mani Ram Sharma, Kathmandu

Sharma CK (1977) Geology XE "Geology" of Nepal. Educational Enterprises, Kathmandu

Tokuoka T, Takeda S, YosHIDA M, Upreti B (1988) The Churia (Siwalik) Group in the western part of the Arung Khola area, west central Nepal. Memoirs of Faculty of Science, Shimane University 22:131–140

Upreti B (1999) An overview of the stratigraphy and tectonics of the Nepal Himalaya XE "Nepal Himalaya" . J Asian Earth Sci 17:577–606

Valdiya K (1964) The unfossiliferous formations of the Lesser Himalaya and their correlation. In: 22nd International Geological Congress. Delhi, pp 15–36

Valdiya K (1984) Evolution of the Himalaya. Tectonophysics 105:229–248

Veh G, Korup O, Roessner S, Walz A (2018) Detecting Himalayan glacial lake outburst floods XE "Floods" from Landsat time series. Remote Sens Environ 207:84–97

Land Use and Land Cover

5

Basanta Paudel and Basanta Raj Adhikari

Abstract

The land is one of the major components of the Earth's surface. Land with anthropogenic activities and interaction is known as land use and land with natural status is recognized as land cover. This chapter deals with the overall land use and land cover (LULC) pattern of Nepal, with the changing trends of major LULC types. Further, it also covers an analysis of the land tenure system and land use policies of the country and discusses the strength and constraints of sustainable land use management and its future direction. The recent land-use pattern of Nepal shows that forest cover is the dominant land cover type, with 44.47% of total land area in 2018. The status of built-up area is low compared to other types, though this is in increasing trend. Long-term analysis (1910–2010) shows that the amount of agricultural land is expanding rapidly; however, it is observed to be slightly declining due to the rapid expansion rate of urban/built-up areas and abandoned farmland after 2010. Most of the agricultural land is being converted to the forest due to successful community forestry practices in Nepal. Similarly, snow/glacial cover changes significantly due to increasing temperature in the Nepal Himalaya. The long-term status of the land tenure system and land use policies in Nepal change in different periods based on different socio-economic and political systems in the country. In addition, for sustainable LULC management, the newly formulated 11-LULC zones by Land Use Policy 2015 need to be implemented in an appropriate way. It is vital to manage sustainable land management and food insecurity in Nepal in the future.

B. Paudel (✉)
Key Laboratory of Land Surface Pattern and Simulation, Institute of Geographic Sciences and Natural Resources Research, Chinese Academy of Sciences, Beijing, China
e-mail: paudelb@igsnrr.ac.cn

B. R. Adhikari
Institute of Engineering, Tribhuvan University, Kathmandu, Nepal

Keywords

Land use • Land cover • Land tenure system • Policies • Nepal

5.1 Introduction

The land is one of the major components of the Earth's surface and is often changeable due to anthropogenic and natural factors, which are recognized as land use and land cover change (Foley et al. 2005; Paudel et al. 2016). The changing phenomenon of land use and land cover (LULC) is therefore also acknowledged as the direct and indirect interaction between anthropogenic activities and environmental factors (Klein Goldewijk et al. 2011). The term "land cover" indicates the physical existence of land types on the Earth surface, whereas the term "land use" refers to how humans interact with and use the land (Chapagain et al. 2018; Lambin et al. 2001).

The long-term LULC status of global, regional, and local scales shows that it is changing on a different scale (Klein Goldewijk et al. 2011; Ramankutty and Foley 1999). In Nepal, the long-term historical changing trend of different LULC types differs from place to place (Adhikari 2016). The agricultural land use trend shows that it has expanded largely over the last 100 years (Paudel et al. 2018); however, the forest cover status has positive- and negative-moving trends between 1978 and 2018 (LRMP 1986; MoFE 2019). Very high deforestation occurred in Nepal before the 1990s, though the forest area began to recover noticeably after the launch of the community forestry program over the ensuing decade (Ghartichhetri et al. 2016). Previous studies reported that the long-term status of snow/glacial cover was in declining order (Bajracharya et al. 2014; Shrestha and Aryal 2011) whereas the amount of built-up area was increasing (Rimal et al. 2020, 2018b).

R. B. Ojha and D. Panday (eds.), *The Soils of Nepal*, World Soils Book Series,
https://doi.org/10.1007/978-3-030-80999-7_5

The recent pattern of LULC is different from previous decades (e.g., the 1990s), with the highest forest cover (44.47%) recorded in 2018 (MoFE 2019). The amount of agricultural land was expanding until 2010 (Paudel et al. 2017), though after 2010 this land-use type declined slightly due to a large rate of urban expansion (Rimal et al. 2018b; Thapa and Murayama 2009, 2010) and farmland abandonment (Chaudhary et al. 2018; Paudel et al. 2020a). Further, the recent spatial land use pattern shows that the urban area is expanding at a higher rate in the Tarai region (Rimal et al. 2020). In addition, the high rate of temperature increase in the Nepal Himalaya (Paudel et al. 2020b) contributes to the reduction of the snow/glacial cover.

The previous overall status of LULC change shows that agricultural land is mainly converted to urban/built-up area and snow/glacial cover is mainly converted into barren land (LRMP 1986; Uddin et al. 2015). Because there is a large gap regarding the implementation of existing agricultural policies for appropriate management of agricultural land in Nepal (Khanal et al. 2020), the land use policy in 2015 introduced different 11 LULC zones in Nepal for better management of the land and land tenure system (MoLRM 2015; Nepal et al. 2020).

The LULC change projection shows that the agricultural land and snow/glacial cover are in declining order, whereas the urban/built-up area and barren land are expanding. In addition, the forest cover is also projected to expand in Nepal due to the increasing trend of farmland abandonment and effective management of the community forestry program. The LULC is essential for humankind and is necessary for conserving the overall environment of Earth's surface. Better management of the LULC is necessary for a sustainable future and development; therefore, this chapter mainly deals with the overall LULC pattern, trends, and land-use policies in Nepal. In addition, it also covers the strengths and constraints of the sustainable land use management system with future direction and perspective on the context of Nepal.

5.2 Land Use and Land Cover Pattern

5.2.1 Agricultural Land

The recent land-use pattern of Nepal shows that the total agricultural land was 29.8% in 2010 (Fig. 5.1) (Uddin et al. 2015), and that it decreased to 28.2% in 2014 (Reddy et al.

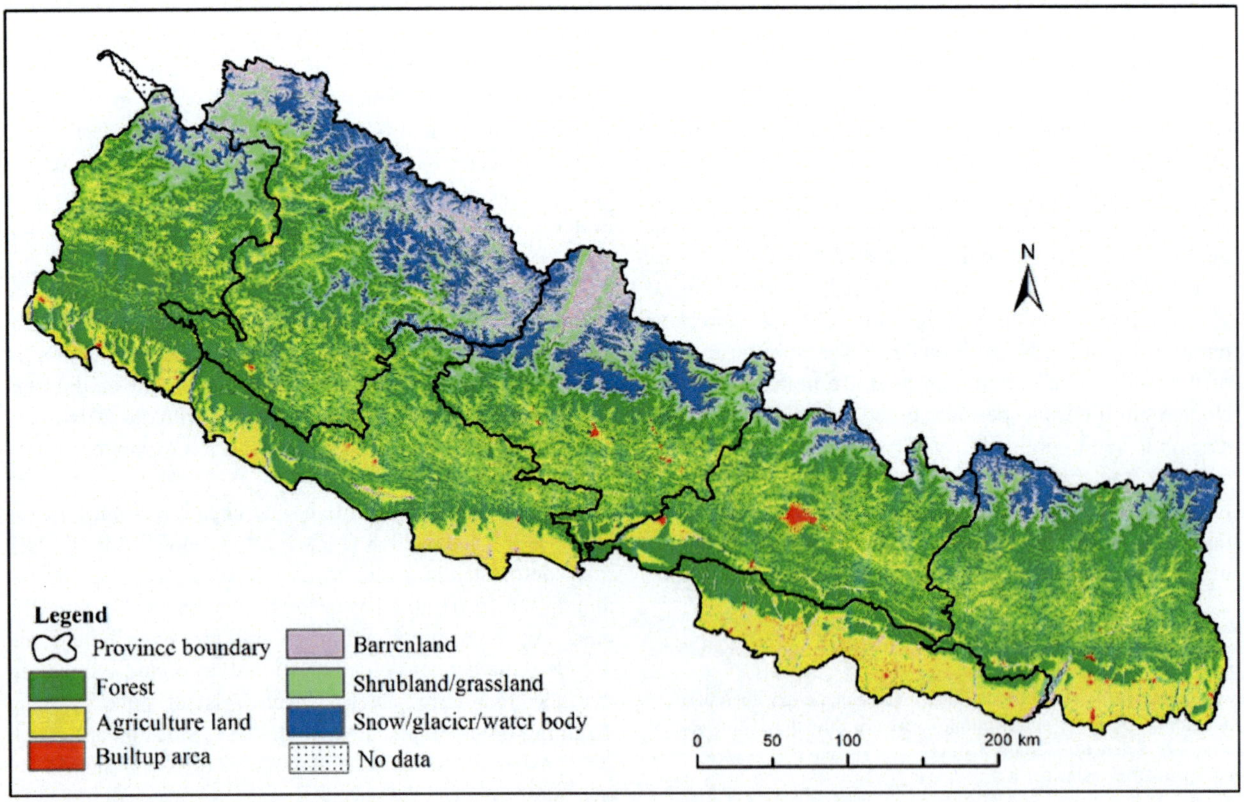

Fig. 5.1 Land use and land cover pattern in Nepal. *Data source* Uddin et al. (2015)

2018). The spatial pattern of the agricultural land in Nepal indicates that about 32.1% of the total agricultural land is distributed within the Tarai region and is useful for agricultural activities (Paudel et al. 2016). Similarly, the agricultural area of Nepal spatially expanded in the river valley and terraces (Fig. 5.2) of the Siwaliks, Middle Mountain, and High Mountain regions (Uddin et al. 2015). Due to its extreme climatic conditions and topography, there is far fewer agricultural area in the High Mountain region, which has only 0.5% of the total agricultural land (Paudel et al. 2016).

5.2.2 Forest

The study carried out by the Ministry of Forest and Environment, Government of Nepal reported that the forest cover made up 44.5% area of the total land area in Nepal in 2018 (MoFE 2019), though that amount was slightly less (39.1%) in a national study in 2010 (Fig. 5.1) (Uddin et al. 2015). The spatial distribution pattern of forest cover in Nepal shows that it has covered area mainly in the Siwalik-Mahabharat range (Fig. 5.3), the Middle Mountain region, and the High Mountain region (Paudel et al. 2016). Similarly, some area in the Tarai region is covered by dense tropical forest, especially in the national parks (e.g., Chitawan National Park, Bardiya National Park) and community forests. Out of the total forest cover (39.1%) in 2010, broadleaved closed forest occupied 14.4% and broadleaved open forest occupied 9.6%, followed by needleleaved closed forest at 9.5%, while the remaining 5.6% was covered by needleleaved open forest (Uddin et al. 2015).

5.2.3 Built-Up Area

Almost 0.32% of Nepal was covered by built-up area in 2010 (Fig. 5.3) (Uddin et al. 2015). The spatial pattern of built-up area in Nepal shows that the large area of the Kathmandu Valley is covered by this land use type (Rimal et al. 2018a). The built-up area has mainly been increasing in the country's emerging cities (Fig. 5.4). Further, recent

Fig. 5.2 Terrace farmland in Ilam district, Nepal. *Photo credit* Zhang Yili/Basanta Paudel/Wu Xue

Fig. 5.3 Forest cover in the Mahabharat range of Dadeldhura district, Nepal. *Photo credit* Basant Raj Adhikari

Fig. 5.4 Built-up area of Phidim, the district headquarters of Pachthar district, Nepal. *Photo credit* Zhang Yili/Basanta Paudel/Wu Xue

studies show that the built-up area in the Tarai region is expanding rapidly (Rimal et al. 2019, 2020, 2018b), as distributed within large cities (e.g., Pokhara and Bharatpur) (Rai et al. 2020).

5.2.4 Snow/Glacial/Waterbodies

Snow/glacial and water bodies covered 8.80% of Nepal's land in 2010, and out of this area, 0.03% was occupied by lakes and 0.57% by rivers (Uddin et al. 2015). This means that almost 8.20% country's area was covered by snow/glaciers (Fig. 5.1). This snow/glacial area is mainly located in the northern part of the country covering the High Himalaya/Himal region (Fig. 5.5a). These regions are mostly covered by snow/glaciers year-round due to the tundra and nival climatic conditions. Most of the melted water from snow cover runs from Nepal's major rivers in the north–south direction (Fig. 5.5b).

5.2.5 Shrubland/Grassland

The land cover pattern in Nepal shows that in 2010 around 11.3% of the country's area was covered by shrubland and grassland. Out of this, around 7.9% was coved by grassland and the remaining 3.4% by shrubland (Fig. 5.1) (Uddin et al. 2015). The distribution of grassland is mainly shown in the areas of the Middle and High Mountain regions (Fig. 5.6), and the valleys, Inner Himalaya, and alpine regions (Paudel et al. 2016). There is a small portion of grassland present in the lowland area.

5.2.6 Others

The other types of land use patterns, including barren land, sand, and boulders, cover the remainder of Nepal (Fig. 5.7). Barren land covers 10.65% of the country's area and is mainly distributed in the High Mountain and trans-Himalaya regions of the country (Uddin et al., 2015), whereas most of the sand and boulders lie in the major river basins such as Koshi, Gandaki, and Karnali.

5.3 Land Use and Land Cover Change

5.3.1 Agricultural Land

Spatio-temporal analysis of LULC change shows that agricultural land occupied 15,120 km^2 in 1910 (Paudel et al. 2018), which was increased to 37,030 km^2 in 1970 (Paudel et al. 2017). Further, a study based on aerial photography found that the country's agricultural land was 40,019 km^2 in 1978 (Fig. 5.8a) (LRMP 1986), which based on satellite data analysis expanded to 43,910 km^2 in 2010 (Fig. 5.8d) (Uddin et al., 2015). Over the period between 1978 and 2010, agricultural land expanded from 27.19% to 29.83% of the total land area of the country (Table 5.1) (Paudel et al. 2016). The spatial distribution of agricultural land in 1978 (LRMP 1986), 1990 (ICIMOD 2014a), 2000 (ICIMOD 2014b), and 2010 (Uddin et al. 2015) is illustrated in Fig. 5.8.

There are increasing trends of agricultural land in Nepal between 1910 and 2010 (Paudel et al. 2019). However, these changes have slightly decreased in recent years due to high

Fig. 5.5 Snow/glacial cover on **a** Mt. Dhaulagiri in Mustang district and the **b** Dudhkoshi River in the border area of Okhaldhunga and Khotang district, Nepal. *Photo credit* Basant Raj Adhikari (**a**) and Basanta Paudel (**b**)

Fig. 5.6 Grassland on Chirdung Mountain in Dolakha district, Nepal. *Photo credit* Basanta Paudel/Wu Xue

Fig. 5.7 **a** Barren land in Mustang district and **b** sand and boulders in the Karnali River in the Bardiya district of Nepal. *Photo credit* Basant Raj Adhikari

levels of farmland abandonment in the Hill and Mountain regions of the country (Chaudhary et al. 2018), and from people migrating from highland to lowland and rural to urban areas (Paudel et al. 2020a), along with a higher rate of urban expansion in the Tarai region (Rimal et al. 2020, 2018b). These changes might create instability in sustainable land management, food insecurity, and vulnerability for peoples' livelihoods.

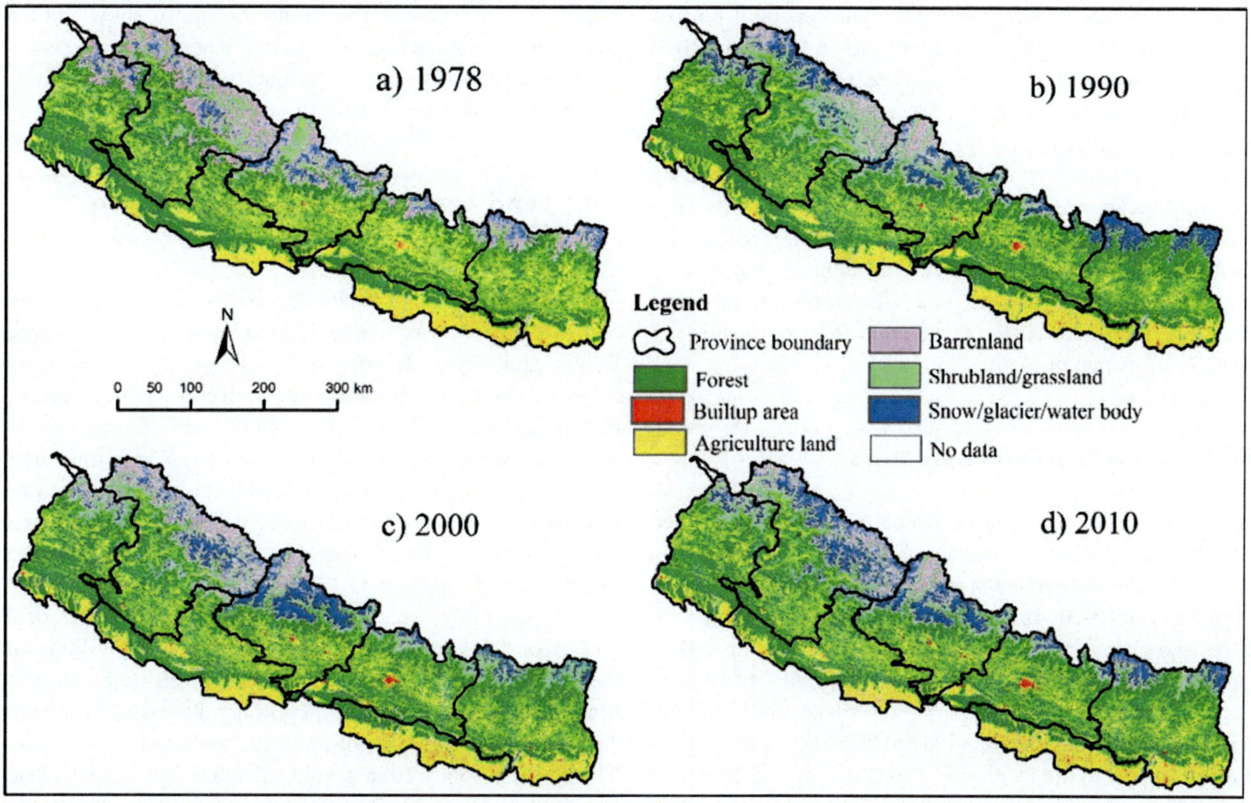

Fig. 5.8 Land use and land cover change between 1978 and 2010. *Data sources* **a** LRMP (1986); **b** ICIMOD (2014a); **c** ICIMOD (2014b); and **d** Uddin et al. (2015) for the years 1978, 1990, 2000, and 2010, respectively

Table 5.1 The conversion statistics of major land use between 1978 and 2010

LULC type	Land cover in 1978, km^2	Land cover in 2010, km^2	Land cover in 1978, %	Land cover in 2010, %	Conversion between 1978 and 2010, %
Agricultural land	40,019	43,910	27.19	29.83	+2.64
Forest cover	55,944	57,537	38.01	39.09	+1.08
Grassland	16,985	11,634	11.54	7.91	−3.63
Built-up area	122	469	0.08	0.32	+0.24
Shrub land	6888	5008	4.68	3.4	−1.28
Other lands	21,939	16,561	14.91	11.25	−3.66

Source Paudel et al. (2016)

5.3.2 Forest

Forest cover holds the highest percentage of land cover type in Nepal. Forest cover areas in Nepal were reported at 43.5% in 1964 and 38.9% in 1978 (Fig. 5.8a) and were reduced to 29% in 1994 (Paudel et al. 2016). The study conducted by Uddin et al. (2015) shows that this land-use type increased to cover 39.1% area of the country in 2010 (Fig. 5.8d). Furthermore, a recent study reported that the amount of forest area has largely increased and covered 44.5% of the area in 2018 (MoFE 2019). These changes can be seen at different rates in different periods; for example, the rate of deforestation was much higher before the 1990s, and this decreased as the amount of forest cover increased due to the country's successful implementation of the community forestry program (Ghartichhetri et al. 2016).

5.3.3 Built-Up Area

The total built-up area in Nepal was 0.32% in 2010 (Uddin et al. 2015) and 0.50% in 2014 (Reddy et al. 2018), a significant change in only a few years. The total built-up area in 1978 was 0.08%, though it increased to 0.32% in 2010 (Table 5.1) (Paudel et al. 2016). The spatial trend of the built-up area shows high rates of increase in the Tarai region (Fig. 5.8). A recent satellite-based recent study of the western Tarai region explains that built-up area increased by 256 km^2 between 1986 and 2016 as more area was converted from agricultural land (Rimal et al. 2020). Similarly, the built-up areas are on an increasing trend in major cities (e.g., Kathmandu, Pokhara, and Bharatpur) as the land around metropolitan cities is converted from farmland (Rai et al. 2020; Rimal et al. 2018a).

5.3.4 Snow/Glacial/Waterbodies

The status of the snow/glacial area has significantly changed over the past decades. For example, the total area occupied by snow/glacial/waterbodies occupied 5168.3 km^2 in 1980, 4506.3 km^2 in 1990, 4210.9 km^2 in 2000, and 3902.4 km^2 in 2010, showing a declining trend (Bajracharya et al. 2014). Further, snow/glacial/waterbodies covered 8.80% of Nepal's area in 2010 (Fig. 5.8d), including 0.03% covered by lakes and 0.57% by rivers (Uddin et al. 2015). These changes are directly related to the increasing trends of temperature in the Nepal Himalaya (Shrestha and Aryal 2011).

5.3.5 Shrubland/Grassland

Long-term spatio-temporal analysis shows that shrubland/grassland in Nepal is on a decreasing trend. The total area coverage by shrubland was 4.68% in 1978, which declined to 3.4% in 2010 (Table 5.1) (Paudel et al. 2016). The grassland area decreased by 3.63% between 1978 to 2010 from 11.54% to 7.91%, respectively (Fig. 5.8a–d) (LRMP 1986; Uddin et al. 2015).

5.3.6 Others

The area of the remaining categories (i.e., barren land, sand, and boulders) also decreased by 3.66% between 1978 and 2010 (Table 5.1) (Paudel et al. 2016) and was reported at 21,939 km^2 in 1978 (Fig. 5.8a) and 16,561 km^2 in 2010 (Fig. 5.8d) (LRMP 1986; Uddin et al. 2015). These changes might be due to climate change in the Nepal Himalaya; for example, erratic rainfall and cloudbursts create landslides in the mountains and flooding in mountain rivers. These rivers transport the sediment generated in the mountains to the flatlands and deposit sand in the floodplain. Such events change the land cover status to the sand and boulder category. Similarly, the increasing trend of temperature in the Nepal Himalaya contributes to the melting of snow and ice and with it the change to barren land. However, such changes are also controlled by micro-climate and geomorphological differences within the country.

5.4 Land Tenure System and Land Use Policies

Historically, the land tenure system in Nepal, especially before the 1950s was mainly divided into two major classes: Raikar and Kipat (Regmi 1978). These two land tenure systems were subdivided into several sub-categories, including Birta, Jagera, Jagir, Guthi, and Rakam. These categories were redefined as Statutory and Non-formal after 1951. The Statutory category is further divided into three types as Raikar (Private), State (Public/Governmental), and Guthi, whereas Slums are included within Non-formal (Khanal 2002; Nepal et al. 2020).

A recent study explains that there is a 409-year historical record of land use policies and legislative provisions in Nepal, including the legal code of Ram Shah in the Gorkha kingdom (1606–1636), Royal Orders between 1768 and 1854, and the land use policy of 2015 (Nepal et al. 2020). The land use policy 2015 developed and categorized the land tenure system into 11 different zones, including agricultural, residential, commercial, industrial, mines and minerals, cultural and archeological, rivers–lakes–reservoirs, forests, public use and open spaces, building materials excavation, and other (MoLRM 2015).

5.5 Strength and Constraints of Sustainable Land Use Management

Sustainable land use focuses on conservation of the natural resources of the terrestrial ecosystem allowing reasonable use of these resources that do not permanently destroy the natural balance of the ecosystem. The United Nation Earth Summit in 1992 defined Sustainable Land Management (SLM) as "The use of land resources, including soils, water, animals, and plants, for the production of goods to meet changing human needs, while simultaneously ensuring the long-term productive potential of these resources and the maintenance of their environmental functions" (Meakin 1992). Therefore, SLM provides the opportunity to enhance the quality of life. Improvement in soil quality using different direct and indirect techniques is an important aspect of SLM. This can be achieved in many ways, such as through climate change adaptation and mitigation. Climate change impacts on land use and land cover change not only destroy the natural ecosystem but also enhance natural hazards (e.g., erosions, floods, landslides, etc.). These natural hazards can destroy fertile soil and lead to loss of property. In contrast, reducing greenhouse gas emissions from different anthropogenic activities aids the C sink and protects the degrading environment.

Land use planning and management helps to reduce land-use conflicts, conserves critical ecosystems, protects the environmentally sensitive habitat, and restores degraded land to ensure a healthy life. SLM integrates the natural, socio-cultural, and socio-economical resources of a particular area. Each area of the Earth's surface has its own unique natural ecosystem and must be dealt with in a unique way. Sometimes a proper method of land use might have negative impacts on others: for example, extensive land use by planting coffee or tea using water can contaminate groundwater in the mountain ecosystem and trigger landslides due to oversaturation of soil. People are often hesitated to adopt new technologies of SLM due to poor interaction between landowners and experts.

5.6 Future Direction and Perspectives

The Land Use Land Cover (LULC) change can be seen throughout the Earth's history in different forms. The effects of climate change, changes in human behavior, the migration of people from rural to urban areas, and improper land use management are major developmental challenges. While such changes can be assumed to occur in the future, different LULC change models exist, which Heistermann et al. (2006) summarized into two groups: Geographical models and Economic models. Geographical models usually consider spatial information for the analysis of drivers of LULC change and discuss how the change is related to the land properties for suitability of use (Heistermann et al. 2006). This model also projects future LULC changing scenarios considering long-term climate change parameters. Economic models, meanwhile, generally consider welfare optimization principles either explicitly or implicitly to model production and consumption patterns. There is a third group of models that considers land use change and its impact on ecological services (Verburg et al. 2004). All of the model groups explained above use historical data to predict future scenarios; however, it is difficult to make accurate predictions due to complex human migration patterns and the changing climate.

Temperature increase and unpredicted rainfall have driven LULC change in the Nepal Himalaya (Paudel et al. 2019). Many land-use plans and policies exist in Nepal based on international conservation treaties, and implementation of these plans and policies is a major challenge due to the low literacy rate and lack of awareness about conservation. The government of Nepal has already reformed the political system and begun practicing federalism while improving land use policy and creating the 11 major land use zones. These federal states have some rights and duties for better management of local LULC and will prepare new plans for better management of LULC at the state level as well as the local level. Moreover, as people will continue to migrate from rural to urban areas the urban population will increase, and the LULC pattern and practice will change significantly.

Therefore, the following recommendations are proposed for the successful implementation of LULC for sustainable development and biodiversity conservation:

1. Strict implementation of LULC plan and policy in the new political environment.
2. Preparation of short-, medium-, and long-term land-use and disaster management plans considering changing climate.
3. Scientific urban planning to encourage compact, mixed-use, and walkable communities.
4. Formulation of an appropriate plan and program to mitigate the increasing trend of farmland abandonment.
5. Development of an appropriate plan and policy for integrated and compact settlement in rural and emerging town areas.
6. Preparation of a risk-sensitive land use plan and implementation.
7. Preservation of open space and greenery in urban areas to combat urban heat islands.

5.7 Conclusion

The spatial pattern of the recent Land Use Land Cover (LULC) shows that forest cover holds the highest land use type followed by agricultural land. There are significantly increasing trends of urban/built-up areas in the major cities and in the Tarai. Agricultural land and forest cover are the most converted land cover types. Agricultural lands are on a decreasing trend of being converted to the built-up area. Forest cover has increased in recent decades due to the successful implementation of community forestry practices in the 1990s. The land tenure system was improved by the Land Use Policy 2015 and categorized the country into 11 different LULC zones. However, this policy must be strictly implemented to aid the advancement of sustainable LULC management in Nepal. The policy will also help reduce the haphazard expansion of urban/built-up areas in agricultural land for the sustainable future.

Acknowledgements We are grateful to the International Centre for Integrated Mountain Development (ICIMOD) for sharing their valuable data and to Prof. Zhang Yili and Dr. Wu Xue for their valuable photos of different land cover.

References

Adhikari BR (2016) Land and soil, compendium of environment statistics. Central Bureau of Statistics, Government of Nepal, pp 47–68

Bajracharya SR, Maharjan SB, Shrestha F, Bajracharya OR, Baidya S (2014) Glacier status in Nepal and decadal change from 1980 to 2010 based on Landsat data. International Centre for Integrated Mountain Development

Chapagain PS, Rai MK, Paudel B (2018) Land use land cover XE "Land cover" change and its pathways in Sidin VDC, Panchthar district, Nepal. Geogr J Nepal 11:77–94. https://doi.org/10.3126/gjn.v11i0.19550

Chaudhary S, Wang Y, Khanal NR, Xu P, Fu B, Dixit AM, Yan K, Liu Q, Lu Y (2018) Social impact of farmland abandonment and its eco-environmental vulnerability in the high mountain XE "High Mountain" region of Nepal: a case study of Dordi River Basin. Sustainability 10:2331

Foley JA, DeFries R, Asner GP, Barford C, Bonan G, Carpenter SR, Chapin FS, Coe MT, Daily GC, Gibbs HK, Helkowski JH, Holloway T, Howard EA, Kucharik CJ, Monfreda C, Patz JA, Prentice IC, Ramankutty N, Snyder PK (2005) Global consequences of land use. Science 309:570. https://doi.org/10.1126/science.1111772

Ghartichhetri DB, Cheng S, Xu Z, Bhandari J, Wang L, Liu X (2016) Community forestry and livelihood in Nepal: a review. J Anim Plant Sci 26:1–12

Heistermann M, Müller C, Ronneberger K (2006) Land in sight?: Achievements, deficits and potentials of continental to global scale land-use modeling. Agriculture XE "Agriculture." Ecosyst & Environ 114:141–158. https://doi.org/10.1016/j.agee.2005.11.015

ICIMOD (2014a) Land cover of Nepal 1990. International Centre for Integrated Mountain Development (ICIMOD), Kathmandu, Nepal. http://rds.icimod.org.

ICIMOD (2014b) Land cover of Nepal 2000. International Centre for Integrated Mountain Development (ICIMOD), Kathmandu, Nepal. http://rds.icimod.org.

Khanal N (2002). Land use and land cover dynamics in the Himalaya: a case study of the Madi watershed, western development region, Nepal. Unpublished PhD dissertation, Tribhuvan University, Nepal

Khanal NR, Nepal P, Zhang Y, Nepal G, Paudel B, Liu L, Rai R (2020) Policy provisions for agricultural development in Nepal: a review. J Clean Prod 261:121241. https://doi.org/10.1016/j.jclepro.2020.121241

Klein Goldewijk K, Beusen A, van Drecht G, de Vos M (2011) The HYDE 3.1 spatially explicit database of human-induced global land-use change over the past 12,000 years. Glob Ecol Biogeogr 20:73-86. https://doi.org/10.1111/j.1466-8238.2010.00587.x

Lambin EF, Turner BL, Geist HJ, Agbola SB, Angelsen A, Bruce JW, Coomes OT, Dirzo R, Fischer G, Folke C, George PS, Homewood K, Imbernon J, Leemans R, Li X, Moran EF, Mortimore M, Ramakrishnan PS, Richards JF, Skånes H, Steffen W, Stone GD, Svedin U, Veldkamp TA, Vogel C, Xu J (2001) The causes of land-use and land-cover change: moving beyond the myths. Glob Environ Chang 11:261–269. https://doi.org/10.1016/S0959-3780(01)00007-3

LRMP (1986) Land utilization report, land resource mapping project. Kenting Earth Science Canada and Department of Topography, Government of Nepal, Kathmandu, Nepal, p 112

Meakin S (1992) The rio earth summit: summary of the United Nations conference on environment and development. Library of Parliament, Research Branch

MoFE (2019) National level forests and land cover analysis of Nepal using google earth images. Kathmandu, Nepal, p 34

MoLRM (2015) Land use policy 2015. Ministry of Land Reform and Management, Government of Nepal, Kathmandu, Nepal, p 20

Nepal P, Khanal NR, Zhang Y, Paudel B, Liu L (2020) Land use policies in Nepal: an overview. Land Degrad Dev. https://doi.org/10.1002/ldr.3621

Paudel B, Wu X, Zhang Y, Rai R, Liu L, Zhang B, Khanal NR, Koirala HL, Nepal P (2020a) Farmland abandonment and its determinants in the different ecological villages of the Koshi River Basin, Central Himalayas: synergy of high-resolution remote sensing and social surveys. Environmental Research, p 109711. https://doi.org/10.1016/j.envres.2020.109711

Paudel B, Zhang Y, Li S, Liu L (2018) Spatiotemporal changes in agricultural land cover XE "Land cover" in Nepal over the last 100 years. J Geog Sci 28:1519–1537. https://doi.org/10.1007/s11442-018-1559-9

Paudel B, Zhang Y, Li S, Liu L, Wu X, Khanal NR (2016) Review of studies on land use XE "Land use" and land cover XE "Land cover" change in Nepal. J Mt Sci 13:643–660. https://doi.org/10.1007/s11629-015-3604-9

Paudel B, Zhang Y, Li S, Wu X (2017) Spatiotemporal reconstruction of agricultural land cover XE "Land cover" in Nepal from 1970 to 2010. Reg Environ Change 17:2349–2357. https://doi.org/10.1007/s10113-017-1164-y

Paudel B, Zhang Y, Yan J, Rai R, Li L (2019) Farmers' perceptions of agricultural land use XE "Land use" changes in Nepal and their major drivers. J Environ Manage 235:432–441. https://doi.org/10.1016/j.jenvman.2019.01.091

Paudel B, Zhang Y, Yan J, Rai R, Li L, Wu X, Chapagain PS, Khanal NR (2020b) Farmers' understanding of climate change XE "Climate change" in Nepal Himalayas: important determinants and implications for developing adaptation strategies. Clim Change 158:485–502. https://doi.org/10.1007/s10584-019-02607-2

Rai R, Yili Z, Paudel B, Khanal NR, Acharya BK (2020) Satellite image-based monitoring of urban land use change and assessing the driving factors in Pokhara and Bharatpur metropolitan cities, Gandaki Basin, Nepal. J Resour Ecol 11(87–99):13

Ramankutty N, Foley JA (1999) Estimating historical changes in global land cover XE "Land cover": croplands from 1700 to 1992. Global Biogeochem Cycles 13:997–1027. https://doi.org/10.1029/1999gb900046

Reddy SC, Vazeed PS, Satish KV, Saranya KRL, Jha CS, Krishna Murthy YVN (2018) Quantifying nationwide land cover XE "Land cover" and historical changes in forests of Nepal (1930–2014): implications on forest fragmentation. Biodivers Conserv 27:91–107. https://doi.org/10.1007/s10531-017-1423-8

Regmi MC (1978) Thatched huts and stucco palaces: peasants and landlords in 19th-century Nepal. Vikas Publishing House, New Delhi, India

Rimal B, Keshtkar H, Sharma R, Stork N, Rijal S, Kunwar R (2019) Simulating urban expansion in a rapidly changing landscape in eastern Tarai, Nepal. Environ Monit Assess 191:255. https://doi.org/10.1007/s10661-019-7389-0

Rimal B, Sloan S, Keshtkar H, Sharma R, Rijal S, Shrestha UB (2020) Patterns of historical and future urban expansion in Nepal. Remote Sens 12:628

Rimal B, Zhang L, Keshtkar H, Haack BN, Rijal S, Zhang P (2018a) Land use/land cover dynamics and modeling of urban land expansion by the integration of cellular automata and Markov chain. ISPRS Int J Geo Inf 7:154

Rimal B, Zhang L, Stork N, Sloan S, Rijal S (2018b) Urban expansion occurred at the expense of agricultural lands in the Tarai XE "Tarai" region of Nepal from 1989 to 2016. Sustainability 10:1341

Shrestha AB, Aryal R (2011) Climate XE "Climate" change in Nepal and its impact on Himalayan glaciers. Reg Environ Change 11:65–77. https://doi.org/10.1007/s10113-010-0174-9

Thapa RB, Murayama Y (2009) Examining spatiotemporal urbanization patterns in Kathmandu valley, Nepal: remote sensing and spatial metrics approaches. Remote Sens 1:534–556. https://doi.org/10.3390/rs1030534

Thapa RB, Murayama Y (2010) Drivers of urban growth in the Kathmandu valley, Nepal: examining the efficacy of the analytic hierarchy process. Appl Geogr 30:70–83. https://doi.org/10.1016/j.apgeog.2009.10.002

Uddin K, Shrestha HL, Murthy MSR, Bajracharya B, Shrestha B, Gilani H, Pradhan S, Dangol B (2015) Development of 2010 national land cover XE "Land cover" database for the Nepal. J Environ Manage 148:82–90. https://doi.org/10.1016/j.jenvman.2014.07.047

Verburg PH, Schot PP, Dijst MJ, Veldkamp A (2004) Land use change modelling: current practice and research priorities. GeoJournal 61:309–324. https://doi.org/10.1007/s10708-004-4946-y

Soil Formation

6

Roshan Man Bajracharya

Abstract

Soil formation in the Himalayan region in general, and in Nepal in particular, is determined primarily by geology, geomorphologic processes, and climate. However, soils in the populated hills and valleys have to some extent been modified by anthropogenic influences. At high elevations above 4000 m, soils tend to be very young and at the initial stages of development. These soils are developed from glacial or glacio-fluvial deposits from the last major glaciations and postglacial events. At intermediate altitudes of about 2500 to 4000 m, soils are mostly formed on glacial outwash plains, alluvial deposits, and colluvial debris from mass movements. In the hills and mountains, soils are generally formed on the residual parent material along the ridges and hill tops, colluvial debris on the hill slopes and foot-slopes, and on alluvial deposits along the valley bottoms. In the lower valleys and plains, soils are mostly developed on depositional materials of alluvial or aeolian origin. Therefore, Entisols and Inceptisols are the dominant soil orders in the high mountains and along river banks. In the middle mountains and valleys, in addition to Entisols and Inceptisols, Alfisols can also be found, while at lower elevations and in the plains region occasional Mollisols, Histosols and Oxisols also occur.

Keywords

Climate • Geology • Geomorphology • Himalaya • Soil genesis • Weathering

6.1 Introduction

Soils are the weathering products of rocks and other geologic materials, broken down through physical, chemical, and biological processes and influenced by local climate. They are a vital component of terrestrial ecosystems and serve to sustain the innumerable forms of life on land, including humans. At first glance, and to the untrained eye, soil appears lifeless and many consider it to be an abiotic component of our physical environment. However, not only are soils home to a wide variety of life forms including plants, animals, and microorganisms, but they can themselves be considered a living entity as they are intricately associated with life and the organic matter originating from living organisms. Soils, along with the land upon which they are formed, carry out numerous essential functions, including biotic and production functions; hydrologic (movement and storage of water); buffering, filtering, and transformation of compounds and wastes; living space and connectivity; archiving (preserving artifacts and records of past civilizations); storehouse of raw materials; and, combined with the global water and carbon cycles, they help to regulate the climate (FAO 1995).

Despite considerable advances in food and agricultural technology, soils continue to serve as the primary means of large-scale crop production and have supported the growth and development of human societies since the dawn of civilization, beginning with the onset of settled agriculture nearly 12,000 yrs ago (Brady 1995). Soils may be regarded as open systems (Buol et al. 1997) with continuous throughput of materials and energy. They are constantly changing and evolving entities comprising the interface among the atmosphere, hydrosphere, lithosphere, and biosphere. Energy in the forms of sunlight and heat, both from solar radiation and from within the Earth, as well as materials of geologic and biologic origin along with water, are transformed through a multitude of intensely interacting processes. These include wetting and drying, heating and

R. M. Bajracharya (✉)
Department of Environmental Science and Engineering, Kathmandu University, Dhulikhel, Kavrepalanchowk, Nepal
e-mail: rmbaj@ku.edu.np

R. B. Ojha and D. Panday (eds.), *The Soils of Nepal*, World Soils Book Series,
https://doi.org/10.1007/978-3-030-80999-7_6

cooling, freezing/thawing, evapotranspiration, leaching, weathering, erosion, and deposition (Buol et al. 1997).

6.1.1 Soil Genesis Concepts

While most soils tend not to be very old according to the geologic time scale (most are younger than the Tertiary or Pleistocene, or 2.58 M yrs, in age), due to their simultaneous formation and destruction, it typically takes several hundreds to thousands of years to form just an inch of soil (Brady 1995; Buol, et al. 1997). Soil formation in the Himalayan region in general, and in Nepal in particular, is governed by the geology, geomorphologic processes, and climate, and, to some extent, has been modified by anthropogenic influences in the populated hills and valleys.

The soil genesis or formation process follows several key concepts as outlined by Buol et al. (1997). They are listed below:

1. Soil formation processes that are active at present have been "in operation over time and varying degrees of expression over space." Hence, the geologic uniformitarian principle, i.e., "the present is the key to the past," can also be applied to soils.
2. The various soil forming, as well as destroying, processes "proceed simultaneously in a soil, and the resulting profile reflects the balance of these processes, present and past."
3. "Distinctive regimes or combinations of processes produce distinctive soils."
4. "Five external factors of soil formation drive the internal pedogenic processes within the soil." These are: climate, organisms, relief (topography), parent material, and time.
5. "Present-day soils may carry the imprint of a combination of pedogenic processes that was active in the geologic past."
6. "A succession of different soils may have taken place at a particular site as soil genetic factors changed, and soil erosion and deposition of soil material proceeded over time, with the soil surface lowered or raised."
7. Soils are not very old on a geologic time scale "because they are either destroyed or buried by geologic events or modified by shifts in climate."
8. Soil genesis tends to be complex rather than a simple process.
9. "Soils are natural clay factories."
10. "Scientific soil classification systems cannot be based entirely on genesis, because genetic processes can seldom be observed." They occur over long timeframes.
11. Knowledge about soil genesis is fundamental for proper use and management of soils.

The formation of soil and the development of the soil profile with its distinctive features begins with the type and nature of the parent material or unconsolidated upper layer of the regolith (weathered loose terrestrial surface). The parent materials from which soils develop originate with rocks occurring at the surface of the Earth's crust. These are typically classified as igneous, sedimentary, or metamorphic rocks. Igneous rocks are derived from molten lava that cools at or near the land surface. Sedimentary rocks result from deposition and re-cementation of weathering products of rocks and often exhibit distinctive layering features. Metamorphic rocks are formed by changes in the form of igneous or sedimentary rocks through intense pressure and heat within the Earth's crust. Common igneous rocks are granite, basalt, diorite, andesite, and peridotite. Important sedimentary rocks include limestone, dolomite, sandstone, and shale. Commonly occurring metamorphic rocks are gneiss, schist, quartzite, slate, and marble. A brief list of rock types and the dominant minerals comprising them are provided in Table 6.1. In the Himalayas, sedimentary and metamorphic rocks are most common, while igneous rocks are only present in minor amounts due to the absence of active volcanoes in the region. Most occur as sedimentary and meta-sedimentary rocks.

The rock weathering process is initiated by heat and cold (freeze/thaw) and the action of water, as well by as the abrasion by wind-blown fine particles, causing physical disintegration of the material. This is accompanied by chemical decomposition of the minerals through reactions such as hydrolysis, hydration, acid corrosion, oxidation–reduction, and dissolution (Brady 1974). Further transformation of the parent materials is ultimately achieved by biophysical and biochemical processes with the appearance and proliferation of living organisms. Initially, lower forms of life such as lichens and mosses colonize rocks and parent material, but eventually this life is succeeded by higher plants, microorganisms, and animals.

6.1.2 Emergence and Uplift of the Himalayas

Soil formation in the 2400 km long and 300 to 400 km wide Himalayan region ensued following the uplift and mountain building process that led to the emergence of the Himalayan mountain ranges. This uplift process began with the collision of the Indian tectonic plate with the Asian continental plate between approximately 50 to 65 M yrs ago (Valdiya 2001). The collision resulted in the heavier geologic substrata of the Indian plate being subducted under the lighter layers of the Asian plate and the subsequent folding and faulting of the rock layers. Most of the Asian plate, along with some of the upper layers of the Indian plate, was uplifted to form a series of hill and mountain ranges (see Table 6.2 for the

Table 6.1 A short list of common rocks found in Nepal and the dominant minerals comprising them

Category	Name of rock	Dominant mineral(s)
Igneous rocks	Granite	Quartz
	Diorite	Feldspars, muscovite
	Gabbro	Muscovite, biotite, horneblende
	Basalt	Muscovite, biotite, horneblende
	Peridotite	Horneblende, biotite, augite
Sedimentary rocks	Limestone	Calcite
	Sandstone	Quartz
	Shale	Clays
	Dolomite	Dolomite
	Conglomerate	Variable
Metamorphic rocks	Gneiss	Variable
	Schist	Variable
	Quartzite	Quartz
	Slate	Clays
	Marble	Calcite

Source Adapted from Brady (1974)

approximate elevation ranges of the series). The ocean between the two tectonic plates gradually shrank to become the Neotethys Sea and eventually disappeared, resulting in the Indus-Tsangpo suture. The highest peaks of the Greater Himalaya (or Himadri) are relatively young (about 10 to 20 M yrs old) and still being uplifted (Valdiya 2001). Following collision and uplift, the ensuing down-cutting and geomorphologic processes became the basis for soil formation and destruction in the region.

6.2 Soil Formation Factors

6.2.1 Geology and Geomorphology

Refer to Chap. 4 for the explanation of geology and geomorphology found in Nepal. Against the backdrop of the orogenic processes leading to formation of the Himalayan mountain ranges, the resulting geologic zones and materials make up the precursors to parent materials from which different soil types result. A number of geologic zones are identified from south to north across Nepal, separated by faults running east–west. Towards the south of the country and extending into India is the Tarai region, which comprises the northern part of the Gangetic Plain. To the north of Tarai and separated by the Main Frontal Thrust (MFT) fault is the narrow, broken belt of the Siwalik range, also known as the Sub-Himalayan Zone. Further north of the Siwaliks lies the Lesser Himalaya (also called the Mahabharat range and the Middle Mountains in Nepal) separated from the former range by the Main Boundary Thrust (MBT) fault. This fault forms the active plate boundary where subduction of the Indian plate beneath the Asian plate continues (Whitehouse 1990). At the northern edge of the Lesser Himalaya lies the Main Central Thrust (MCT) fault, to the north of which is the Greater Himalayan zone. Still further north of the Greater Himalaya, extending into the Tibetan plateau, is the Tethys Himalaya, bounded on the south by the South Tibetan Detachment System (STDS) and on the north by the Indus-Tsangpo suture (Valdiya 2001; Dahal 2006).

The Tarai region of Southern Nepal is predominantly a depositional belt that merges into the Gangetic Plains of India. At the southernmost reaches bordering India, the soils are mainly derived from alluvial deposits of fine materials. In the middle of Tarai zone lies a marshy belt of about 10 to 12 km in width consisting of unconsolidated brown clay

Table 6.2 Elevation ranges of the cross-section from South to North of the Himalayan Region

Physiographic region	Elevation range (m)	Geographic location
Tarai plains	~80 to 250	Along the India border
Siwaliks range	250 to 800	East–west across Nepal
Lesser Himalaya	800 to 3000	East–west across Nepal
Greater Himalaya	3000 to 8000+	Northern Nepal/Southern Tibet
Tethys Himalaya	3600 to 5000+	Along Tibet–Nepal border

Source Valdiya (2001)

sediments with gravel or pebble contents. Numerous springs and Artesian wells emerge in this zone forming shallow lakes and swamps in low-lying areas. Further north, near the foothills of the Siwaliks range, lies the Bhabar zone, which is made up largely of pebbles, cobble stones, and boulders derived from rocks (mostly sandstone) of the Siwaliks and Lesser Himalayan ranges. This zone forms the aquifer recharge area for the Tarai region with streams and rivers disappearing into the permeable and highly porous substrate (Dahal 2006).

Just north of the Tarai region is the Siwaliks or Sub-Himalayan zone occupying an east–west belt ranging from 8 to 50 km in width. The region is made up mostly of fossil-rich fluvial deposits ranging from 1.6 to 23 M yrs old (Neogene age). The rock beds of the Siwaliks generally dip northwards with several east–west running thrusts. The lower Siwaliks are made up of laminated beds of greenish fine-grained sandstone and siltstone with alternating beds of variegated, multi-colored mudstone. The middle Siwaliks consists of medium- to coarse-grained salt-and-pepper sandstone inter-bedded with mudstone. The upper reaches of the Siwaliks range are mainly comprised of conglomerate and boulder beds of Lesser Himalayan rocks as well as lower amounts of sand and silt beds. Also present are massive and irregular mudstone beds rich in invertebrate fossils (Dahal 2006).

The Lesser Himalayan region consists of rock beds that are thrust southward over the rocks of the Siwaliks. The rocks of this zone are predominantly sedimentary and meta-sedimentary ranging from Precambrian to Eocene in age, namely, schist, phyllite, slate, quartzite, limestone, dolomite, etc., with some granitic intrusions. The Greater Himalayan range, north of the Lesser Himalaya, is comprised mainly of thick (about 10 km) layers of coarse-grained metamorphic rocks. These include kyanite-sillimanite mineral-bearing gneiss, marble, schist, and some granite in the upper layers. North of the Greater Himalaya, and extending across the Tibetan plateau, lies the Tethys-Himalayan range. In Nepal, this region occupies only limited portions of the north, in the west-central part of the country (the Manang, Mustang, and Dolpa areas). Here, the rocks range from Cambrian to Eocene in age and consist mostly of sedimentary rocks such as shale, limestone, and sandstone (Dahal 2006). Table 6.3 provides a brief list of the dominant rocks and landform processes in different physiographic zones across Nepal.

6.2.2 Climate and Vegetation

Refer to Chap. 3 for the explanation of different types of climate found in Nepal. Nepal is a narrow, elongated country

Table 6.3 Dominant rocks and landform processes in different physiographic zones of Nepal

Zone	Extent and elevation range	Dominant rock types	Major geomorphological processes
Tarai plains	20 to 50 km; 60 to 200 m	Alluvium, coarse gravel, conglomerate	Stream deposition, erosion, tectonic uplift
Churia range (Siwaliks)	10 to 50 km; 200 to 1300 m	Sandstone, mudstone, shale, conglomerate	Tectonic uplift, erosion, slope failure
Dun valleys (Churia)	5 to 30 km; 200 to 300 m	Fine to coarse alluvial sediments	Erosion–deposition, slope failure, tectonic uplift
Mahabharat and Middle Mountains (Lesser Himalaya)	10 to 35 km; 1000 to 3000 m	Schist, phyllite, gneiss, quartzite, limestone, granite	Tectonic uplift, erosion, slope failure, weathering
Midland valleys	40 to 60 km; 300 to 2000 m	Schist, phyllite, gneiss, marble	Tectonic uplift, erosion, slope failure, weathering
High Himalaya (Greater Himalaya)	10 to 60 km; 5000 + m	Gneiss, schist, marble, migmitites	Tectonic uplift, glacial erosion, slope failure
Inner/trans Himalaya (Tethys-Himalaya)	5 to 50 km; 2500 to 5000 m	Gneiss, schist, marble, and Tethyan sediments (shale limestone, sandstone, etc.)	Tectonic uplift, wind and glacial erosion, slope failure

Source Adapted from Dahal (2006) and Valdiya (2001)

wedged between India to the south, east, and west, and China (Tibet) to the north. Within its short latitudinal span ranging from 150 to 200 km in width, there are dramatic variations in climatic regimes. A key factor leading to such climatic variation is the drastic elevation difference across a short distance from the lowest point of about 60 m above mean sea level to the highest peak at 8848 m of Mount Everest. Another factor is the influence of two major weather systems, namely, the summer monsoon circulation during June to September and the westerly circulation from November to May. While the summer monsoons bring about 80 to 85% of annual precipitation, total amounts are higher toward the southeastern areas of the country, whereas the westerly derived winter precipitation has higher amounts in the northwestern part (Karki et al. 2015; Shrestha 2000).

The mean annual temperature variation across Nepal ranges from a high of about 28 °C in the southern Tarai belt to a low of less than −12 °C in the High (Greater) Himalayan Region. While maximum temperatures in the Tarai exceed 40 °C during the summer, winter lows typically remain at near 0 °C for much of the country (at elevations below 1000 m asl). In the hills and mountains, however, at elevations above 1000 m, winter low temperatures dip well below 0 °C. Apart from a few areas in the northwestern part of the country, mean annual total precipitation is greater than 1000 mm. For much of the country, mean total precipitation is in the range of 1500 to 2000 mm yr^{-1}. However, there are a few pocket areas in the east, central, and western development regions of the country with a perhumid moisture regime having a mean precipitation of 4500 to 5500 mm yr^{-1} (Karki et al. 2015).

Based on a modified Koppen–Gieger climate classification, Karki et al. (2015) proposed eight different climatic types across Nepal (Table 6.4). These include, from warmest to coldest: Tropical Savannah, Temperate with dry winter and hot summer, Temperate with dry winter and warm summer, Cold with dry winter and warm summer, Cold with dry winter and cold summer, Cold Arid Steppe, Polar Tundra, and Polar Frost climate. Temperate climate types cover the largest area across the country, followed by Tropical Savannah. These broad climatic regimes lead to the occurrence of dominant vegetation patterns and distinctive soil types with further variations brought about by localized differences in micro-climatic conditions.

The prevailing climate determines the dominant vegetation found in each region, and altitude plays an important role due to its effect on mean annual temperatures. At low elevations of the Tarai region, tropical deciduous forest occurs, with the dominant tree species being *Shorea robusta*, along with a variety of savannah grasses and shrubs. *Shorea robusta* is also widely found in the lower Siwalik region (Churia range and Bhabar zone), along with *Acacia catechu* and *Dalberigia sissoo*. In the western part of the country, *Adina cordifolia* and *Terminalia tomentosa* are also commonly seen along with an infinite variety of grasses and shrubs. At elevations of about 1000 m and higher, *Pinus roxbourghii* dominates in the west, while *Schima walichii* and *Castanopsis indica* occur in the eastern part of the country. Further up in elevation between about 2000 and 3000 m, Oak–Rhododendron forests dominate, with *Quercus semicarpifolia*, *Q. lamellose*, *Q. pachyphylla*, and *Rhododendron arboretum* being most common. In the sub-alpine altitudinal zone between 3000 and 4000 m, coniferous forests are most common in the west, while deciduous broadleaf forests can be found in the eastern areas. Species such as *Abies pindrow*, *A. densa*, *Betula utilis*, and *R. arboretum* commonly occur. At still higher elevations between 4000 and 5500 m, alpine vegetation can be found, including *Delphinium spp.*, *Rhododendron anthropogon*, Juniper, and various grasses and herbs (Hagan 1972; Valdiya 2001).

Table 6.4 Climate types and area coverage across physiographic regions of Nepal

Physiographic region	Climate type	Area, km^2	Area, %
Tarai, Lower Siwalik	Tropical Savannah	31,078	21.1
Siwalik/Churia valleys	Temperate with dry winter and hot summer	46,011	31.3
Upper Siwalik and Midland valleys	Temperate with dry winter and warm summer	33,133	23.5
Lower Mahabharat	Cold with dry winter and warm summer	6,232	4.2
Upper Mahabharat and Middle Mountains	Cold with dry winter and cold summer	669	0.5
Inner/trans Himalaya	Cold Arid Steppe	864	0.6
High Himalaya	Polar Tundra	26,797	18.2
High Himalaya	Polar Frost climate	2,399	1.6

Source Modified from Karki et al. (2015), Dahal (2006), and Whitehouse (1990)

6.2.3 Land Systems and Major Landform Types

The landform types across a landscape reflect the dominant geomorphologic processes shaping the land and soil types in each area. Across Nepal, a series of different land systems and corresponding landform types have been identified for each of the five major physiographic regions of the country (LRMP 1986). These clearly influence the soil formation (and destruction) processes active in each region. In the Tarai region, alluvial plains and fans make up most of the land area and deposition processes dominate. Active alluvial plains, present river channels, sand and gravel bars, and near-level flood plains that are frequently flooded are the main landforms. At slightly higher landscape positions, less frequently flooded, gently sloping to undulating slopes, as well as low-lying depressional areas and flat high ground can be seen. On the upper piedmonts, alluvial fans and apron complexes occur with slight to gentle slopes and relatively non-dissected to highly dissected gently rolling topography (LRMP 1986).

The Siwalik region to the north of the Tarai has, aside from active and recent alluvial plains, fans and aprons, ancient river terraces (Tars), and depositional basins or valleys (Duns). These landforms have areas ranging from very gently sloping to undulating and rolling terrain. Also present are moderate to steeply sloping hills (15–30° slopes), as well as steeply to very steeply sloping mountainous terrain (30–60° slopes).

In the Middle Mountain region, which includes the Mahabharat and part of the Lesser Himalaya, alluvial plains and fans occur in valleys and areas adjacent to major rivers and streams with steep gradients. Commonly occurring are ancient river terraces that may be dissected or non-dissected. However, much of the land area consists of moderate to steeply sloping mountainous terrain (30–60° slopes), and steep to very steep mountainous terrain (>60° slopes).

The High Mountain region, comprising parts of the Lesser Himalaya and Greater Himalaya, is predominantly past-glaciated, with moderate to very steeply sloping mountain terrain categorized based on elevation range below and above the limit of arable agriculture. Here, the active and recent alluvial plains, fans, and ancient alluvial terraces are confined to narrow areas adjacent to rivers and streams with steep gradients that make deep incisions into the mountainous terrain. The High Himal region comprises much of the Greater Himalaya and parts of the Tethys Himalaya bordering Tibet. It consists mostly of very steep and rugged terrain as well as glaciated areas with a preponderance of rocks and boulders with little or no soil formation (LRMP 1986).

6.2.4 Humans as a Factor Influencing Soil Development

Anthropogenic influences on soil formation occur because humans have manipulated soil from the inception of settled agriculture more than 10,000 yrs ago. The extent and intensity of impacts on the land progressively grew as civilizations expanded and global population increased along with the advancement of technology enabling tremendous exploitation of natural resources to meet the needs of humans. In Nepal, agriculture has been the mainstay of economic development for many generations. For centuries, farmers have been cultivating the land and re-shaping hill slopes through terracing and impounding water in valleys for paddy. Over many decades of tillage, addition of soil amendments, and irrigation, the physical, chemical, and biological properties of the soil and even the profile features are altered. Hence, farming practices that involve over-turning the soil, exposing subsoil, or burying topsoil, along with other artifacts of human civilization, over prolonged periods of time, lead to soils that develop distinctive characteristics that are a consequence of human influence. Such characteristics have, in recent years, become recognized as commonly occurring in human-impacted soils the world over. These soils are classified as having an "anthropic epipedon" (surface layer) within the USDA soil taxonomic system (Soil Survey Staff 2014).

6.3 Key Soil Formation Processes by Physiographic Region

Refer to Chap. 4 for the explanation of physiographic regions of Nepal. In Nepal, there are two key processes that dominate in soil genesis influenced by altitudinal gradients from north to south, and geomorphologic effects of (yet) young and active mountain ranges. In the high mountains, glacial and glacio-fluvial processes dominate (Baulmer and Zech 1994), while in the mountain valleys and plains, fluvial and aeolian processes are most prevalent. Hence, soils at high altitudes (above about 4000 m) tend to be very young soils at the initial stages of development and influenced mainly by orographic position and elevation, which has the effects of integrating climatic and biophysical factors (Baulmer and Zech 1994). These soils are developed from glacial or glacio-fluvial deposits from the last major glaciations and postglacial events. At intermediate elevations (~2500 to 4000 m), soils are mostly formed on glacial outwash plains, alluvial deposits, and colluvial debris from slips, slumps, and landslides (Reddy and Singh 1993). In the hills and mountains, soils tend to be formed on residual parent material (ridges and hill tops), colluvial debris (hill slopes and foot-slopes), and on alluvial deposits along the valley bottoms. In the lower valleys and plains, soils are mostly developed on depositional materials of alluvial or aeolian origin. Likewise, the soil types that occur across Nepal are consistent with the soil genesis processes and reflect the influence of altitude and monsoonal variation, which determine the overall climatic and biophysical characteristics across the physiographic regions.

6.3.1 High Himal Region

At elevations above about 5500 m in the High Himal region of the Greater and Tibetan-Tethys Himalayas, the annual average temperatures are almost always below freezing. The landscape is covered with snow and ice and dominated by boulders and rocks. Therefore, in this region, there is practically no soil development as there is a lack of vegetation and very little biological activity.

6.3.2 High Mountains Region

This region includes the Alpine and Sub-Alpine climatic zones of the Greater Himalayan range situated at altitudes of 3000 to 5000 m asl. Here, glacial and glacio-fluvial processes dominate, with parent material being mainly glacial till and alluvial sediment from glacial rivers on outwash plains. Therefore, soils are mostly very young to slightly developed. Commonly found soil orders include Entisols, Inceptisols, and Spodosols, of which Dystrochrepts, Haplumbrepts, and Cryumbrepts (some of which have calcareous materials). Great-groups are most common according to the United States Department of Agriculture (USDA) taxonomic classification (Soil Survey Staff, 2014). In the Alpine zone, Entisols are very shallow (8 to 12 mm), rocky soils with minimal soil development lacking in B horizons, whereas in the lower reaches of the Alpine zone, Spodosols with their characteristic ashen subsoil horizon can be found. Only in a few locations, with favorable temperature and moisture conditions, are Inceptisols observed, having somewhat more subsoil development than Entisols. In the Sub-Alpine zone, apart from Inceptisols, Alfisols, which have well-developed soil profiles and features, can be found, of which Haplustalfs are most common. A markedly contrasting situation is observed in the arid zone of the Trans-Himalayan region (parts of the Tibetan-Tethys-Himalaya), where Aridisols predominate. Due to the rain-shadow effect of the High Himalayan range, low annual total precipitation occurs here, typically in the range of 250 to 500 mm yr^{-1} or less and aeolian processes of wind erosion and deposition dominate.

6.3.3 Middle Mountains and Midland Valleys

In the Middle Mountains and valleys therein, fluvial processes predominate, with water erosion being the main weathering agent along with the action of gravity (colluvial deposition) and mass movements of steep slopes. Wind erosion is limited here but can have a seasonal impact during the prolonged dry season (October to May). Thus, soils in this region tend to be moderately developed and fall mostly in the orders of Alfisols, Inceptisols, and Entisols. Alfisols dominate on the upper slopes and ridges and are characterized by deep soil profiles with moderately developed subsoil horizons but low base saturation. Of this order, the Great-groups Haplustalfs and Hapludalfs are most common. Inceptisols and Entisols (Dystrochrepts and Eutrochrepts) are found mainly on the lower slopes and valleys where depositional processes dominate, particularly those of streams and rivers that are known to carry seasonally high sediment loads owing to the active geology of the region (Chalise and Khanal 1997; Sharma 1997). Occasionally, Mollisols (Argiudolls) and Vertisols (with high shrink-swell clay contents) form on lacustrine deposits in valleys that were once inundated by glacial lakes, such as Kathmandu and Banepa.

6.3.4 Siwalik Region and Dun Valleys

In this region, comprised of alluvial fans and gentle to moderately sloping terrain, both alluvial and aeolian deposits can be found due to gentle to moderate slopes with many streams and rivers leading to water erosion, as well as a prolonged dry season during which wind erosion can also occur. On the upper slope positions and level summit areas of hills, mature soils with well-developed profiles are commonly found, whereas along the lower slopes and near stream or river banks, younger soils with less profile development tend to occur. The lower hills and plains region have mature, well-developed soils due to high temperatures and seasonal moisture contents. The predominant soil order found is Alfisol (Rhodustalfs, Haplustalfs, and Hapludalfs) with occasional pockets of Oxisols and Histosols. Soils tend to be reddish-brown in color, indicating high contents of Fe and Al oxides as typified by significantly weathered and mature soils. In the floodplains of active streams and rivers, however, Inceptisols and Entisols may also occur on recently deposited sediment.

6.3.5 Tarai Region

This region is made up of level to gently sloping plains with mostly southward flowing rivers and streams, as well as depressional (low-lying) areas that form swamps and shallow wetlands. Areas adjacent to rivers with active seasonal flooding have frequent deposition of sand and silt where young soils (Entisols) form, of which, Udorthents, Ustorthents, and Psamments are common. Also, frequently found are Inceptisols (Eutrochrepts, Udochrepts, and Ustochrepts), which are relatively young but with weakly developed subsoil features and that are found on upper, non-flooded parts of the plains, and the lower reaches of alluvial fans. In depressional and frequently flooded areas, both Inceptisols and Entisols are found, including Haplaquepts, Haplaquents, and Psammaquents. Oxisols are likely to be found only in a few localized areas in the southeastern Tarai near the Indian border, while small areas of Histosols occur in and around wetland areas of the southern plains. Only rarely do Mollisols occur where high organic matter, deep profiles, and high base saturation result from a combination of rapid vegetation turnover, suitable landscape position, and calcareous parent material. In a few similar areas that are frequently wet or regularly flooded, Haplaquents with calcareous materials can also be found. A short list of major soil types in each of the main physiographic subdivisions of the country is provided in Table 6.5.

Table 6.5 Major soil orders and dominant soil types in different physiographic regions of Nepal

Physiographic region	Major soil orders	Main suborders	Dominant great-groups	Remarks
High Himal	–	–	–	Rocks and ice only
High Mountains	Inceptisols	Ochrepts, Umbrepts	Dystrochrepts, Haplumbrepts, Cryumbrepts*	*with calcareous materials
	Alfisols	Ustalfs	Haplustalfs	South-facing upslope areas
	Entisols	Psamments	Ustipsamments	Along riverbanks
Middle Mountains	Inceptisols	Ochrepts, Umbrepts	Dystrochrepts, Haplumbrepts	Back-slope areas
	Alfisols	Ustalfs, Udalfs	Hapustalfs, Rhodustalfs, Hapludalfs	Ridge tops and south-facing areas
	Entisols	Aquents, Arents	Haplaquents, Haplarents	Along riverbanks
Inner Tarai	Entisols	Aquents, Orthents	Haplaquents, Udorthents, Ustorthents	Depositional areas
	Inceptisols	Ochrepts	Dystrochrepts, Eutrochrepts	Lower slope areas
	Mollisols	Udolls	Argiudolls	Limited areas
Siwalik/Churia	Inceptisols	Aquepts, Ochrepts	Haplaquepts, Eutrochrepts	Lower slope areas
	Alfisols	Ustalfs, Udalfs	Rhodustalf, Hapludalfs	Level upslope areas
	Entisols	Aquents, Psamments, Orthents	Haplaquents, Psammaquents, Udipsamments, Ustorthents	Along riverbanks and depositional areas
Tarai plains	Entisols	Aquents, Psamments, Orthents	Haplaquents,* Psammaquents, Eudorthents, Ustorthents	*with calcareous materials
	Inceptisols	Aquepts, Ochrepts	Haplaquepts, Eutrochrepts	Slightly developed
	Alfisols	Ustalfs, Udalfs	Hapludalfs, Haplustafs	Mature, well-developed
	Histosols, Oxisols	Folists, Hemists; Udox, Ustox	Medifolists, Luvihemists; Hapludox, Haplustox	Limited/small areas

Source Modified from data of Soil Science Division, Nepal Agricultural Research Council

6.4 Imperatives for Sustainable Production and Management of Land in Nepal

6.4.1 Key Considerations

While the plains and mid-hills regions of Nepal are densely populated and have limited productive agricultural lands, demands on the soil and land-based resources continue to steadily increase. With nearly two-thirds of the population directly or indirectly dependent on agriculture accounting for about 33% of the national gross domestic product (GDP), food production at present falls short of domestic requirements (MOAC 2014). This is largely due to the nearly stagnant production levels of major food crops and suboptimal yields caused by the diminishing fertility of cultivated land from a lack of replenishment of soil organic matter and inadequate application of fertilizers (Bajracharya and Sherchan 2009; Karki 2006).

Soils have, in recent decades, gained recognition for their ability to serve as sinks or sources of greenhouse gases, namely, carbon dioxide (CO_2), methane (CH_4), and nitrous oxide (N_2O). The accelerated global warming phenomenon attributed to human activities such as the burning of fossil fuels and clearing of vast areas of forest could potentially be mitigated through carbon sequestration in soils. This is because soils contain about 75% of the terrestrial carbon pool ($\sim$1500 pg organic C and $\sim$ 1700 pg inorganic C), that is, nearly three times more than the amount of carbon stored in living organisms and the atmosphere combined (FAO 2001; Lal and Kimble 1994). It is therefore essential to conserve soils and enhance their productive potential to maintain a global climatic balance, as well as to nurture the ever-growing population.

6.4.2 Future Priorities and Outlook

The consequence of the increasing pressures upon the land is a steady degradation of the soil through erosion, fertility loss, and reduced depth. The forms of degradation are interrelated and reinforce one another usually initiated by the clearing and destruction of vegetation. Once the protective and stabilizing vegetative cover is removed, surface runoff and overland flow of rain water increases while infiltration and percolation into the ground is reduced. Thus, soils become susceptible to various forms of erosion, which removes the fertile topsoil and subsequently reduces the effective plant rooting depth (Lal 1997). Furthermore, the eroded sediment and nutrients washed away into rivers and streams often have adverse consequences downstream (Merz 2004). Hence, the cycle of degradation, once set-in motion, builds up momentum, and if un-reversed, could exceed the critical threshold beyond which recovery of the agro ecosystems may become difficult (Bajracharya et al. 2007). Therefore, it is imperative that policymakers, conservation workers, and farmers all work in a coordinated manner to urgently take the requisite precautions to safeguard our life-giving soils.

References

Bajracharya RM, Sitaula BK, Sharma S, Jeng A (2007) Soil quality in the Nepalese context - an analytical review. Int J Ecology Environ Sci 33(2–3):143–158

Bajracharya RM, Sherchan DP (2009) Fertility status and dynamics of soils in the Nepal Himalaya: a review and analysis. IN: Lucero DP, Boggs JE (eds) Soil fertility. Nova Science Publisher Inc., New York, pp 111–135

Baumer R, Zech W (1994) Soils of the high mountain region of Eastern Nepal: classification, distribution and soil forming processes. CATENA 22:85–103

Brady NC (1974) The nature and properties of soils, 8th edn. MacMillan Publisher Co., Inc. New York, pp 277–288, 303–312

Brady NC (1995) The nature and properties of soils, 10th edn. Prentice Hall of India Pvt. Ltd., New Delhi, India

Buol SW, Hole FD, McCracken RJ, Southerd RJ (1997) Soil genesis and classification, 4th edn. Lowa State University Press, Ames, Iowa, pp 3–41

Chalise SR, Khanal NR (1997) Erosion processes and their implications in sustainable management of watersheds in Nepal Himalayas. IAHS Publ 246:325–334

Chamlagain D, Gautam D (2015) Seismic hazard in the Himalayan intermontane basins: an example from Kathmandu Valley, Nepal. In: Shaw R, Nibanupudi HK (eds) Mountain hazards and disaster risk reduction, disaster risk reduction, pp 75–103

Dahal RK (2006) Geology for technical students. Bhrikuti Academic Publications, Kathmandu, Nepal, p 756p

FAO (1995) Planning for sustainable use of land resources: towards a new approach. FAO land and water bulletin 2. Rome, 60p.

FAO (2001) Soil carbon sequestration for improved land management. Food and Agriculture Organization of the United Nations, Rome, 62p

Hagen T (1972) Nepal: the Kingdom in the Himalayas. Publ. Robert Hale and Co., 264p

Karki R, Talchabhadel R, Aalto J, Baidya SK (2015) New climatic classification of Nepal. Theor Appl Clim (Springer-Verlag Wien, Published online), 10p

Karki KB (2006) Impact of cropping intensification on nutritional balance in Nepalese soils. In: Proceedings of international seminar on environmental and social impacts of agricultural intensification in Himalayan Watersheds. Held Oct. 15–17, 2006, Kathmandu, Nepal, pp 27–33

Lal R, Kimble JM (1994) Soil management and the greenhouse effect In: Lal R, Kimble JM , Levine E (eds), Soil processes and greenhouse effect, USDA, SCS, National Soil Survey Center, Lincoln, NE, USA, pp 1-5

Lal R (1997) Degradation and resilience of soils. Phil Trans R Soc Lond B 352:997–1010

LRMP (1986) Land resources mapping project: summary report. HMG Nepal and Govertment of Canada, Kenneth Earth Sciences Ltd. Publ. 104p

Merz J (2004) Water balances, floods and sediment transport in the Hindu Kush-Himalayas. Data analyses, modeling and comparison of selected meso-scale watersheds. Ph.D. Dissertation, University of Berne, Publ. Int'l. Center for Integrated Mountain Dev., Kathmandu, Nepal, 339p

MOAC (2014) Statistical information on Nepalese agriculture 2013/2014, Ministry of Agriculture and Co-operatives, Agri-business Promotion and Statistics Division, Agri-statistics Section, Kathmandu, Nepal, 212p

Reddy VS, Singh JS (1993) Changes in vegetation and soil during succession following landslide disturbance in the Central Himalaya. J Environ Manag 39:235–250

Sharma CK (1997) A treatise on water resources of Nepal. Kathmandu, Nepal, 493p

Shrestha ML (2000) Interannual variation of summer monsoon rainfall over Nepal and its relation to Southern Oscillation. Meteorol Atmos Phys 75:21–28

Soil Survey Staff (2014) Keys to soil taxonomy, 12th edn. United States Department of Agriculture, Natural Resource Conservation Service, 372p

Valdiya KS (2001) Himalaya: evolution and emergence. University Press, Hyderabad, India, 139p

Whitehouse IE (1990) Geomorphology of the Himalaya: a climato-techtonic framework. NZ Geogr 46(2):75–85

Soil Types, Soil Classification, and Mapping

7

Subhasha Nanda Vaidya, Dil Prasad Sherchan, Krishna Raj Tiwari, Subash Subedi, Krishna Bahadur Karki, Dinesh Panday, and Roshan Babu Ojha

Abstract

Nepal exhibits unique topographic features with great biodiversity variation within a short gradient. The variation, along with the gradient, is coupled with heterogeneous soil types. Entisols, Inceptisols, Alfisols, and Mollisols are commonly found soils in Tarai and the Middle Mountain physiographic regions. Similarly, the Siwaliks and High Mountain physiographic regions are dominated with Entisols and Inceptisols. The High Himalayan physiographic region is dominated with Inceptisols and Spodosols with rock outcrops. However, a different topo-sequence of soil is observed even in different aspects of the same topography. Soil variation with a microclimatic difference manifests the possibility of growing a wide range of crops that can contribute to the country's gross domestic product. Additionally, different soil types are promising for maintaining soil biodiversity which can be nurtured with proper identification and conservation of the soils in the region. As such, proper management of soil is necessary to keep up diversity to secure and foster the Nepalese economy.

Keywords

Alfisols • Entisols • Inceptisols • Mollisols • Physiography • Topography

S. N. Vaidya (✉) · D. P. Sherchan · K. B. Karki · R. B. Ojha
Nepal Agricultural Research Council, Kathmandu, Nepal

K. R. Tiwari
Institute of Forestry, Tribhuvan University, Pokhara, Kaski, Nepal

S. Subedi
Genesis Consultancy Pvt. Ltd./Geoinformation and Earth Observation Services, Lalitpur, Nepal

D. Panday
Department of Biosystems Engineering and Soil Science, The University of Tennessee-Knoxville, Knoxville, TN, USA

7.1 Introduction

From the time cultivation began, humans noticed the differences in soils and classified them according to their suitability for different uses. In general, farmers have important knowledge for soil classification based on physical soil properties and qualify soil quality to produce crops (Nath et al. 2015). Farmers used descriptive names such as *black soils*, *red soils* or *yellow soils* based on soil color. Similarly, they classify soil as clay, loamy, or silty based on textural classes indicating different levels of workability during tillage. Other soil names such as *limestone soils, piedmont soils, sandy soil*, and *alluvial soils,* are still in common use today and suggest the parent materials from which the soil was formed. These terms can convey valuable meaning to local users but they are inadequate for helping us organize our scientific knowledge of soils or defining relationships among the soils of the world (Barrera-Bassols et al. 2006). So, a system of soil classification was developed to create a common understanding.

Soil classification systems vary in different areas and soil is classified using traditional to scientific approaches. A single soil classification system is essential for fostering global communication about soils not only among soil scientists, but for all people concerned with the management of land and the conservation of soil resources (Brady et al. 2008). Classification is the process of ordering soils into groups with similar properties and potential uses (Fitzpatrick 2013). Each country has adopted either a soil taxonomy system (Soil Survey Staffs, 2010) or the IUSS Working Group WRB (2007) for the general purpose of broad soil classification that communicates soil information at international scales. These classification systems provide valuable broad conceptual understanding of soils in terms of the morphological, physical, and chemical properties of soil profiles (Fitzpatrick 2013).

The purpose of any classification is to organize our knowledge so that the properties of objects may resemble

R. B. Ojha and D. Panday (eds.), *The Soils of Nepal*, World Soils Book Series,
https://doi.org/10.1007/978-3-030-80999-7_7

one another and be easily understood in terms of their relationship for a specific purpose. Soil classification is often referred to as a taxonomy, since the system follows a procedure and criteria that forms well-defined groups and exists within a set of rules for allocation (Mishra, 2016). Soil is classified using multiple approaches because of its complexity and diversity. Through such systems, we can take advantage of research and experience at one location to predict the behavior of similarly classified soils at another location. Therefore, through a classification system, we can create a universal language of soils that enhances communication among users of soils around the world.

Located between 26°22' to 30°27' N latitude and 80°4' to 88°12' E longitude with a total area of about 147,181.0 km^2, Nepal is a country of extremes, both in its landscape and its climate. It has vast, a flat to almost flat plain in the south to very highly rugged mountainous terrain in the north. Its elevation above mean sea level ranges from about 55 m in the south to the world's highest peak, the Mount Everest (8848.86 m, the new height of Mt. Everest) in its extreme north. Its climate ranges from hot and humid Subtropical to Arctic. Though very small in size, so extreme is the country's diversity that one can easily see bananas and apples ripening in the same season just by driving a few km from south to north. This indicates that Nepal is diverse in its landscape and climate (Vaidya and Maskey 2000). Naturally, these diversities have their marked impact on the occurrence of varied soils in the country. Climate, Organisms (Biota), Relief, Parent materials, and Time (ClORePT) have universally been recognized as the most important factors that influence the formation and occurrence of soils in a given place (Chap. 6). Hence, a summarized description of these important features of Nepal will be relevant in understanding its soils.

7.2 Physiographic Regions of Nepal

Based on the grouping of the repeating patterns of landforms (Nelson; FAO 1980), the existence of the following five physiographic regions has been recognized in Nepal (Fig. 7.1). These regions represent well-defined geographical areas, different land features, bedrock geology, geomorphology, and climate (Carson et al. 1986).

Tarai: Bordered to the south by the Indo-Gangetic plains and to the north by the Siwalik physiographic region, the Tarai physiographic region is generally flat with a minor relief of approximately 2–10 m km^{-2}. The elevation ranges between 60 and 330 m asl from south to north. The region is also called the northern edge of the Gangetic Plain, whose geology consists primarily of recent and post-Pleistocene alluvium materials. It is built by alluvial deposits from rivers, and its major deposits are formed in recent periods or reworked by water. Four major rivers (Koshi, Gandaki, Karnali, and Mahakali) and their tributary rivers are the major sources of the alluvial deposit in the region. It consists mainly of recent to post-Pleistocene quaternary alluvial deposits, predominantly loamy textured and stone free soil (LRMP 1986).

Within this physiographic region, three different land systems such as Active alluvial plains, Recent Alluvial plains (Lower piedmont), and Upper piedmont plains have been recognized (LRMP 1986). The active alluvial plains constitute the lowest level plains in this region because it lies in the proximity of rivers that are subjected to annual flooding during the monsoon season. The recent alluvial plains (the lower piedmonts) are of intermediate level and are mostly separated from the intermittent flooding during monsoons. The upper piedmont plains, which forms the highest level of the plains, lie near the Siwalik Hills. Within each of these various level plains, land types vary based on their proximity to rivers and their land position. This region constitutes about 14.6% of the country's total area.

Siwaliks: The Siwaliks physiographic region is located between the Tarai upper piedmont plains to the south and the Middle Mountain region to the north. It is generally made up of tertiary, interbedded sandstones, shale conglomerates, and quaternary alluvium, subtropical. The elevation ranges from about 300 to 700 m asl. The landscapes are quite rugged with high relief. The region has plains as well as steep mountainous terrains. The region consists of young hills and mountains and is one of the most fragile and vulnerable due to soil erosion and landslides (Ghimire et al. 2013). Soil formation is mainly dominant as active alluvial deposit in this region.

Within the plains, three different types of landforms occur (a) Active and recent alluvial plains (b) Fans, aprons, and ancient river terraces (c) Depositional basins or Dun valley (LRMP 1986). The mountainous terrain is generally moderately steep to very steep. The plain portions are generally termed as inner Tarai and or Dun valleys. They are structurally controlled valleys whose outlets at some points were blocked by the rapid tectonic uplift (LRMP 1986) of the Siwalik range to the south. Most agriculture and settlements in the Siwaliks region are centered on these plains. The major inner Tarai/dun valleys of the country are Chitwan, Surkhet, Dang-Deokhuri, Kamala, Trijuga, and northern Nawalparasi. The region constitutes predominantly north dipping, semi-consolidated interbedded tertiary to quaternary sandstones, mudstones, siltstones, and conglomerates

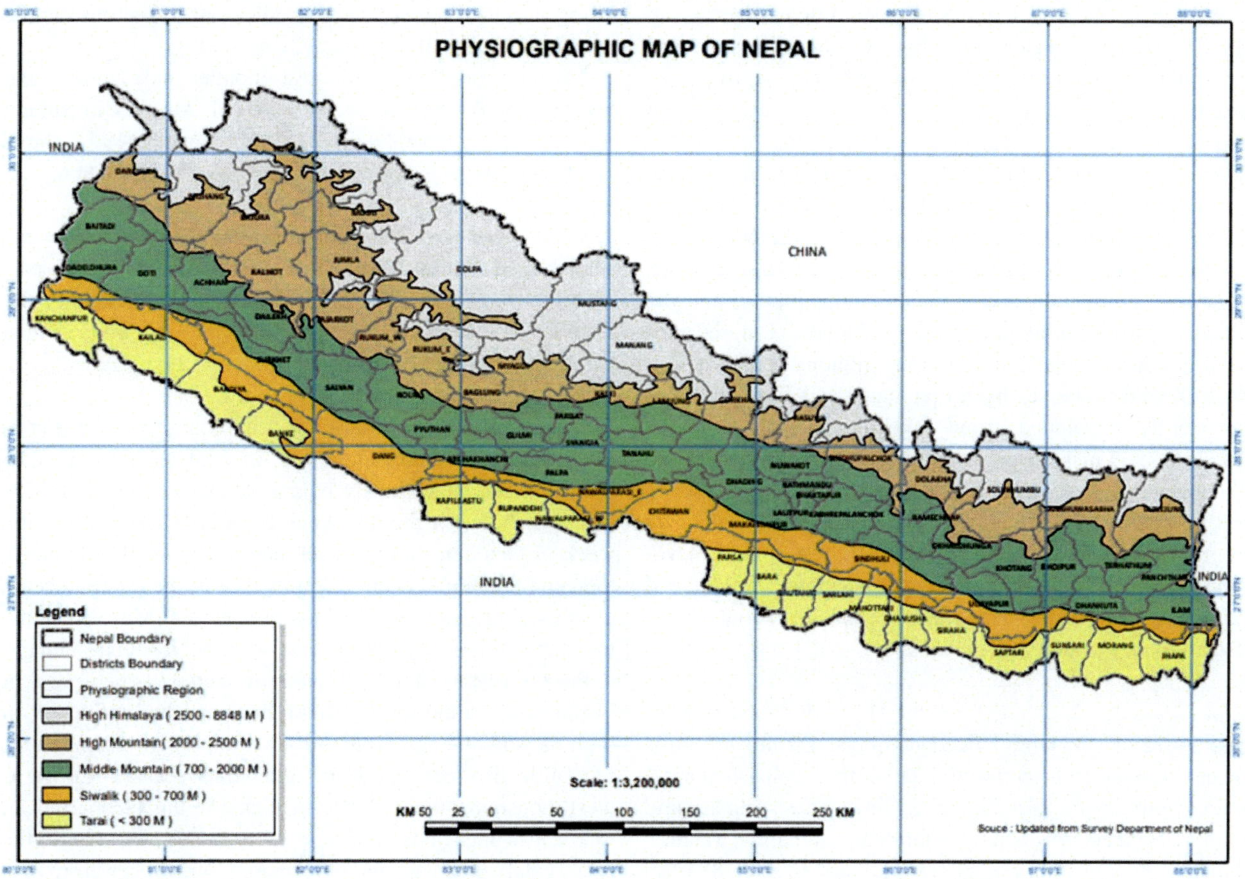

Fig. 7.1 Map of Nepal showing different physiographic zones

(LRMP 1986). This region constitutes about 12.7% of the total land area of the country.

Middle Mountains: The Middle Mountain region is located between the Siwaliks region to the south and the High Mountain region to the north. The elevation ranges from 700 to 2000 m asl. The region consists of Precambrian to Cambrian phyllites, schists, quartzites, granites, and limestones (LRMP 1986). These rocks and minerals are generally deeply weathered and formed in the subtropical to warm temperate climate. Within this region, the Kathmandu valley, a fine example of a tectonic valley, can be found. During the tectonic uplift of the Mahabharat Mountain (the middle mountain), the Bagmati River of the Kathmandu valley was dammed and remained as a lake. Later, it was successful in breaking the geological fault line at Chobhar, draining the water. The famous Pokhara valley is also located in this region. The landscapes of this middle mountain region are generally rugged with high relief. The region contains both nearly level plains and steep to very steep mountainous terrain. The proximity of the river and the land position have marked influence on the nature of alluvium, and hence the soils. The elevation ranges from 700 m to about 2000 m asl. Diverse climate and landscapes support year-round agricultural production on river valley plains and on the sloping and level terraces of the mountainous terrain. Soil formation in this region is mainly dominated by residual parent materials (ridge and hilltops), colluvial debris (hill slopes and foot-slopes), and alluvial deposit of the ancient terrace along the river valley bottoms (Bajracharya 2013). Soils of this region tend to be moderately developed in the valleys to mature in the plain/tar areas. This region constitutes about 29.5% of the country's total land area.

High Mountains: The High Mountain physiographic region is located between the Middle Mountain physiographic region in the south and the High Himal physiographic region in the north. The elevation of this region ranges from 2000 to 3000 m asl. It consists mainly of Precambrian to Eocene gneiss, quartzites, and schists. Phyllites and limestones are generally not deeply weathered, glaciated, and are warm temperate to alpine. The landscapes in the region are very rugged with very high relief (>3000 m asl). As in the Middle Mountain region, the High Mountain region contains both areas that are nearly plains and steep mountainous terrain. However, in this region, the strips of nearly plain area along the major rivers are relatively smaller. All the valleys in this region have been glaciated. The region has highly metamorphosed Precambrian to Eocian gneisses, quartzites, schists, phyllites, and limestones (LRMP 1986). The High Mountain region also has limited land area for agriculture production due to its steep topography, and the climate is also limited for crop production. Limited soil information is available in this region. Glacial and glacio-fluvial processes have dominated soil formation and are mainly developed from deposits of the last main glaciations, which are predominated by para-gneisses (Baulmer and Zech 1994). It is reported that soil formation in the High Mountain region is strongly influenced by orographic position (Righi and Lorphelin 1986) and elevation (Carson 1992), which essentially integrates climatic and bioclimatic parameters. This region constitutes about 19.7% of the country's total land.

High Himalaya: Located above the High Mountain physiographic region, the High Himalayan physiographic region is the last frontier of the country in the north. It consists mainly of Precambrian to Eocene gneiss, limestones, schists, and granites, with active glaciations formed in subalpine to arctic climate. The elevation ranges from around 3000 to 8849 m asl. The etymological meaning of Himalaya in the Nepali language means "house of snow," and hence, most of the country's snow peak are in this region. The High Himalaya region has very small plain areas along the major rivers where agricultural land and major settlements are located. The region has Precambrian to Eocene gneisses, limestones, schists, and granites, along with shales. Over 86% of the region has bedrock at or near its surface. The region constitutes about 23.7% land area of the country.

7.3 Climate of Nepal

Though small, Nepal contains extremely diverse climates, subtropical to arctic temperature regimes, and arid to per-humid moisture regimes. High relative relief and pronounced wet and dry monsoon seasons are characteristic features of the country that can be considered responsible for the present state of diversity in ecosystem, flora, and fauna.

Temperature: The temperature decreases as elevation increases in Nepal. With each 200 m gain in elevation, the temperature drops by 1 °C. Over 90 percent of the variation in mean annual temperature for all weather stations can be explained by elevation alone (Table 7.1).

May to June are the hottest months and January and February are the coldest months in Nepal. Absolute maximum and minimum temperatures range approximately 15 °C above and below the mean annual temperature for any station.

Soil Temperature: Data on soil temperature are rarely available in Nepal. Mean annual soil temperature is estimated as mean annual air temperature plus 1 °C based on the assumption that soil warms up before and cools off later than the air temperature immediately above (Table 7.2).

Precipitation: Eastern monsoons originating from the Bay of Bengal have a marked influence on climate of Nepal. The monsoon season generally begins in the last week of May and ends between the last week of September and the first week of October, during which about 85% of the country's total annual rainfall occurs. About 10% of the rainfall occurs during the pre-monsoon months of March to May. Since the monsoons originate from the east, eastern Nepal has a longer monsoon season than the western Nepal; hence, eastern Nepal receives more rain, which has a marked influence on both the soil and agricultural production. A small amount of rainfall is also received during the winter season. Known as west monsoons, these storms originate in the Mediterranean Sea, and winter rains are active more in the west than in the east. Total measured annual precipitation ranges from less than 300 to 5,000+ mm. There exists major seasonal as well as local variation in precipitation. The orientation of the mountain range seems to have a remarkable influence on the distribution of rainfall. For example, the Nepalgunj area located at the southern border with India receives about 1,200 mm rainfall annually, while Chisapani, which is located about 20 km north at the foothill of Siwalik, receives about 2,200 mm rainfall. Such variations are quite common throughout the country. Some of the high mountain valleys by virtue of being located at the rain shadow receive even less rainfall annually (annual precipitation < 300 mm).

7.4 Early Work in Soil Classification and Mapping

The local system of soil classification is an important and easy to understand tool that could be applied for improvement. It does not encompass all soil types found in the country and is not sufficient for explaining variations. Soil is classified differently in different regions of Nepal, and soil color and clay content-based soil classification is widely used (Tables 7.3 and 7.4).

Table 7.1 Relationship between elevation, climate zone, and temperature

Elevation, m	Mean annual air temperature, °C	Climate zone
<1000	20 to 25	Subtropical
1000 to 2000	15 to 20	Warm temperate
2000 to 3000	10 to 15	Cool temperate
3000 to 4500	3 to 10	Sub-alpine, alpine
>4500	<3	Arctic like

Source LRMP (1986)

Table 7.2 Relationship between soil temperature regime and elevation

Elevation range, m	Mean annual soil temperature, °C	Soil temperature regime
<800	>22	Hyperthermic
800–2200	15–22	Thermic
2200–3600	8–15	Mesic
3600–5200	0–8	Cryic
>5200	< 0	Pergelic

Source LRMP (1986)

Table 7.3 Local soil classification based on color that indicates soil productivity (Shah 1995)

Local terminology	Soil Colour	Levels of soil fertility
Kalo mato (Black soil)	Black soil to very black soils	High to very high fertile
Rato mato (Red soil)	Dark brown to reddish brown	Moderate to high fertile
Kamero mato (White soil)	Pinkish white, light gray	Low fertile
Fusro mato (Gray soil)	Very dark grayish and gray	Marginal

Table 7.4 Local terminology used to classify soils by farmers based on clay content (Shah 1995)

Class	Soil classification local term used	Major properties	Remarks
1	*Chimte*	High clay content	35–45% clay content
2	*Dumat*	Loam	Approximately 20% clay
3	*Balute*	Less than 10% clay	Coarse, rough feeling
4	*Pango*	Silty	70% silt content, 20% clay

7.4.1 Soil Survey Work

Bidur Kumar Thapa initiated survey work and travelled across the Kavre and Dolakha districts collecting soil samples. M.L. Pradhan is the first agriculturist who dug a soil profile in Nepal and led a soil survey team. Several agriculture projects such as the UNDP/FAO UNDP cotton development project conducted soil surveys of the Tarai district, including the Sunsari, Saptari, Jhapa, Palpa, Kanchanpur, and Kailali districts. Comprehensive work on soil classification was completed in 1986 under the Land Resource Mapping Project (LRMP). The Soil Science Division (SSD) of the Nepal Agricultural Research Council supported soil survey work under LRMP. The project is multidisciplinary and includes the mapping of all land resources according to soil type and land capability.

Soil surveys with the following scale is carried out in Nepal:

1. General reconnaissance, or exploration type surveys: These surveys are carried out to identify potential areas and extensive work at provincial level as they occupy five to six districts. The scale of mapping is 1:1,000,000 for regional-/provincial-level planning.
2. Semi-detailed surveys: These surveys are carried out for district-level planning. The scale of soil mapping used in these surveys is 1:50,000 or 1:125,000. Medium intensity of sampling or checking is required.
3. Detailed survey: These surveys are conducted with a 1:10,000 scale. The target area includes irrigation projects and watershed studies, usually with high-intensity soil sampling. National land use planning projects use

this scale to determine the local scale of land suitability for the entirety of Nepal.

7.5 Soil Mapping and Classification System in Nepal

Soils of Nepal are described according to different geomorphic mapping units within each physiographic region. A soil map has been prepared based on the SOTER database and the Land Systems Report of the Land Resources Mapping Project (LRMP 1986). This report gives extensive and widely covered soil information for Nepal, and hence, the information from this report has been extensively used in preparing this text. Soils have been classified per the USDA soil taxonomy. Since the time of the project, significant modifications and revisions have been made in the USDA soil taxonomy; however, the soil map and descriptions follow the same system. A few examples applicable to Nepal's context are stated here. The most widely found soil in the alluvial plains of Nepal is the Haplaquepts great group of the Aeric and Typic subgroups. However, in the revised taxonomy, there is no Haplaquepts. The presently used term Endoaquepts seems to resemble Haplaquepts. Similarly, while Dystrochrepts and Ustochrepts soils are found extensively in Nepal, in the revised version of the USDA Soil taxonomy, there is no Ochrepts suborder. The present-day soil taxonomic suborders Udepts and Ustepts resemble Ochrepts. In this present text, references have also been taken from recent soil studies conducted by the National Land Use project (NLUP 2020) of the Survey department of the government of Nepal. In these reports, recent revisions have been considered. As a result, the Endoaqupts of both aeric and typic subgroups and Haplustepts and Dystrudepts and Eutrudepts are extensively described in place of Haplaquepts, Ustochrepts, Dystrochrepts, and Eutrochrepts. Profile descriptions of some dominant soils of Nepal have been given. Those taken from the LRMP (1986) reports show the older version while those taken from the NLUP (2020) show the revised system as discussed above.

The soils of Nepal are also classified according to the World Reference Base (WRB) soil classification system (Vaidya and Sah 2014). Here, we will describe soils found in different land systems within each physiographic region using both USDA (Fig. 7.2) and WRB (2007) classification systems.

7.5.1 Land System Units and Soil Types

7.5.1.1 Tarai

Active alluvial plains: These lowest level plains in the Tarai cover about 1.6% of the total land area of Nepal, including the present river channels. The areas are annually flooded, with a slope gradient less than one degree. Depending on the proximity to the river and the position, different land types occur, and this land has marked influence on the occurrence of floods and the nature of alluvial deposits.

Soil Types (USDA): Ustifluvents, Fluvaquents, and Ustochrepts, which are at certain places calcareous, are the major soils occurring in this unit, with some Psamments adjacent to rivers and Fluventic Haplaquepts at the higher positions commonly found in this region.

Soil Types (WRB): Calcaric Fluvisol followed by some Gleyic Fluvisols and Eutric Cambisols are commonly found in this region.

Recent alluvial plains (Lower Piedmont plains): This intermediate-level alluvial plains of the Tarai covers about 7.9% of the total land area of Nepal. The area is plain, with a slope gradient of less than one degree. Within this intermediate level of plains, slight variations in land positions are also found.

Soil Types (USDA): Except for the land occurring at higher positions where moderately well-drained Ustochrepts and some Haplustolls occur, almost all areas have poorly drained Haplaquepts of typic subgroups in the depressional land and imperfectly drained Haplaquepts of the aeric subgroup occurring in the rest of the plains.

Soil Types (WRB): Eutric Gleysols, Eutric Cambisols, and some minor Haplic Phaezome soils are found in this land unit.

Upper piedmont Plains: Located near the Siwaliks hills, the highest-level plains in Tarai (upper piedmont plains) covers about 4.9% of the total land area of Nepal.Made up of an alluvial fan and apron complex, the area has high relative relief with increasing slopes.

Soil Types (USDA): Rapid to well-drained Haplustolls, Dystrochrepts, and Ustochrepts are the major soils occurring on the highest-level plains of Tarai.

Soil Types (WRB): Phaezomes (Calcaric and Haplic) and Cambisols (dystric and Eutric) are commonly found soils in this land unit.

7.5.1.2 Siwaliks

Active and recent alluvial plains: The active and recent alluvial plains including the river channels and sandbars of the Siwaliks covers about 1.01% of the total land area of the

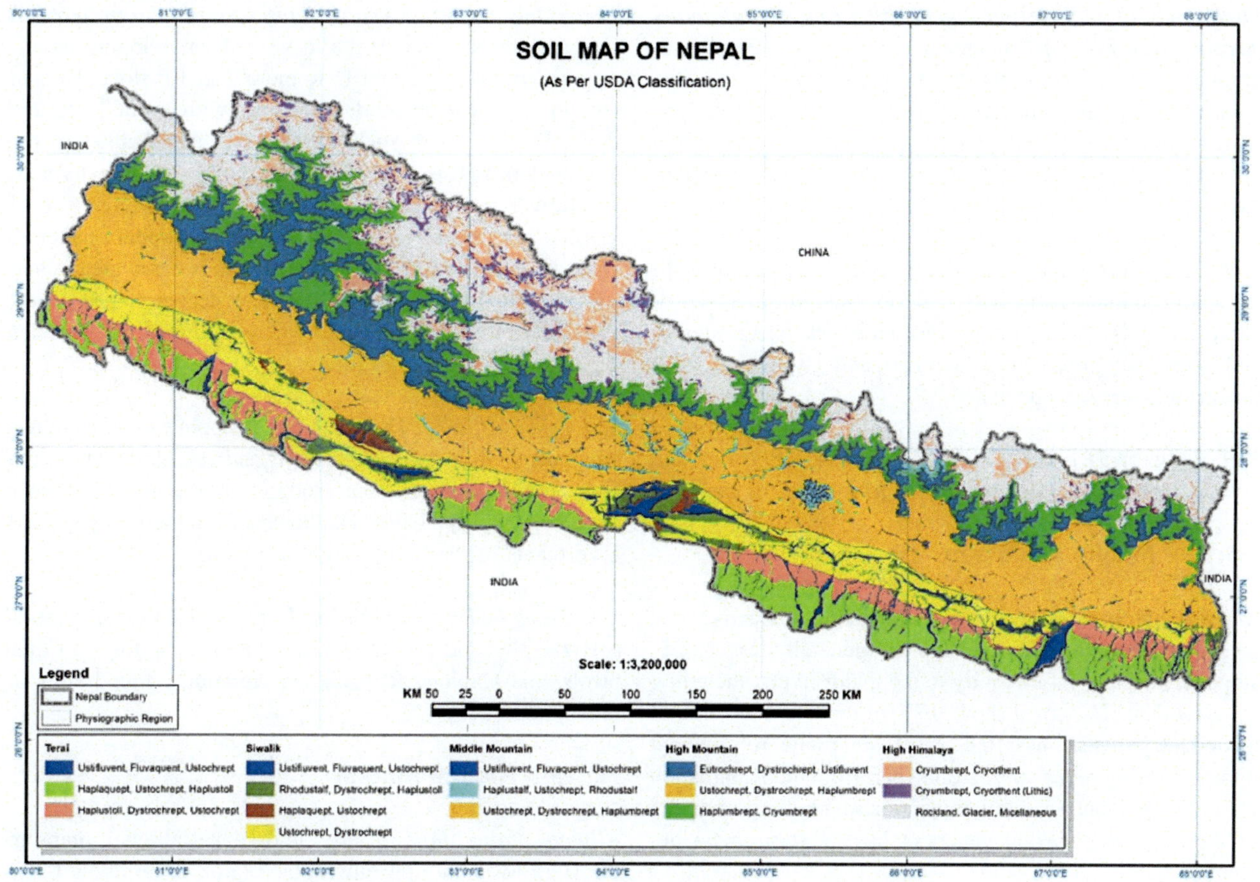

Fig. 7.2 Soil map of Nepal based on the United States Department of Agriculture classification system

country. As in the Tarai region, variations occur in land type based on proximity to the rivers and position. The lower- and intermediate-level plains are subject to annual flooding, whereas the higher-level plains are generally removed from frequent flooding events.

Soil Types (USDA): Ustifluvents, Fluvaquents, and Ustochrepts are the major soils found in the region, with some Haplaquepts and Ustochrepts.

Soil Types (WRB): Fluvisols (Calcaric and Gleyic), with some Eutric Cambisols and Eutric Gleysols are found in this land unit.

Fans, aprons, and ancient river terraces: The ancient river terraces, including fans and aprons, of this physiographic region cover about 2.2% of the total land area of the country. It consists of nearly plain land to gently sloping land with about 5° slope.

Soil Types (USDA): Moderately well- to well-drained Rhodustalfs, Dystrochrepts, and Haplustolls are the major soils occurring in these areas.

Soil Types (WRB): Rhodic Alisols, Phaeozems (Haplic and Luvic), and Eutric Cambisols.

Depositional basins: The depositional basins of this physiographic region cover about 0.37% of the total land area of the country. They consist of almost level plains, with some depressional to some gently rolling areas.

Soil Types (USDA): Haplaquepts and Ustochrepts are the major soils found in the region, with some areas consisting of Haplustalfs.

Soil Types (WRB): Eutric Gleysols and Eutric Cambisols, with some Humic Acrisols, are commonly found in the region.

Moderately to very steeply sloping hilly and mountainous terrains: The hilly and mountainous terrain of this physiographic region covers about 9% of the total land area of the country. Major areas are very steep with more than 30° slope with minor areas of about 15 to 20° slopes. This hilly and mountainous terrain is made up mainly of poorly consolidated sandstones and mudstones with some conglomerates.

Soil Types (USDA): The soils found in this region are well- to rapidly drained Typic and Anthropic subgroups of Ustochrepts and Dystrochrepts in the moderately steep slopes, and Lithic subgroups of Ustochrepts and Dystrochrepts in the steep to very steep slopes.

Soil Types (WRB): Cambisols (Chromic and Dystric) and Dystric Regosols are commonly found soils in the region.

7.5.1.3 Middle Mountains

Alluvial plains and fans: Made up mainly of active and recent alluvium, the almost level plain (generally less than 5° slope) areas lying in proximity to the major rivers, including the river channels, cover about 0.7% of the total land area of the country. Areas near the rivers are prone to frequent flooding during monsoon season. Apart from the Kathmandu valley, which has a warm temperate climate, the remainder of the areas in this unit generally have a subtropical hot climate.

Soil Types (USDA): As in the Tarai and Siwaliks, soils lying near the rivers are well-drained Psamments, Ustifluvents, and Fluvaquents, while those lying in a slightly higher elevation are Ustochrepts/Haplustepts with minor Haplustalfs.

Soil Types (WRB): Fluvisols (Gleyic and Eutric) and Cambisols (Eutric and Dystric), and minor Gleyic Luvisols are found in this region.

Ancient lake and river terraces: Made up mainly of older alluvium, these almost level second-level terraces cover about 0.8% of the total land area of the country. Parts of the area are dissected. Areas lying within the Kathmandu valley have a warm temperate climate.

Soil Types (USDA): Moderate- to well-drained Haplustalfs, Rhodustalfs, and Haplustepts are the major soils occurring in these areas.

Soil Types (WRB): Luvisols (Chromic and Rhodic), in association with Haplic Cambisols, are commonly found soils in these areas.

Moderately to very steeply sloping mountainous terrain: This area includes moderate to very steeply sloping mountainous terrain from about 15 to more than 30° slope. Human settlements occur on moderate slopes with about 15 to 25° slope. Cultivation in such areas is conducted either in the sloping terraces or on level terraces. The sloping terraces generally have maize-based cropping patterns while the level terraces are generally used for rice-based cropping patterns. With the increase in slope, terrace risers become greater and terrace widths become narrower. The steeper slopes generally have forest land use with varying forest species depending upon the elevation.

Soil Types (USDA): The moderately sloping terrains generally have well-drained Haplustepts and Haplumbrepts of the typic, rhodic, and anthropic subgroups of soil. The steeper slopes generally have Haplustepts, Haplumbrepts, and Ustorthents of the lithic subgroup of soil.

Soil Types (WRB): Cambisols (Eutric and Dystric), Umbrisols (Haplic and Rhodic), Leptic Regosols, Leptic Cambisols, and Haplic Arenosols are commonly found soils in this region.

7.5.1.4 High Mountain

Alluvial plains, fans: These almost level alluvial plains in the High Mountain physiographic region cover about 0.2% of the total land area of the country. The climate is warm to cool temperate.

Soil Types (USDA): Soils with well- to moderately well-drained Eutrochrepts and Dystrochrepts (Haplustepts, Hapludepts), with some minor Ustifluvents, are found in this region.

Soil Types (WRB): Cambisols (Eutric, Gleyic, and Dystric) and Fluvisols (Haplic and Eutric) are commonly found soils in the region.

Post-glaciated mountainous terrain below the upper limit of arable agriculture: About 3500 m above mean sea level has been considered the upper limit for arable agriculture. Comprised of moderate to steep sloping mountainous terrain (generally less than 30° slope) below this upper limit, this area covers about 10% of the total land area of the country. Cultivation generally occurs on the sloping terraces. The climate is generally warm to cool temperate.

Soil Types (USDA): Soils with well- to moderately well-drained Eutrochrepts, Dystrochrepts, and Haplumbrepts of the anthropic and typic subgroups and the lithic subgroup on the steeper slopes are found in this region.

Soil Types (WRB): Cambisols (Eutric and Dystric), Leptic Cambisols, Leptic Regosols (RGle), and Humic Umbrisols are commonly found in this region.

Post-glaciated mountainous terrain above the upper limit of arable agriculture: Comprised of steep to very steeply sloping mountainous terrain (generally more than 35° slope) above the upper limit of arable agriculture, this region covers about 9.4% of the total land area of the country. The climate is generally subalpine to alpine.

Soil Types (USDA): Soils with well- to moderately well-drained typic and lithic subgroups of Haplumbrepts, Cryumbrepts, and Cryorthents are found in this region.

Soil Types (WRB): Umbrisols (Haplic, Cryic, and Leptic) and Arenosols (Cryic and Leptic) are commonly found soils in this region. However, in steeply to very steeply sloping mountainous terrain, mostly in rock land, several areas are covered with Leptic Regosols (RGle).

7.5.1.5 High Himalaya

Glacio alluvial, colluvial, and morainal depositional surfaces: Covering about 3.2% of the total land area of the country, these areas have very high relief with minor gently sloping to steeply sloping lands.

Soil Types (USDA): Soils with well- to imperfectly drained Cryumbrept and Cryorthents are the major soils occurring in these areas. Some Typic Calciorthids are also found in this region.

Soil Types (WRB): Umbrisols (Cryic and Leptic), Regosols (Cryic and Leptic), and Podzols (Cryic and Calcaric) are commonly found in this region.

Steeply to very steeply sloping mountainous terrain: Covering about 2% of the total land area of the country, these areas are very steep, with shallow till colluviums over bedrock.

Soil Types (USDA): Soils with well-drained fragmental loamy lithic Cryumprepts and Cryorthents are the major soils, with very minor Typic Cryorthods found in this region.

Soil Types (WRB): Umbrisols (Cryic and Leptic), Cryic Regosols, and Cryic Podzols are commonly found in this region.

Rocky land: Covering about 18.5% of the total land area of the country, this very steep mountainous terrain constitutes hard rock with no surficial material. They remain snow-covered most of the time.

7.6 Description of Some Typical Soil Profiles

7.6.1 Tarai

Typic Ustifluvents

Location: 1 km west of Fatepur Banke.

Horizon	Depth, cm	Description
C1	0 to 30	Light olive brown (2.5Y5/4) silt loam; single grain, loose: common fine roots; inter-bedding of silts and very fine sand throughout; abrupt and smooth boundary to
C2	30 to 120+	Light yellowish brown (2.5Y6/4) loamy fine sand; single grained; loose; few fine roots, no stones, pH 8.1

Source LRMP (1986)

Typic Haplaquepts

Location: 1 km north of Nepalgunj Airport, Banke.

Horizon	Depth, cm	Description
Apg	0 to 10	Light brownish gray (2.5Y6/2); many fine distinct strong brown mottles; silty clay loam; coarse subangular blocky; very hard, many fine roots; pH 7; abrupt smooth boundary to
Bpan	10 to 15	Light brownish gray (2.5Y6/2); many fine distinct strong brown mottles; silty clay loam; massive (platey); very hard; few fine roots; pH 7.7; clear smooth boundary to
Bg	15 to 35	Very dark grayish brown (10YR4/2); common fine faint dark yellowish brown mottles; silty clay loam; strong medium subangular blocky; very firm; few fine roots; pH 8.0; gradual smooth boundary to
BCg	35 to 75	Light olive brown (2.5Y5/4); common fine distinct red mottles; silty clay loam; weak medium subangular blocky; very firm; few fine roots; pH 7.6; clear smooth boundary to
Cg1	75 to 120	Olive (5Y5/3); common prominent red mottles; silty clay loam; massive; very firm; pH 7.8; slight reaction with HCl;
Cg2	120 to 175+	Light yellowish brown (2.5Y6/4); common fine prominent red mottles; silty clay loam; massive very firm pH 8; moderate reaction with HCl

Source LRMP (1986)

Typic Ustochrepts

Location: Suklaphanta reserve forest, Kanchanpur 180 m asl.

Horizon	Depth, cm	Description
Ah	0 to 22	Very dark grayish brown (10YR3/2); fine sandy loam; weak medium granular; friable, common fine roots; pH 7.7; gradual smooth boundary to
Bm	22 to 50	Dark grayish brown (10YR3/2); fine sandy loam; weak medium subangular blocky; friable; common fine roots; pH 7.4; clear smooth boundary to
Bc	50 to 120	Brown (10YR4/3); loamy fine sand; single grain; loose; pH 7.8; clear smooth boundary to
C	120 to 150	Brown(10YR4/3); fine sand; single grain; loose; pH 8.3

Source LRMP (1986)

Epiaquic Eutrudepts

Location: Nidiahawa, Rupandehi 93 m asl.

Horizon	Depth, cm	Description
Ap	0 to 15	Dark grayish brown (2.5Y4/2); silt loam; moderate fine and medium subangular blocky; firm, plastic, sticky; many fine roots; pH 6.4; clear smooth boundary to
AB	15 to 40	Dark grayish brown (2.5Y4/2); common fine faint strong brown mottles; silty clay loam; medium fine subangular blocky; friable; sticky, plastic; pH 7.4; clear smooth boundary to
B11	40 to 60	Light olive brown (2.5Y5/4); many medium prominent strong brown mottles; silty clay loam; weak fine subangular blocky; friable, sticky, slightly plastic; pH 7.3; clear smooth boundary to
B12	60 to 100	Light olive brown (2.5Y5/4); many fine distinct strong brown mottles; silty clay loam; weak fine subangular blocky; friable, sticky, slightly plastic; pH 7.2

Source National Land Use Project (2017)

Typic Dystrochrepts

Location: 2 km north of Anarmani birta, Jhapa.

Horizon	Depth, cm	Description
Ap	0 to 10	Dark grayish brown (10YR3/2); sandy loam; strong medium; firm; slightly sticky; slightly plastic; pH 5.3; clear smooth boundary to
AB	10 to 15	Very dark grayish brown (10YR4/2); sandy loam; strong medium subangular blocky; firm; slightly sticky, slightly plastic; pH 5.3; clear smooth boundary to
Btj	15 to 40	Very dark grayish brown (10YR3/2); sandy loam; strong medium subangular blocky; firm; slightly sticky, slightly plastic; pH 5.4; clear smooth boundary to
BC	40 to 85	Dark grayish brown (10YR4/2); loamy sand; moderate medium subangular blocky; loose; non-sticky, non-plastic; pH 5.3; clear smooth boundary to
C	85 to 150 +	Dark brown(10YR4/3); sand; single grain; loose; pH 5.53

Source LRMP (1986)

Typic Haplustolls

Location: Near Babai river, 190 m asl.

Horizon	Depth, cm	Description
Ah	0 to 15	Dark gray (10YR4/1); silt loam; moderate medium granular; friable; many fine and medium roots; pH 7.5; clear smooth boundary to
Bm	15 to 40	Dark brown (10YR3/3); silt loam; weak medium subangular blocky; slightly firm; common medium roots; pH 7.12; clear smooth boundary to
BC1	40 to 65	Brown (10YR4/3); loam; weak medium subangular blocky; slightly firm; few coarse roots; pH 7; clear smooth boundary to
BC2	65 to 95	Brown (10YR5/2); loam; massive; slightly firm; pH 6.8; clear smooth boundary to
C	95 to 150	Yellowish brown (10YR5/5); very fine sandy loam; massive; slightly firm; few coarse roots; pH 6.8

Source LRMP (1986)

Typic Dystrochrepts

Location: 2 km north of Anarmani birta, Jhapa.

Horizon	Depth, cm	Description
Ap	0 to 10	Dark grayish brown (10YR3/2); sandy loam; strong medium; firm; slightly sticky; slightly plastic; pH 5.3; clear smooth boundary to
AB	10 to 15	Very dark grayish brown (10 YR4/2); sandy loam; strong medium subangular blocky; firm; slightly sticky, slightly plastic; pH 5.3; clear smooth boundary to
Btj	15 to 40	Very dark grayish brown (10YR3/2); sandy loam; strong medium subangular blocky; firm; slightly sticky, slightly plastic; pH 5.4; clear smooth boundary to
BC	40 to 85	Dark grayish brown (10YR4/2); loamy sand; moderate medium subangular blocky; loose; non-sticky, non-plastic; pH 5.3; clear smooth boundary to
C	85 to 150+	Dark brown (10YR4/3); **sand;** single grain; loose; pH 5.5

Source LRMP (1986)

Typic Haplustolls

Location: Near Babai river, 190 m asl.

Horizon	Depth, cm	Description
Ah	0 to 15	Dark gray (10YR4/1); silt loam; moderate medium granular; friable; many fine and medium roots; pH 7.5; clear smooth boundary to
Bm	15 to 40	Dark brown (10YR3/3); silt loam; weak medium subangular blocky; slightly firm; common medium roots; pH 7.1; clear smooth boundary to
BC1	40 to 65	Brown (10YR4/3); loam; weak medium subangular blocky; slightly firm; few coarse roots; pH 7; clear smooth boundary to
BC2	65 to 95	Brown (10YR5/2); loam; massive; slightly firm; pH 6.8; clear smooth boundary to
C	95 to 150	Yellowish brown (10YR5/5); very fine sandy loam; massive; slightly firm; few coarse roots; pH 6.8

Source LRMP (1986)

7.6.2 Siwaliks

Udic Rhdustalfs

Location: 8 km east of Surkhet, Surkhet 660 m asl.

Horizon	Depth, cm	Description
Ap	0 to 10	Dark reddish brown (5YR3/4); silty clay loam; strong medium granular; friable; common fine roots; pH 6.2; clear smooth boundary to
AB	10 to 40	Dark reddish brown (2.5YR3/5); clay; weak fine subangular blocky; friable; common fine roots; pH 6.5; gradual smooth boundary to
Bt	40 to 120+	Dark red (2.5YR3/6); clay; strong medium subangular blocky; slightly firm; few fine roots; pH 6.7; clear smooth boundary

Source LRMP (1986)

Typic Haplustepts

Location: Madhyapur Pithuwa, Chitwan, 203 m asl.

Horizon	Depth, cm	Description
Ap	0 to 18	Dark grayish brown (2.5Y4/2); common fine distinct strong brown mottles; loam; massive; firm; pH 5.6; clear smooth boundary to
AB	18 to 28	Dark brown (10YR4/3); few fine faint strong brown mottles; loam; massive; firm; pH 6.1; clear and smooth boundary to
B1	28 to 38	Dark brown (10YR4/3); silt loam; moderate subangular blocky; friable; pH 6.1; gradual and smooth boundary to
B2	38 to 65	Dark yellowish brown (10YR4/4); silt loam; moderate subangular blocky; friable; pH 7; gradual and smooth boundary to
B3	65 to 90 +	Brown (10YR5/3); loam; weak subangular blocky; friable; pH 6.8

Source National Land Use Project (2011)

Udic Haplustalfs

Location: West of Tulsipur, Deukhuri, Dang 700 m asl.

Horizon	Depth, cm	Description
Ah	0 to 15	Dark reddish brown (5YR4/3); loam; strong fine granular; firm; slightly sticky, non-plastic; many fine roots; pH 5.5; gradual smooth boundary to
Bt	15 to 40	Reddish brown (5YR4/3); loam; strong fine subangular blocky; firm, slightly sticky, non-plastic; many fine roots; pH 5.3; gradual smooth boundary to
BC	40 to 75	Reddish brown (5YR4/4); clay loam; moderate fine subangular blocky; firm, sticky, slightly plastic; many fine roots; pH 5.2; gradual smooth boundary to
C	75 to 100	Reddish brown to yellowish red (5YR4/4-4/6); clay loam; massive; firm, sticky, slightly plastic; pH 5.2

Source LRMP (1986)

Aridic Haplustalfs

Location: Rayamajhitole, Pithuwa, Chitwan 198 m asl.

Horizon	Depth, cm	Description
Ap1	0 to 11	Brown (10YR4/3); common fine distinct strong brown mottles; loam; massive; firm; pH 5.5; clear and smooth boundary to
Ap2	11 to 28	Dark brown (10YR3/3); loam; massive; firm; pH 7.1; clear and smooth boundary to
B1	28 to 42	Yellowish brown (10YR5/4); loam; moderate subangular blocky; firm; pH 7.1; gradual and smooth boundary to
B21	42 to 131	Strong brown (7.5YR5/6); loam; moderate subangular blocky; firm; pH 6.9; clear and smooth boundary to
B22	131 to 143	Yellowish red (5YR4/6); clay loam; moderate/strong subangular blocky; very firm

Source National Land Use Project (2011)

Oxyaquic Haplustepts

Location: Dhamala tole, Pithuwa, Chitwan 199 m asl.

Horizon	Depth, cm	Description
Ap1	0 to 10	Olive brown (2.5Y4/4); loam; massive; firm; pH 6.0; clear and smooth boundary to
Ap2	10 to 20	Olive brown (2.5Y4/4); common fine distinct strong brown mottles; silt loam; massive; firm; pH 6.3; gradual and smooth boundary to
B11	20 to 69	Dark yellowish brown (10YR4/4); silt loam; moderate subangular blocky; firm; pH 6.6; gradual and smooth boundary to
B12	69 to 94+	Dark brown (7.5YR5/6); loam; moderate subangular blocky; firm; pH 6.9; clear and smooth boundary to
B22	131 to 143	Yellowish red (5YR4/4); silt loam; moderate/strong subangular blocky; friable

Source National Land Use Project (2011)

Epiaquic Haplustepts

Location: Panitanki danda, Dumkibas, Nawalparasi 182 m asl.

Horizon	Depth, cm	Description
Ap	0 to 19	Olive brown (2.5Y4/4); few fine faint strong brown mottles; sandy clay loam; moderate subangular blocky; firm; slightly sticky, slightly plastic; pH 7.3; clear and smooth boundary to
B11	19 to 38	Dark brown (10YR3/3); sandy loam; moderate fine and medium subangular blocky; friable; slightly sticky, slightly plastic; pH 6.8; clear and smooth boundary to
B12	38 to 50	Dark brown (10YR3/3); sandy loam; moderate fine and medium subangular blocky; friable; slightly sticky, slightly plastic; pH 6.7 gradual and smooth boundary to
B3	50 to 62	Brown (10YR5/3); sandy loam; weak subangular blocky; friable; pH 6.7

Source National Land Use Project (2012)

7.6.3 Middle Mountain

Typic Rhodustalfs

Location: Pakalwa Tansen, Palpa 734 m asl.

Horizon	Depth, cm	Description
Ap	0 to 28	Strong brown (7.5YR5/6); silt loam; weak fine medium subangular blocky; firm; slightly sticky, slightly plastic; very few fine roots; pH 6.6; clear and smooth boundary to
B21t	28 to 70	Yellowish red (5YR5/6); silty clay loam; moderate medium subangular blocky; very firm; sticky, plastic; pH 6.9; clear and wavy boundary to
B22t	70 to 100	Reddish brown (5YR4/4); silty clay loam; moderate fine and medium subangular blocky; friable; sticky, plastic; about 25% gravel coarse fragments; pH 6.9

Source National Land Use Project (2018)

Aquic Hapludalfs

Location: Arkhale, Tansen, Palpa 700 m asl.

Horizon	Depth, cm	Description
Ap	0 to 11	Very pale brown (10YR7/4); loam; weak fine medium subangular blocky; very hard; sticky, plastic; few fine roots; about 10% gravel coarse fragments; pH 6.7; clear and smooth boundary to
B21t	11 to 22	Brown(7.5YR5/6); silty clay loam; moderate fine and medium subangular blocky; firm; sticky, plastic; pH 6.9; clear and wavy boundary to
B22t	22 to 59	Strong brown (7.5YR5/6); silty clay loam; moderate medium subangular blocky; friable; sticky, plastic; about 25% gravel coarse fragments; pH 6.9
B3	59 to 100	Dark brown (7.5YR3/4); silt loam; weak fine and medium subangular blocky; friable; slightly sticky, slightly plastic; about 60% gravel coarse fragments; pH 6.9

Source National Land Use Project (2018)

Anthropic Dystrochrepts

Location: Madhi, Achham, 610 m asl.

Horizon	Depth, cm	Description
Ah	0 to 10	Brown (10YR4/3); loam; moderate medium subangular blocky; firm; common fine roots; 15% gravel and stones; pH 5; clear smooth boundary to
AB	10 to 50	Brown (10YR4/3); loam; moderate medium subangular blocky; firm; common fine roots; pH 5; 15% gravel and stones; clear and smooth boundary to
Be	50 to 100	Dark yellowish brown (10YR4/4); loam; moderate fine subangular blocky; firm, pH 5; gradual smooth boundary to
BC	100 to 170	Strong brown (7.5YR4/4/6); clay loam; massive; slightly firm; pH 5.2; abrupt and smooth boundary to
R	170+	Rock and saprolitic materials

Source LRMP (1986)

Paralithic Ustorthents

Location: Chechikot, Tansen, Palpa 1420 m asl.

Horizon	Depth, cm	Description
Ap	0 to 15	Brownish yellow (10YR4/3); loam; weak fine subangular blocky; very friable; many very fine roots; 5% gravel and stones; pH 6.9; abrupt smooth boundary to
c	15 to 30	Yellowish brown (10YR5/6); sandy loam; weak fine subangular blocky and granular; very friable; very fine roots; slightly sticky and slightly plastic; pH 6.6; 10% gravel and stones
R layer	30+	Rock and saprolitic materials

Source National Land Use Project (2018)

Typic Haplustepts

Location: Srinagar danda, Tansen, Palpa 1408 m asl.

Horizon	Depth, cm	Description
Ah	0 to 10	Dark grayish brown(10YR4/2); loam; weak fine granular; loose; slightly sticky, slightly plastic; many very fine roots; 5% gravel and stones; pH 6.8; gradual and irregular boundary to
B11	10 to 62	Dark brown (10YR4/3); silt loam; weak fine granular; loose; very fine roots in common; pH 5; 15% gravel and stones; gradual and irregular boundary to
B12	62 to 100	Dark brown (10YR4/3); silt loam; weak fine granular; loose; about 10% gravel coarse fragments; few very fine roots; pH 5.2 gradual smooth boundary

Source National Land Use Project (2018)

7.6.4 High Mountain

Arenic Fluvaquents

Location: Bharagaun, Chandannath, Jumla 2349 m asl.

Horizon	Depth, cm	Description
A1	0 to 20	Very dark grayish brown (10YR3/2); loamy sand; weak fine crumb; loose; non-sticky, non-plastic; abundant fine roots; 1% gravel and stones; pH 5.7; gradual and irregular boundary to
C1	20 to 37	Very dark grayish brown (10 YR3/2); loamy sand; weak fine crumb; loose; non-sticky, non-plastic; common fine roots; pH 7.4; abrupt and smooth boundary to
C2	37 to 60	Very dark grayish brown (10YR3/2); loamy sand; weak fine crumb; loose; non-sticky, non-plastic; common fine roots; pH 7.4; abrupt and smooth boundary to
	60+	Water level

Source Department of Survey, Topographic Survey and Land Management Division (2020)

Humic Eutrudepts

Location: Gaurigaun, Chandannath, Jumla 2344 m asl.

Horizon	Depth, cm	Description
Ap	0 to 30	Very dark grayish brown (10YR3/2); sandy loam; weak fine granular; loose; very friable; non-sticky, non-plastic; many fine roots; 30% gravel and stones; pH 5.1; clear and irregular boundary to
B	30 to 57	Dark grayish brown (10 YR4/2); sandy loam; weak fine subangular blocky and granular; very friable; non-sticky, non-plastic; few fine roots; pH 7.1; clear and smooth boundary to
C2	57 to 68	Dark grayish brown (10YR4/2); loamy sand; weak fine subangular, granular; very friable; non-sticky, non-plastic; very few fine roots; pH 7.2; abrupt and smooth boundary to
	68+	Stones and gravel

Source Department of Survey, Topographic Survey and Land Management Division (2020)

Anthropic Cumulic Haplumbrepts

Location: Bhero, Bajhang; 2300 m asl.

Horizon	Depth, cm	Description
Ap	0 to 14	Grayish brown (10YR5/2); rubbly loam; weak medium subangular blocky; friable; many fine roots; 15% stones; pH 5.8; clear smooth boundary to
Bm	14 to 42	Yellowish brown (10YR5/4); rubbly loam; weak coarse subangular blocky; friable; few fine roots; pH 5; 20% stones; clear and smooth boundary to
IIa burried	42 to 64	Brown (10YR4/3); rubbly loam; moderate fine subangular blocky; friable: 40% stones; pH 5.5; abrupt irregular boundary to colluvial boulders

Source LRMP (1986)

Typic Haplustepts

Location: Lamasima, Chandannath, Jumla 2790 m asl.

Horizon	Depth, cm	Description
Ap	0 to 20	Brown (10YR4/3); sandy loam; moderate fine medium subangular blocky; friable; many fine roots; slightly sticky and slightly plastic; 15% stones; pH 5.8; clear smooth boundary to
Bm	14 to 42	Yellowish brown (10YR5/4); rubbly loam; weak coarse subangular blocky; friable; few fine roots; pH 5; 20% stones; clear and smooth boundary to
IIa burried	42 to 64	Brown (10YR4/3); rubbly loam; moderate fine subangular blocky; friable: 40% stones; pH 5.5; abrupt irregular boundary to colluvial boulders

Source Department of Survey, Topographic Survey and Land Management Division (2020)

Lithic Cryumbrepts

Location: 4 km south west of Bumra, Jumla 3700 m asl.

Horizon	Depth, cm	Description
Ah	0 to 16	Dark brown (10YR3/3); silty clay loam; moderate medium granular; friable; many medium and fine roots; pH 5.1; smooth boundary to Dark brown (10YR4/3); silty clay loam; weak fine subangular blocky; friable; common fine roots; pH 5.1; irregular gradual boundary to
Bm	16 to 27	Olive brown (2.5Y5/4); fine sandy loam; massive friable; few fine roots; pH 6; irregular abrupt boundary to
C	27 to 45	Saprolite/bedrock
S	45 to 70+	

Source LRMP (1986)

Humic Cryustepts

Location: Futa, Chandannath, Jumla 3984 m asl.

Horizon	Depth, cm	Description
A1	0 to 30	Dark brown (10YR3/3); sandy loam; moderate fine subangular blocky and crumby; very friable; slightly sticky, slightly plastic; common fine roots; pH 4.2; clear and smooth boundary to
B11	30 to 52	Dark brown (10YR3/3); sandy loam; moderate fine subangular blocky and crumb; friable; slightly sticky, slightly plastic; common fine roots; pH 5.6; abrupt and smooth gradual boundary to
B12	57 to 100	Dark yellowish brown (10YR4/4); sandy loam; moderate fine subangular blocky and crumb; friable; slightly sticky, slightly plastic; few fine roots; pH 5.8

Source Department of Survey, Topographic Survey and Land Management Division (2020)

Typic Calciorthids

Location: 1 km south of Tungbe village, Mustang 3000 m asl.

Horizon	Depth, cm	Description
–	0 to 7	Stone pavement olive brown (2.5Y5/4); fine sandy loam; single grain; loose; non-sticky, non-plastic wet; 60% stone gravel, pH 8
AC		
C	7 to 40 +	Light brownish gray (2.5Y6/2); gravelly loam; single grain; loose, non-sticky, non-plastic, 65% stone gravel; pH 8

Source LRMP (1986)

7.6.5 High Himalaya

Typic Cryorthods

Location: 1 km north of Tengboche village, Solukhumbu 3700 m asl.

Horizon	Depth, cm	Description
Ahe	0 to 8	Very dark gray (7.5YR3/0); silt loam; angular blocky (frozen); very hard; few fine roots; pH 4.7; abrupt smooth boundary to
Ae	8 to 18	Pinkish gray (7.5YR7/2); loam; platy (frozen); very hard; few fine roots; pH 7.7; clear irregular boundary to
Bfh	18 to 45	Yellowish red (5YR4/6); sandy loam; weak medium angular blocky; medium and coarse roots; 20% stone gravel; pH 4.9; gradual smooth boundary to
BC	45 to 70	Yellowish red (5YR5/8); rubbly sandy loam; massive; medium and coarse roots; 10% stone and gravel; pH 5.8

Source LRMP (1986)

7.7 Illustration of Dominant Soil Types of Nepal

Soil type: Epiaquic Eutrudepts
Physiographic Region: Tarai
Location: Nidiahawa, Rupandehi
Elevation: 93 m asl

Horizon	Depth, cm	Description
Ap	0 to15	Dark greyish brown (2.5Y4/2); **Silt loam**; moderate fine and medium subangular blocky; firm, plastic, sticky; many fine roots; pH 6.4; clear smooth boundary to
AB	15 to 40	Dark grayish brown (2.5Y4/2); common fine faint strong brown mottles; **Silty clay loam**; medium fine subangular blocky; friable; sticky, plastic; pH 7.4; clear smooth boundary to
B11	40 to 60	Light olive brown (2.5Y5/4); many medium prominent strong brown mottles; Silty **clay loam**; weak fine subangular blocky; friable, sticky, slightly plastic; pH 7.3; clear smooth boundary to
B12	60 to 100	Light olive brown (2.5Y5/4); many fine distinct strong brown mottles; Silty **clay loam**; weak fine subangular blocky; friable, sticky, slightly plastic; pH 7.2

Source National Land Use Project (2011)

Soil type: Typic Haplustepts
Physiographic Region: Siwaliks
Location: Madhyapur Pithuwa, Chitwan
Elevation: 203 m asl

Horizon	Depth, cm	Description
Ap	0 to 18	Dark greyish brown (2.5Y4/2); common fine distinct strong brown mottles; **Loam**; massive; firm; pH 5.6; clear smooth boundary to
AB	18 to 28	Dark brown (10 YR4/3); few fine faint strong brown mottles; **Loam**; massive; firm; pH 6.1; clear and smooth boundary to
B1	28 to 38	Dark brown (10 YR4/3); **Silt loam**; moderate subangular blocky; friable; pH 6.1; gradual and smooth boundary to
B2	38 to 65	Dark yellowish brown (10 YR4/4); **Silt loam**; moderate subangular blocky; friable; pH 7.0; gradual and smooth boundary to
B3	65 to 90+	Brown (10 YR5/3); **Loam**; weak subangular blocky; friable; pH 6.8

Source National Land Use Project (2011)

Soil type: Aridic Haplustalfs
Physiographic Region: Siwaliks
Location: Rayamajhitole, Pithuwa Chitwan
Elevation: 198 m asl

Horizon	Depth, cm	Description
Ap1	0 to 11	Brown (10YR4/3); common fine distinct strong brown mottles; Loam; massive; firm; pH 5.5; clear and smooth boundary to
Ap2	11 to 28	Dark brown (10YR3/3); Loam; massive; firm; pH 7.1; clear and smooth boundary to
B1	28 to 42	Yellowish brown (10YR5/4); Loam; moderate subangular blocky; firm; pH 7.1; gradual and smooth boundary to
B21	42 to 131	Strong brown (7.5YR5/6); Loam; moderate subangular blocky; firm; pH 6.9; clear and smooth boundary to
B22	131 to 143	Yellowish red (5YR4/6); Clay **loam;** moderate/strong subangular blocky; very firm

Source National Land Use Project (2011)

Soil type: Oxyaquic Haplustepts
Physiographic Region: Siwaliks
Location: Dhamala tole, Pithuwa Chitwan
Elevation: 199 m asl

Horizon	Depth, cm	Description
Ap1	0 to 10	Olive brown (2.5Y4/4); Loam; massive; firm; pH 6.0; clear and smooth boundary to
Ap2	10 to 20	Olive brown (2.5Y4/4); common fine distinct strong brown mottles; Silt **Loam;** massive; firm; pH 6.3; gradual and smooth boundary to
B11	20 to 69	Dark yellowish brown (10YR4/4); Silt **loam**; moderate subangular blocky; firm; pH 6.6; gradual and smooth boundary to
B12	69 to 94+	Dark brown (7.5YR5/6); Loam; moderate subangular blocky; firm; pH 6.9; clear and smooth boundary to
B22	131 to 143	Yellowish red (5YR4/4); Silt **loam**; moderate/strong subangular blocky; friable

Source National Land Use Project (2011)

Soil type: Epiaquic Haplustepts
Physiographic Region: Siwaliks
Location: Panitanki danda, Dumkibas Nawalparasi
Elevation: 182 m asl

Horizon	Depth, cm	Description
Ap	0 to 19	Olive brown (2.5Y4/4); few fine faint strong brown mottles; Sandy **clay loam;** moderate subangular blocky; firm; slightly sticky, slightly plastic; pH 7.3; clear and smooth boundary to
B11	19 to 38	Dark brown(10YR3/3); **Sandy loam;** moderate fine and medium subangular blocky; friable; slightly sticky, slightly plastic; pH 6.8; clear and smooth boundary to
B12	38 to 50	Dark brown (10YR3/3); **Sandy loam;** moderate fine and medium subangular blocky; friable; slightly sticky, slightly plastic; pH 6.7 gradual and smooth boundary to
B3	50 to 62	Brown (10YR5/3); Sandy **loam;** weak subangular blocky; friable; pH 6.7

Source National Land Use Project (2012)

Soil type: Aquic Hapludalfs
Physiographic Region: Middle Mountain
Location: Arkhale, Tansen, Palpa
Elevation: 700 m asl

Horizon	Depth, cm	Description
Ap	0 to 11	Very pale brown (10YR7/4); **Loam;** weak fine medium subangular blocky; very hard; sticky, plastic; few fine roots; about 10% gravels coarse fragments; pH 6.7; clear and smooth boundary to
B21t	11 to 22	Brown (7.5YR5/6); **Silty clay Loam;** moderate Fine and medium subangular blocky; firm: sticky, plastic; pH 6.9; clear and wavy boundary to
B22t	22 to 59	strong brown (7.5YR5/6); **Silty clay loam;** moderate medium subangular blocky; friable; sticky, plastic; about 25% gravel coarse fragments; pH 6.9
B3	59 to 100	Dark brown (7.5YR3/4); **Silt loam;** weak fine and medium subangular blocky; friable; slightly sticky, slightly plastic; about 60% gravel coarse fragments; pH 6.9

Source National Land Use Project (2018)

Soil type: Paralithic Ustorthents
Physiographic Region: Middle Mountain
Location: Chechikot, Tansen, Palpa
Elevation: 1420 m asl

Horizon	Depth, cm	Description
Ap	0 to 15	Brownish yellow (10YR4/3); **Loam**; weak fine subangular blocky; very friable; many very fine roots; 5% gravels and stones; pH 6.9; abrupt smooth boundary to
c	15 to 30	Yellowish brown (10YR5/6); **Sandy Loam**; weak fine subangular blocky and granular; very friable; very fine roots; slightly sticky and slightly plastic; pH 6.6; 10% gravels and stones
R		Rock and saprolitic materials

Source National Land Use Project (2018)

Soil type: Typic Haplustepts
Physiographic Region: Middle Mountain
Location: Srinagar danda,Tansen, Palpa
Elevation: 1408 m asl

Horizon	Depth, cm	Description
Ah	0 to 10	Dark grayish brown (10YR4/2); **Loam**; weak fine granular; loose; slightly sticky, slightly plastic; many very fine roots; 5% gravels and stones; pH 6.8; gradual and irregular boundary to
B11	10 to 62	Dark brown (10 YR4/3); **Silt Loam**; weak fine granular; loose; common very fine roots; pH 5.0; 15% gravels and stones; gradual and irregular boundary to
B12	62 to 100	Dark brown (10 YR4/3); **Silt Loam**; weak fine granular; loose; about 10% gravel coarse fragments; few very fine roots; pH 5.2 gradual smooth boundary to

Source National Land Use Project (2018)

Soil type: Arenic Fluvaquents
Physiographic Region: High Mountain
Location: Bharagaun, Chandannath, Jumla
Elevation: 2349 m asl

Horizon	Depth, cm	Description
A1	0 to 20	Very dark grayish brown (10YR3/2); **Loamy sand**; weak fine crumb; loose; non-sticky, non-plastic; abundant fine roots; 1% gravels and stones; pH 5.7; gradual and irregular boundary to
C1	20 to 37	Very dark grayish brown (10YR3/2); **Loamy sand**; weak fine crumb; loose; non-sticky, nonplastic; common fine roots; pH 7.4; abrupt and smooth boundary to
C2	37 to 60	Very dark grayish brown (10YR3/2); **Loamy sand**; weak fine crumb; loose; non-sticky, nonplastic; common fine roots; pH 7.4; abrupt and smooth boundary to
	60+	Water

Source Department of Survey, Topographic Survey and Land Management Division (2020)

Soil type: Humic Eutrudepts/Eutrochrepts
Physiographic Region: High Mountain
Location: Gaurigaun, Chandannath, Jumla
Elevation: 2344 m asl

Horizon	Depth, cm	Description
Ap	0 to 30	Very dark grayish brown (10YR3/2); **Sandy Loam** week fine granular; loose; very friable; non-sticky, non-plastic; many fine roots; 30% gravels and stones; pH 5.1; clear and irregular boundary to
B	30 to 57	Dark grayish brown (10YR4/2); **Sandy loam**; weak fine subangular blocky and granular; very friable; non-sticky, nonplastic; few fine roots; pH 7.1; clear and smooth boundary to
C2	57 to 68	Dark grayish brown (10YR4/2); **Loamy sand**; weak fine subangular, granular; very friable; non-sticky, nonplastic; very few fine roots; pH 7.2; abrupt and smooth boundary to
	68+	Stones and gravels

Source Department of Survey, Topographic Survey and Land Management Division (2020)

Soil type: Typic Haplustepts
Physiographic Region: High Mountain
Location: Lamasima, Chandannath, Jumla
Elevation: 2790 m asl

Horizon	Depth, cm	Description
Ap	0 to 20	Brown (10YR4/3); **Sandy loam**; moderate fine medium subangular blocky; friable; many fine roots; slightly sticky and slightly plastic 15% stones; pH 5.8; clear smooth boundary to
Bm	14 to 42	Yellowish brown (10YR5/4); **Rubbly loam**; weak coarse subangular blocky; friable; few fine roots; pH 5.0; 20% stones; clear and smooth boundary to
IIa burried	42 to 64	Brown (10YR4/3); **Rubbly loam**; moderate fine subangular blocky; friable: 40% stones; pH 5.5 abrupt irregular boundary to colluvial boulders

Source Department of Survey, Topographic Survey and Land Management Division (2020)

Soil Type: Humic Cryustepts
Physiographic Region: High Mountain
Location: Futa, Chandannath, Jumla
Elevation: 3984 m asl

Horizon	Depth, cm	Description
A1	0 to 30	Dark brown (10YR3/3); **Sandy loam**; moderate fine subangular blocky and crumby; very friable; slightly sticky, slightly plastic; common fine roots; pH 4.2; clear and smooth boundary to
B11	30 to 52	Dark brown (10YR3/3); **Sandy loam**; moderate fine subangular blocky and crumb; friable; slightly sticky, slightly plastic; common fine roots; pH 5.6; abrupt and smooth gradual boundary to
B12	57 to 100	Dark yellowish brown (10YR4/4); **Sandy loam**; moderate fine subangular blocky and crumb; friable; slightly sticky, slightly plastic; few fine roots; pH5.8

Source Department of Survey, Topographic Survey and Land Management Division (2020)

7.8 Conclusion

A comprehensive soil map of Nepal was prepared showing the extent and distribution of different soils occurring in different land systems within each of the five physiographic regions of Nepal based on the land resource mapping project database. Soils identified have been classified based on both the USDA and WRB system. The major soils dominant across the zones are Entisols, Inceptisols, Alfisols, Mollisols, Ultisols, and Spodosols, as mentioned in this chapter. However, several studies also reported the presence of Gelisols and Aridisols. Interestingly, Baulmer and Zech (1994) reported Andisols of non-volcanic origin from the Dudh Koshi Valley (86°43′10" E, 27°39′10" N) in Eastern Nepal. They argued that restricted microbial degradation of organic materials in colder climates coupled with aluminium leads to insolubilization and stabilization resulting in Al–humus complexes. These complexes do not exhibit volcanic material properties (i.e., short-range order minerals). Other soil types such as Histosols, Oxisols, and Vertisols have not been reported in Nepal to date. Recently, Technosols (a new order in the WRB classification) or Anthrosols have been reported in the cultivating terraces of Nepal, which are known as "human influenced" soils. These terraces have existed since cultivation began >300 yr ago. The soil types available in Nepal are of relatively young age and our cultivation practices are affecting them. As a result, the soils of Nepal require intensive care and management to retain productivity for a longer period.

References

Bajracharya RM (2013) Soils. In: Jha PK, Neupane FP, Shrestha ML, Khanal IP (eds) Environment and natural resources. Pulb. Nepal Academy and Science and Technology, Khumaltar, Lalitpur, pp 447–452

Barrera-Bassols N, Alfred Zinck J, Van Ranst E (2006) Symbolism, knowledge and management of soil and land resources in indigenous communities: ethnopedology at global, regional and local scales. Catena 65:118–137; Brady, NC and Weil, RC. 1996. The Nature and Properties of Soils

Baulmer R, Zech W (1994) Soils of the high mountain region of Eastern Nepal: classification, distribution and soil forming processes. Catena 22(2):85–103

Brady NC, Weil RR, Weil RR (2008) The nature and properties of soils. Upper Saddle River, NJ: Prentice Hall, USA

Carson B (1992) The land, the farmer and the future: a soil management strategy for Nepal. ICIMOD Occasional Paper No. 21. ICIMOD, Kathmandu, Nepal.

Carson B, Maharjan PL, Shah PB (1986) Soil landscapes of Nepal; land systems report; land resources mapping project Nepal. Kenting Earth Sciences Ltd. Ottawa, Canada

Fitzpatrick RW (2013) Demands on soil classification and soil survey strategies: special-purpose soil classification systems for local practical use. In: Shahid SA et al (eds) Developments in soil classification, land use planning and policy implications: innovative thinking of soil inventory for land use planning and management of land resources. Springer Science+Business Media, Dordrecht, pp 51–80

Ghimire SK, Higaki D, Bhattarai TP (2013) Estimation of soil erosion rates and eroded sediment in a degraded catchment of the Siwalik Hills. Nepal. Land 370–391. https://doi.org/10.3390/land2030370

IUSS Working Group WRB (2007) World reference base for soil resources 2006, first update 2007. World Soil Resources Reports No 103. FAO, Rome

Land Management Division, Department of Survey (2020) Soils of Chandannath, Jumla, Nepal

LRMP (1986) Land capability report. Land resource mapping project, Kenting Earth Sciences Limited, Ottawa, Canada

Mishra BB (2016) Indian system of soil classification: a way forward. Agri Res Tech: Open Access J 3(2). https://doi.org/10.19080/ARTOAJ.2016.03.555606 (0042)

Nath AJ, Lal R, Das AK (2015) Ethnopedology and soil quality of bamboo (Bambusa sp.) based agroforestry system. Sci Total Environ 521–522:372–379. https://doi.org/10.1016/j.scitotenv.2015.03.059

National Land Use Project (2011) Soils of Pithuwa Village Development Committee, Chitwan, Nepal

National Land Use Project (2012) Soils of Dumkibas Village Development Committee, Nawalparasi, Nepal

National Land Use Project (2017) Soils of Nidahawa Village Development Committee, Rupandehi, Nepal

National Land Use Project (2018) Soils of Tansen, Palpa, Nepal

Righi D, Lorphelin LP (1987) The soils of a typical slope in the Himalayas (Nepal): their main *characteristics* and distribution. CATENA 14:533–551

Shah PB (1995) Indigenous land and soil classification. In: Challenges in mountain resource management in Nepal: processes, trends and dynamics in middle mountain watersheds; Proceedings of a workshop held in Kathmandu, Nepal. 10–12 April, 1995, pp 203–210. International centre for Integrated Mountain Development, Kathmandu, Nepal

Soil Survey Staff (2010) Keys to soil taxonomy. 11th ed. USDA/NRCS. U.S. Government Printing Office, Washington, DC. Available online at ftp://ftpfc.sc.egov.usda.gov/NSSC/Soil_Taxonomy/keys/2010_Keys_to_Soil_Taxonomy.pdf (verified January 24, 2011)

USDA (2014) Key to soil taxonomy.12th edn. USDA, Washington, DC

Vaidya SN, Maskey SL (2000) Soils of the middle mountain region of Nepal; improved soil fertility management for sustainable maize production; proceeding of a working group meeting of the Hill Maize Research Project, Lumle, Nepal

Vaidya SN, Sah K (2014) Harmonized world soil data (HWSD) upgrading work Nepal, Final report; soil science division, Nepal Agricultural Research Council, Khumaltar, Lalitpur, Nepal

Soil Properties

8

Shree Prasad Vista, Krishna Bahadur Karki, Yam Kanta Gaihre, Sonisa Sharma, and Bandhu Raj Baral

Abstract

Nepalese soils are developed dominantly from micaceous parent materials such as phyllites, schists, gneisses, and granites; thus, most of the soil contains a higher proportion of mica. Nepalese soils are mostly friable in hills and mountains because of the continuous use of organic inputs such as farmyard manure and compost, while soils are hard and compacted in the Terai region because of the continuous use of machinery and less use of organic inputs. Soils in hills exhibit higher soil plasticity but the intensity of soil plasticity decreases in lower belts where erosion deposits prevail and most of the soils are sandy in nature. The soils of Nepal also show wide variability in chemical properties, and thereby respond differently to the same crop across ecological zones and soil types. In general, the soils of Nepal are acidic in nature. About 53% of soils are in the acidic range, while 33% are neutral, and 13% are in the alkaline range. In general, soils in hills and mountains are richer in major plant nutrients and organic matter than soils in the Terai region. However, the content of micronutrients such as zinc, boron, and molybdenum are low. Overall, soil nutrients are being mined throughout the nation, with the fastest decline in eastern areas and in the Terai region. Soil erosion, imbalanced use of fertilizers, low or no use of organic inputs, and adoption of intensive cropping systems are some of the major causes for the decline in soil nutrients.

Keywords

Biofertilizers • Nepal • Organic matter • Soil acidity • Soil fertility

8.1 Introduction

Soil properties are comprised of physical, chemical, and biological properties. These properties affect various soil characteristics including soil production capacity and determine suitability for crop husbandry practices. Soils are living, complex, and dynamic in nature, and their properties vary across time and space as they are composed of minerals, organic matter, and air. Soil is the medium for the natural growth of plants and is usually characterized by different horizons governed by various processes of soil formation such as addition, losses, transformation, and translocation. Soil characteristics such as water holding capacity, density, moisture content, drainage, aeration, compactness, retention of plant nutrients vis-à-vis their availability to plants are related to both the physical and chemical properties of soil. Chemical properties are mostly the result of parent material composition. Soil biological properties are often active when organic matter in the soil is adequate. Chemical properties are rich when the soil has enough nutrients derived from its parent materials. Hence, all soil properties are equally important for sustaining crop production.

8.2 Soil Physical Properties

Soil properties are determined by five major factors: parent material, climate, organism, topography, and time. The role of the above-mentioned factors is expressed in the equation developed by V. V. Dokuchaev:

$$S = f(Cl, O, R, P, T) \tag{8.1}$$

S. P. Vista (✉) · K. B. Karki · B. R. Baral
Nepal Agricultural Research Council (NARC), Kathmandu, Nepal

Y. K. Gaihre
International Fertilizer Development Center (IFDC), Kathmandu, Nepal

S. Sharma
Department of Plant and Soil Sciences, Oklahoma State University, Stillwater, OK, USA

R. B. Ojha and D. Panday (eds.), *The Soils of Nepal*, World Soils Book Series,
https://doi.org/10.1007/978-3-030-80999-7_8

where S = soil or any soil property, Cl = climate, O = organism (e.g., vegetation), R = relief or topography, P = parent material, and T = time.

To simplify and understand soil physical properties, we need to understand the constituents of soil as the solid, liquid, and air phases. The solid phase is usually characterized by the physical properties of soil texture and soil structure.

8.2.1 Soil Constituents and Their Relationship

Ideally, by volume soil consists of 45% inorganic matter (solid), 5% organic matter (solid), 25% water (liquid), and 25% air. The percentage of air and water varies with the state of soils. When soil is saturated, all porous space is occupied by water, whereas in dry soils most pores would be occupied by air. Organic matter is composed of dead and living microorganisms and plants whereas inorganic matter is composed of minerals. The mineral particles are of different sizes and include sand, silt, and clay. These particles are the foundational materials for soil.

Some of the terminology commonly used in soil physical properties are as follows:

(a) Bulk density: The mass of oven-dried solid per unit volume of soil. It has a unit of g cm^{-3}, and it usually ranges from 0.8 to 2.0 g cm^{-3}.
(b) Particle density: The mass of oven-dried soil particles per unit volume of soil particles. A value of 2.65 g cm^{-3} is used if the true value of particle density is not known.
(c) Porosity: The volume of porous space per unit volume of soil. It is unitless but expressed as a percentage. The relationship between the three is as follows:

$$\text{Porosity} = 1 - \frac{\textit{Bulk Density}}{\textit{Particle Density}} \quad (8.2)$$

(d) Gravimetric water content (GWC): The mass of water per unit of oven-dried soil. It is mass versus mass. For mineral soil, the value ranges from 0 to 0.6 g g^{-1}.
(e) Volumetric water content (VWC): The volume of water per unit of oven-dried soil. It is volume vs volume. Similarly, the value for volumetric water content ranges from 0 to 0.5 cm^3 cm^{-3}.
If the gravimetric water content is known, then the volumetric water content can be calculated based on soil bulk density and the density of water using Eq. (8.3)

$$\text{VWC} = \text{GWC} * \frac{\textit{soil bulk density}}{\textit{density of water}} \quad (8.3)$$

(f) Available soil water: The water content available between field capacity (water managed by irrigation scheduling) and the permanent wilting point (water unavailable to the plant) is called available soil water.

Soil Texture

The relative proportions of sand, silt, and clay in the soil are referred to as soil texture. There are various ways to identify soil texture, for example, soil textural angel calculator, feel method, texture-based on size, sedimentation method, etc. The properties of soil texture are important in determining the soil's physical, chemical, and biological properties.

Nepalese soils are developed from micaceous parent materials such as phyllytes, schists, gneisses, granites, and others, and most of the soil contains a higher proportion of mica. Loam and sandy loam are the most dominant soil textures in both hills and alluvial terraces (Carson 1992). Many studies show that the different textures in different parts of Nepal, but dominated by loam soil. A study conducted in a watershed in the Likhu river, a tributary of the Trisuli river in the northern Kathmandu valley, shows that soil textures in the soil surface are dominated by loam, silty loam, and silty clay loam (Gardner and Gerrard 2003). Similarly, another study in the Pokhare Khola watershed shows sandy loam soil (Shrestha et al. 2007). Another study from two different sal (*Shorea robusta*) forests found that the soil was sandy loam (60.12–50.58% sand, 28.59–35.24% silt, and 11.12–22.41% clay) (Paudel and Sah 2003). In 50 soil samples collected at 0–20 cm depth in the rice fields at the National Rice Research Program at Hardinath, Dhanusha, the soil texture was dominated by loam types (Khadka et al. 2017b), with sand ranging from 5.1 to 67.6%, silt ranging from 20.5 to 64.5%, and clay ranging from 9.6 to 40.8%. The soil texture for 81 soil samples collected at Regional Agricultural Research Station in Tarahara, Sunsari in 2016 were observed to be loam (34%), clay loam (10%), sandy loam (4%), silt loam (32%), and silty clay loam (20%), respectively (Khadka et al. 2017a). Likewise, 76 soil samples collected at Regional Agricultural Research Station found that most soil texture belongs to the silt loam category, with sand ranging from 15.6 to 41.1%, silt ranging from 40.8 to 63.2%, and clay ranging from 14.8 to 26.2% (Khadka et al. 2018).

Soil Structure

Soil structure is the arrangement of individual particles of sand, silt, and clay into a larger soil particle. Smaller particles are called peds, which are influenced by the interactions of soil particles. The soil structure assists in the movement of air and water in soil (Brady et al. 2008).

Regarding the soils in Nepal, the structure is different in various studies. For example, soils at the National Rice Research Program at Hardinath, Dhanusha, and Regional

Agricultural Research Station in Tarahara, Sunsari had sub-angular blocky structures (Khadka et al. 2017a). Likewise, soils at Regional Agricultural Research Station in Parwanipur, Bara had an angular blocky structure (Khadka et al. 2018). These and the studies above show that there is variability in soil structure and texture within different parts of Nepal.

Soil Consistency

The properties of soil such as friability, plasticity, stickiness, ease or resistance to compression, and others are all related to the moisture and organic matter content of the soil and form some examples of soil consistency. In general, the soils are mostly friable in hills and mountains because of the continuous use of organic inputs such as farmyard manure (FYM) and compost (SSD 2016,2017). However, in the Tarai region where the use of organic inputs is low, hard consistency dominates. The Tarai areas are intensively cultivated and farmers use relatively higher amounts of chemical fertilizers compared to the hill and mountain regions. Application of organic manure in Tarai soils is low, and even crop stubbles from fields are removed to feed animals or as fuel wood.

Soil Plasticity

Plasticity is exhibited by soil containing more than 15% clay and is exhibited over a range of moisture contents referred to as plasticity limits. It is also considered the upper limit of moisture content for tillage operation for most crops excluding rice. Soils in the hills exhibit higher plasticity but the intensity of plasticity decreases in lower belts where erosion deposits prevail and most of the soils are sandy in nature.

Soil Color

Due to the temperate climate, the weathering of the rocks and minerals is slow in higher altitudes (mountains), whereas it is faster in the lower altitudes of the Tarai region. In the same manner, soil color also ranges from brown and dark brown in high altitudes to yellowish to reddish-brown, olive-brown, and white to grayish brown in lower altitudes.

Nepalese farmers always consider black soil to be fertile and brownish-white soil to be unfertile. Even in red soil, the surface soil color changes to black or another color due to physical, chemical, and biological reactions, including organic matter. Because of organic matter, the application of other mineral fertilizers continues biological and chemical reactions that accelerate root decay and increase soil carbon stock in the rhizosphere (Angst et al. 2018).

Soils from the old river terraces known as *Tars* (flat land in the foothills with relatively warm micro-climate) in Nepal that drain to the rivers are red. The same soils when cultivated with seasonal irrigation, and the addition of organic matter gradually turns brown, reddish-brown, and/or yellowish-brown (Walker and Lin 2008). River silt with low organic matter added to cultivated fields changes the soil color to olive and olive-brown. Upland soils, when a sufficient amount of FYM or compost is added, look black. Though the topsoil remains black the subsequent soil horizon remains a mix of dark and brown (Kaiser and Guggenberger 2005). This is mainly due to the movement of humus and clay colloids leaching with the irrigation water. Nepalese Tar soils are very porous, and hence the colloidal particles leach down to the subsoil and even deep into the bottom horizon when there is water to move downwards. These soils have original reddish-brown color. In some conditions, a concretion of manganese oxide develops and is black (Ivarson and Heringa 1972; Zhang and Karathanasis 1997). Such cases are observed in some parts of Terai.

Waterlogged soils are saturated with the groundwater and gleyic condition developed where the soil color is gray to light gray (Motomura 1969; Rabenhorst 1990). This condition is common in most rice cultivated areas including large irrigation command areas of Nepal. In most seasonal irrigated rice fields, the topsoil is generally dark brown to brown and the subsoils are mostly light brown to olive-brown. This olive-brown color is mainly a result of the leaching of soil colloidal particles that are deposited in verticals and/or horizontal ped faces.

8.2.2 Physical Constraints in Nepalese Soils

Soil physical constraints generally refer to the physical degradation of soils, including structural degradation that hinders crop yield (Singh et al. 2014). These properties mainly consist of soil texture, soil erosion, bulk density, soil structure, soil sealing/crusting, soil temperature, water holding capacity, water infiltration capacity, and aeration. When these properties degrade, it is difficult to bring them back to favorable conditions (Sen 2003). The physical constraints of Nepalese soils should be grouped into two categories, as their setting and development differ.

Soil texture: The physical properties of soil play an important role in improving soil fertility. Hill soils are light and developed in phyllite, schists, and quartzite, and hence are coarse (Baumler and Zech 1994), whereas valley and Tarai soils are heavier (Pal et al. 2001; Srivastava et al. 1998). There are patches of calcite soils, along with valleys and foothills of colluvial deposits (Mücher et al. 2018). Glacial deposit soils in Nepal are mostly coarse and may be alkaline, as in the case of the soils of the Pokhara Valley. Light-textured soils are low in nutrient holding, water-holding, high water infiltration, and permeability (Ogban and Babalola 2003). Soil consisting of less than 10% clay particles is said to be light and sandy and have high water permeability; crop productivity of these soils is low (Singh et al. 2014). Tarai soils mostly contain a higher percentage of clay and are fertile. With the addition of 5–10 t ha^{-1} of well-decomposed, organic manure will help to increase water and nutrient-holding capacity in addition to other physical properties especially in the soils of the hills and mountains.

Soil erosion: Erosion is greater in the cultivated fields of Nepal and especially in upland conditions in the hills and mountains where soils are light and susceptible to both water and wind erosion. The eroded finer materials are deposited elsewhere, leaving a coarse skeleton on the surface and making poor-quality soil (Gardner and Gerrard 2003). When soil erosion takes place, soil particles are first detached from the soil aggregates in the surface soil, after which they are transported and the organic matter and other nutrients associated with these surface soil particles are lost. Thus, eroded soils are infertile and have low productivity (Gardner and Gerrard 2003; Schreier et al. 1994). There have been very few studies measuring soil loss from cultivated fields. Schreier et al (2000) reported the loss of up to 20 t ha^{-1} fertile soils annually. Sediment estimation on a watershed basis is 1–9 t ha^{-1} yr^{-1} as the result of erosion. Wind erosion is high in the High Mountain and Himalayan regions where wind speed is very high, though because agriculture production in this area is minimum its effects are not discussed in detail. Eroded sediments are deposited in the valleys and sometimes mix with landslide debris. Landslides and heavy floods deposit sediment-loaded debris in the valleys and plain areas, which are also poor in fertility (Karki 1986; Shrestha 1985).

Soil sealing/crust: Soil crusting generally occurs when there is heavy rain. When raindrops hit the soil surface, fine particles are splashed and transported away leaving the coarse particles on the surface. As a result of raindrops continuously hitting the surface, the aggregates are disintegrated into small particles and in some cases are puddled. After a day or night of rain, the small soil particles form a layer that is hard, brittle, and friable. This layer can range from a fraction of an inch up to 1.5 in thick. In Nepal, growers generally notice this type of crust in dry seedbed nurseries for rice. Even in wet rice nurseries after puddling, the water is cut off and the finer particles are left to settle; in a day or two, the puddled layer forms crust that splits to form cracks. When this seal or crust on the soil surface is formed, water infiltration is lower and it is harder for seeds to germinate. Likewise, soil crust increases bulk density and decreases the porosity percentage resulting in lower root proliferation. The amount of crust depends on the raindrops and the intensity of the rain, which increases soil compactness and bulk density (Bajracharya and Lal). Soil crust formed due to beating by raindrops is prone to soil erosion (Morin and Van Winkel 1996), and the formation of soil crust depends on soil texture, where light textures are prone to crusting (Gabriels et al. 1997; Lui et al. 1996). Since Nepalese soils are light in texture and raindrop intensity is high (Gardner and Gerrard 2003; Gilmour et al. 1987), they are more susceptible to crusting and soil erosion.

Soil temperature: Soil temperature is latitude dependent, as tropical soils are always warmer than temperate soil. Soils on north-facing slopes in the southern hemisphere and soils on south-facing slopes in the northern hemisphere are warmer than soils on southern slopes in the southern hemisphere and soils on northern slopes in the northern hemisphere (McLaren and Cameron 1996). As it is in the northern hemisphere, the soil temperature regime of Nepal is the mesic type. However, other soil temperature regimes can be found in Nepal, such as cryic and frigid in the High Himalayan region and thermic in the Tarai region. In high mountains where snowfall is dominant during winter, seeds are sown during October and remain inside the soil without germination, though these seeds begin germinating soon after the snow starts melting because of the rise in soil temperature. The snow acts as insulation and the soil temperature rises, ultimately favoring seed germination and rapid crop growth.

Soil structure: Soil structure indicates soil pores and soil bulk density (BD). There are several types of soil structure such as crumb and granular, which are the best for infiltration and air movement and thus the best for seed germination. The subangular blocky structure is a medium type of structure that is available in fertile soils. The angular blocky structure is somewhat harder and generally available in the harder plow pan of rice soils. It indicates that this structure is compact, with higher BD and a lower porosity percentage, and that it is unfavorable for root growth. Mohammed et al. (2020) argues that the climate controls structure formation as warmer climate is humid and lead to finer soil aggregates whereas in temperate regions soil structures are harder. The columnar and prismatic structures are much harder and are found in salt-affected and sodic soils (Horn et al. 1964). These soils have higher BD, low porosity, low infiltration, and low permeability. Plant roots do not grow deeper and concentrate only on the surface soil, which is made favorable by intercultural operation. Because Nepalese farmers in the hills and river valleys apply a sufficient amount of organic manure to their field crops, the structure of surface soil is mostly granular or crumb. Due to its highly porous nature and the organic acids formed during the degradation of organic matter leaching down to the subsurface horizon, the subsoil remains as a subangular blocky structure.

Leaching of organic acids from the top horizon favors biological activities and mineralization of nutrients and promotes plant biomass. Paddy soils may have some disturbances in their soil structure that are generally hard, as puddling for rice cultivation destroys soil structure but rotating soybean into the cropping system has been found to improve soil structure (Cass et al. 1994) because the legume roots can break the plow pan and extract nutrients from subsoil while biologically fixing nitrogen. Therefore, scientists advise farmers to include legumes in their crop rotation. Chickpeas, lentils, field peas, and lathyrus are cultivated in a rotation with rice. As the Tarai soils of Nepal are lower in organic matter, the surface soil structure is getting more compact.

Bulk density (BD): Bulk density is the ratio of mass by volume and is generally measured in g cm^{-3}. For good crop growth, the BD should be between 1.0 and 1.5 g cm^{-3}. If the BD is below 1.0 g cm^{-3}, there will not be any anchorage for the plant to stand, whereas above 1.5 g cm^{-3} there will be lower aeration for the proper growth of the plant roots. Meki et al. (2013) argued that conventional tillage increased soil bulk density, whereas no-tillage treatment lowered bulk density while increasing biomass. Begum et al. (2013) pointed out that there is a seasonal difference in bulk density in the hills of Nepal, indicating 1.19 g cm^{-3} in the pre-monsoon season and 1.31 g cm^{-3} in the post-monsoon season, which is optimum bulk density in agricultural soils. Regmi and Zoebisch (2004) showed a slight difference between khet (rice cultivated lowland) soils at 1.38 g cm^{-3} and upland soils at 1.23 g cm^{-3}. Both studies show that bulk density in the hills and mountains is favorable for cultivation of crops. Bulk density in the Tarai soil is a bit higher than in the mountains. Gami et al. (2001) reported that a control plot where no organic manure was added showed 1.69 g cm^{-3} and plots where only fertilizer was applied showed 1.49 g cm^{-3}, whereas other organic manure-treated soils showed 1.40 g cm^{-3}. Similar results (1.69 g cm^{-3} from control plots) were also reported from the Indo-Gangetic plain of India (Gangwar et al. 2006). Both the results are from experimental plots, though in farmers' fields the results can be slightly different.

Water-holding capacity and water-infiltration capacity: The infiltration of water refers to the movement of water through micropores. If the micropores are sealed water movement is also sealed (Ela et al. 1992). In some upland soils in Nepal, there are more micropores, and water moves through the soil at higher speeds; sometimes this movement is so rapid that it is unmeasurable. In irrigated river valleys soil contains a higher amount of silt and fewer micropores result in low water infiltration. No-till cultivation practices in rice-wheat cropping systems in the plains region lower water infiltration more than tillage practices (Jat et al. 2013). Water infiltration in the hills and mountain soils is normally high; the soil moisture index of different physiographic regions is depicted in Fig. 8.1.

Organic matter and clay content in soils are the two main factors affecting water-holding capacity (WHC). When there is high organic matter and clay content in the soil, WHC will also be high (Khaleel et al. 1981; Vengadaramana and Jashothan 2012). Similarly, clay content also increases as WHC increases (Lund 1959). Agvise (2020) showed examples of WHC under different soil textures, with sandy loam at 29.4%, loam at 85.0%, silt loam at 71.9%, and clay at 73.7%. The hill and mountain soils are rich in soil organic matter and hence WHC is high, whereas Terai soil, though low in organic matter, contains high silt and clay content, whereas WHC is also high.

Soil aeration: Soil aeration is related to compaction and bulk density, in addition to soil hydraulic properties, nutrient movement, and root respiration (Grable 1966). When the soil is fully saturated with water, plant roots cannot breathe and plants (with the exception of rice) turn yellow. In Nepal, this problem appears mostly in winter crops, especially wheat and lentils, due to unexpected heavy winter rains. Some vegetable crops such as tomatoes also suffer from heavy rain and flooding conditions. Farmers generally drain water as quickly as they can or wait for the rain to stop and apply urea fertilizer.

8.3 Soil Chemical Properties

Soils of Nepal show wide variability in chemical properties, and thereby respond differently to the same crop across ecological zones and soil types. Variation in topography, soil-forming parent materials, climatic conditions, soil management practices, and organic matter content are major factors that determine the variations in soil chemical properties. The soils of Nepal are acidic in nature (53% of soils are in the acidic range, 33% are neutral, and about 13% are alkaline). About 44% of soils are low in organic matter, 56% in nitrogen, 49% in potassium, and 42% in phosphorus. However, potassium content was high (68%) a decade ago but is also on a decreasing trend (Dawadi and Thapa 2015). Soil fertility status (i.e., the nutrient content in soils) is declining throughout the nation, with the most rapid decline taking place in eastern Nepal. Deficiencies of many nutrients have been observed in different patches that often limit crop production; therefore, improvements in plant nutrient

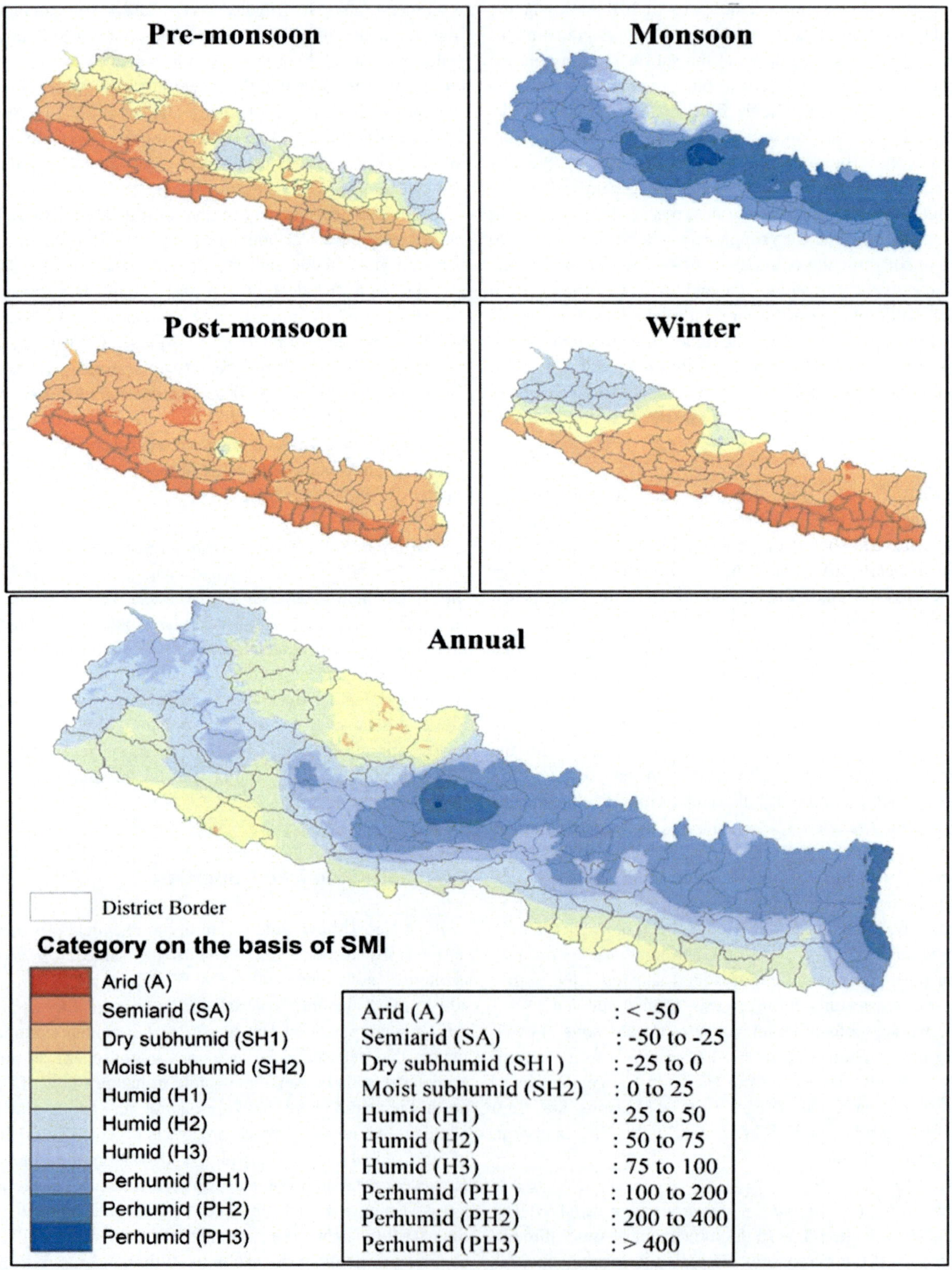

Fig. 8.1 Soil moisture index (SMI) variation in different seasons and physiographic regions of Nepal (maps re-drawn) (Talchabhadel et al. 2019; written permission granted from Talchabhadel on 08 February 2021)

management systems are crucial for enhancing soil fertility and crop productivity. In Nepal, studies on other chemical properties such as cation exchange capacity, base saturation, electrical conductivity, and the status of heavy metals are lacking.

8.3.1 Soil Acidity

Soil pH (activity of H^+ and OH^- ions) is an important soil chemical property that determines the availability of soil microbial activities and nutrient availability to plants. All acid soils contain hydrogen ions, and the strength of the acid depends on the degree of ionization (i.e., the release of hydrogen ions) of the acid. The more hydrogen ions are held by the exchange complex of soil in relation to the basic ions (Ca, Mg, K, and Na), the greater the soil acidity. In Nepal, soil acidification is one of the major challenges for the sustainable improvement of soil fertility.

Status of soil pH in Nepal

Out of 16,585 soil samples analyzed by the Soil Management Directorate (SMD), Department of Agriculture, Ministry of Agriculture and Cooperative, Government of Nepal, about 53% of samples were acidic, followed by 34% neutral and 13% alkaline (SMD, DoA 2013). In general, soils of eastern Nepal are relatively more acidic than soils in the central and western parts of the country (DOA/STSS 1999, Fig. 8.2). Similarly, soils in the hill and mountain regions are more acidic compared to soils in the Tarai region. Soil analyses over the past decade show that soil acidity is increasing due to faulty soil management practices, including imbalanced use of chemical fertilizers (more urea, but lower or no use of phosphorous, potassium or other secondary and micronutrients), lack of awareness among farmers in managing acidic soils, and a lack of availability of agricultural lime on the market. Soil pH determines soil quality as described in Table 8.1 Soil quality rating in Nepal based on soil pH value.

Causes of Soil Acidity in Nepal

Rainfall and irrigation

Rainfall contributes to soil's acidity due to the leaching of basic cations. Water (H_2O) combines with carbon dioxide (CO_2) to form a weak acid: carbonic acid (H_2CO_2). The weak acid ionizes, releasing hydrogen (H^+) and bicarbonate (HCO_2). The released hydrogen ions replace the calcium ions held by soil colloids, causing the soil to become acidic. The displaced calcium (Ca^{++}) ions combine with the bicarbonate ions to form calcium bicarbonate, which, being soluble, is leached from the soil. The net effect is increased soil acidity. High-rainfall areas, especially in eastern Nepal (1,000–2,000 mm), are associated with acidic soils due to increased leaching of base cations (Ca^{2+}, Mg^{2+}, K^+, etc.) (Lucas et al. 2011).

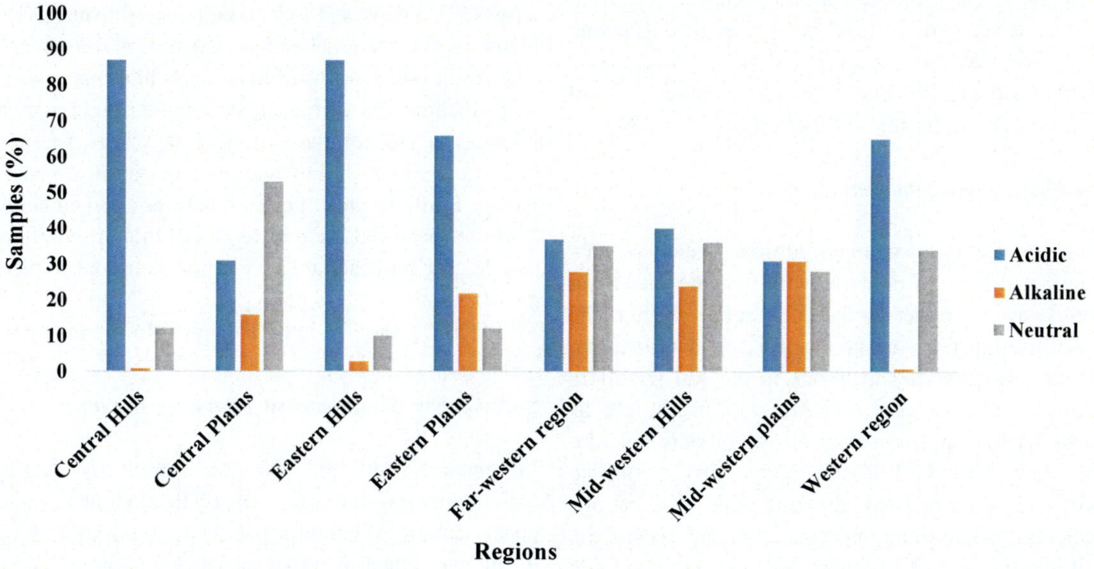

Fig. 8.2 Soil pH status of different regions of Nepal (SMD, DoA 2013, n = 6442)

Table 8.1 Soil quality rating in Nepal based on soil pH value

Soil pH	Ranking value	Soil quality rating
<4, >8.5	0.2	Very poor
4.1–4.9	0.4	Poor
5–5.9	0.6	Fair
6–6.4, 7.6–8.5	0.8	Good
6.5–7.5	1.0	Best

Based on DoA/STSS (1999), Sitaula et al. (2004)

Acid rain

Rain with a pH value less than five is referred to as acid rain. The pH of acid rain is usually between 4 and 4.5 but can sometimes be as low as two. It is formed due to the emission of oxides of nitrogen, sulfur, and carbon from industry, automobiles, and other sources. Acid rain serves as a source of soil acidity, especially in soil with low buffering capacity. The soil in industrial and high-density traffic areas has generally developed acidic soil due to acid rain.

$$SO_2 + {}^1/_2O_2 + H_2O \rightarrow H_2SO_4 \rightarrow SO_4^{2-} + 2H^+$$

Limited studies on acid rain in Nepal have been conducted and as there has been little major industry in Nepal there has been a minimal chance of the country experiencing acid rain. However, it is possible that heavy traffic and low-quality transport fuel have aggravated acid rain especially in the Kathmandu Valley (Shrestha and Malla 1996).

Parent material

The soils of Nepal have developed on Himalayan residuum and alluvium derived from shale, sandstone, and siltstone and have a low buffering capacity (FAO 2006). These weakly buffered minerals facilitate the development of soil acidity, particularly in the mid-hill region.

Imbalanced use of inorganic fertilizers

The continuous use of nitrogenous fertilizers such as urea and ammonium sulfate, particularly when they are used in an imbalanced way, increases soil acidity. The higher the nitrogen fertilization rate, the greater the soil acidification. As ammonium is converted to nitrate in the soil (nitrification), H ions are released. For each kilogram of nitrogen as ammonium, it takes approximately 1.8 kg of pure calcium carbonate to neutralize the residual acidity. Also, the nitrate that is provided or formed can combine with basic cations such as calcium, magnesium, and potassium and leach from the topsoil into the subsoil. As these bases are removed and replaced by H ions, soils become more acidic. Most farmers in Nepal apply only urea as a chemical source of fertilizer, which has an acidic residual effect. The imbalanced use of chemical fertilizer over the long term has increased soil acidity, particularly in the Tarai region where less organic manure is applied.

Carson (1992) reported that factors that contribute to increased soil acidity in hilly areas are the increasing use of nitrogenous fertilizer and the government reforestation program's emphasis on pine tree plantations, both of which have increased acidity in hilly areas. Nevertheless, less acidic soil in agricultural lands compared to forestland suggest that fertilizer-driven acidification is not a major problem in the mountains of Nepal. Per capita fertilizer use in Nepal is very low compared to other south Asian countries. Previous studies revealed that soil acidity induced by fertilizer application was observed only after 10 years of continuous fertilizer use in experimental plots at Bhairahawa (Carson 1992).

Plants

Crop selection is also responsible for creating soil acidity. For example, leguminous crops such as soybeans, alfalfa, and clover tend to take up more cations in proportion to anions. This causes H ions to be released from plant roots to maintain the electrochemical balance within their tissues. The result is net soil acidification. While this process is not well documented in Nepal, continuous legume systems in hilly areas are common practice in places where farmers have no options due to marginal irrigation or lack of irrigation. Similarly, the effects of Sal trees on soil acidification have not been extensively reported, though coniferous species (e.g., Chir-pine) are known for their contribution to soil acidification by releasing organic acids from their roots and by absorbing base cations (Rigueiro-Rodriguez et al. 2012).

Soil Acidity Management Practices in Nepal

In general, soil pH can be raised by adding lime (calcium/magnesium carbonate). The amount to add depends on the cation exchange capacity (nutrient-holding capacity) of the soil, which is based on its clay content and buffering capacity. Soil higher in clay will have a higher cation exchange capacity and requires more materials to raise the soil pH.

Karki and Dacayo (1990) recommended 6–9 t ha^{-1} of lime on various soils, whereas Adhikari et al. (2007) recommended 4 t ha^{-1} lime for western mid-hill regions under maize-based cropping systems. However, considering the remoteness and lack of supportive logistics especially in the Hill regions, there exists the need for alternative methods for correcting soil acidity. For example selection of local and exotic germplasms of maize, wheat, upland rice, soybean, and black gram suitable for acidic environments. Tripathi (2000) recommended applying lime in combination with compost or organic manure. The author also recommended 2 t ha^{-1} lime for maize and wheat that would increase yield up to 35% compared to a non-limed plot. Thus, this is one area in nutrient management that needs to be studied systematically. In addition to liming other management options to overcome soil acidity problems have also been reported; for example, diversified cropping was reported to improve residue quality and quantity by supplying a variety of above- and below-ground biomass and residue inputs, which has the potential to improve soil pH and soil fertility (Umiker et al. 2009). Increasing soil organic matter in soil can increase soil pH and its buffering capacity in highly acidic soil (Umiker et al. 2009).

The government of Nepal has recommended lime requirements based on soil pH value for different soil texture groups and geographic regions (Table 8.2). Although lime recommendations are made by the government, they are rarely practiced by farmers due to the lack of availability of lime on the market. Moreover, the use of lime is limited by its high transport cost, lack of farmers' awareness, lack of soil testing facilities, and other factors.

8.3.2 Organic Matter Content in Nepalese Soil

Organic matter is the core constituent of agricultural soils. Organic matter incorporation improves soil fertility resulting in a good crop harvest. Improved Soil Organic Matter (SOM) forms the basis of soil health and is required for improved soil health and fertility. Due to the unbalanced use of chemical fertilizer, lower use of organic inputs, and a poor strategy of organic matter management, SOM in Nepalese agricultural lands have deteriorated substantially, affecting the soil's productive capacity. The Agriculture Development Strategy (ADS), a 20 year-long term vision of the Nepal government, has set a target of enhancing SOM from the current 1.96% to 4% by 2035. The ADS includes the promotion of integrated soil fertility management (ISFM) and improvement of agricultural practices such as crop rotation, incorporation of crop residue (including direct incorporation and composting), and integrated crop nutrition as the main strategies for achieving 4% SOM. Recent soil analyses carried out by the Land Use Project in the Terai region show a mean organic matter content of 2.24% (minimum 0.3% to a maximum of 7.3%), though this analysis included forest soil as well. The highest OM is found particularly in forest soils, not agricultural soils, where soils were found to have mostly medium organic matter content. In many parts of Nepal, soil organic matter content is less than 2%, and in some areas, it is below 1%, especially in cultivated Terai soils. Soil organic C content is generally higher in higher elevation soil due to cooler climate and slow decomposition, but with shallow soil depth. Soil organic matter depends on the land use and season. Agricultural land has low SOM due to tillage and a poor management system. In areas where sustainable soil management practices are followed, higher SOM was observed compared to soil where conventional practices were followed.

Soil organic matter varies with ecological zones and land-use types. It is higher in grasslands and forests (e.g., in the mountains) and lower in agricultural lands (e.g., in Tarai) as shown in Fig. 8.3. The eastern part of the Tarai region has lower SOM compared to the western part of the Tarai, while the Mid-Hills and Higher Hills have similar ranges across the country that are higher than the Tarai region.

Soil organic matter shows a declining trend, and the causes behind this vary between agro-ecological zones and management practices. The use of FYM has decreased due to a decrease in livestock population and the incorporation of crop residue. Moreover, in Tarai dung cakes are used as fuel for cooking. Intensive and repetitive monocropping, the shift from traditional to conventional agriculture, and mechanization are some of the leading factors contributing to the decreased organic matter in agricultural soils. Moreover, abandoning traditional agro-ecological practices such as the retention of crop residue, fallow cropping, crop rotation, and intercropping coupled with the unbalanced use of chemical fertilizer and unsustainable practices adopted by commercialized agriculture has further resulted in the loss of soil organic matter (MoALD and Practical Action 2020).

8.3.3 Status of Plant Nutrients in Nepalese Soils

The status of plant nutrients in Nepalese soils varies widely between agro-ecological zones, cropping systems, and crop management practices. In general, both macro- and micronutrients are on a declining trend because of increasing crop intensification, decreased use of organic inputs, and removal of crop residues from agricultural land, among other factors. With the intensification of the cropping system, farmers are removing crop residue from their fields and allowing shorter fallow periods, while at the same time use of other organic inputs such as FYM, compost, green manure, and cover crops are declining, resulting in the mining of plant nutrients from soils.

Table 8.2 Recommended lime requirements by the Government of Nepal based on soil type and physiographic region

Soil pH	Hills (t ha^{-1})			Terai (t ha^{-1})		
	Sandy loam	Loam	Clay loam	Sandy loam	Loam	Clay loam
6.4	0.30	0.40	0.48	0.16	0.28	0.44
6.3	0.58	0.80	0.96	0.30	0.48	0.88
6.2	0.86	1.20	1.44	0.46	0.68	1.28
6.1	1.16	1.56	1.96	0.60	0.88	1.72
6.0	1.42	1.84	2.40	0.76	1.04	2.12
5.9	1.70	2.20	2.92	0.90	1.24	2.56
5.8	1.94	2.56	3.32	1.04	1.44	2.92
5.7	2.16	2.84	3.76	1.16	1.64	3.32
5.6	2.38	3.16	4.16	1.28	1.80	3.68
5.5	2.60	3.40	4.60	1.40	2.00	4.00
5.4	2.80	3.76	5.04	1.52	2.20	4.40
5.3	3.00	4.08	5.48	1.62	2.36	4.76
5.2	3.20	4.36	5.88	1.72	2.52	5.08
5.1	3.38	4.56	6.28	1.82	2.72	5.40
5.0	3.52	4.80	6.68	1.92	2.84	5.72
4.9	3.68	5.04	7.04	2.02	3.00	6.04
4.8	3.82	5.24	7.48	2.12	3.16	6.32
4.7	3.98	5.44	7.80	2.22	3.32	6.60
4.6	4.10	5.60	8.12	2.30	3.48	6.80
4.5	4.20	5.80	8.40	2.40	3.60	7.00

Source Agriculture Information and Communication Center (AICC), Krishi Diary (2076), MoAD

Status of macronutrients

Among the primary (NPK) nutrients, Nitrogen (N) is the most limiting nutrient in Nepalese soils. However, the N content in soils varies widely across agro-ecological zones. Most Terai and eastern Mid-Hill soils have a total N content of less than 0.15%, while western Mid-Hill soils are rich in N (Fig. 8.4). This suggests that the Tarai plains and the eastern Mid-Hills require higher amounts of nitrogen fertilizers to achieve optimum crop yields, while in the western Mid-Hills and mountains the N requirement could be smaller compared to Terai soils.

Phosphorus content in soils also varies across agro-ecological zones. As with N content, the phosphorus content of most eastern Terai soils are lower than 55 kg P_2O_5 ha^{-1}, though it is higher than 55 kg P_2O_5 ha^{-1} in most western Terai soils (Fig. 8.5). Hill and mountain soils are richer in phosphorus compared to Terai soils. Different studies conducted by Nepal Agricultural Research Council (NARC) show that the soil phosphorus level is increasing in Nepal as most farmers use Diammonium Phosphate fertilizer to cultivate their crops.

Unlike their nitrogen content, Nepalese soils are rich in potassium. Although Terai soils have relatively lower potassium content compared to hill and mountain soils, the nutrient level falls within the medium range (110–280 kg K_2O ha^{-1}) (Fig. 8.6). As Nepalese soils are rich in potassium, most farmers do not use potassium fertilizers to cultivate their land. NARC has recommended a very low amount of potassium (10–50 kg K_2O ha^{-1}) as a maintenance dose. However, due to intensive crop cultivation coupled with residue removal from farmland and the lower use of organic inputs, potassium mining has begun in Nepalese soils (Ojha et al. 2021). A SMD report based on an analysis of 15,000–20,000 soil samples from different parts of the country shows rapid decline in potassium over the last two decades. So, Ojha et al. (2021) recommend increasing potassium content to 1.5–2 times for rice and 2–2.5 times for wheat to the current rate of potassium recommendation (30 kg K_2O) per hectare in rice–wheat cropping system of Nepal.

Information on available sulfur, calcium, and magnesium in Nepal is lacking. In Nepal, most analyses are conducted to determine soil pH, organic matter, nitrogen, phosphorous, and potassium. It is reported that soils from eastern Nepal contain lower amounts of calcium and magnesium (which is associated with higher rainfall) compared to western Nepal.

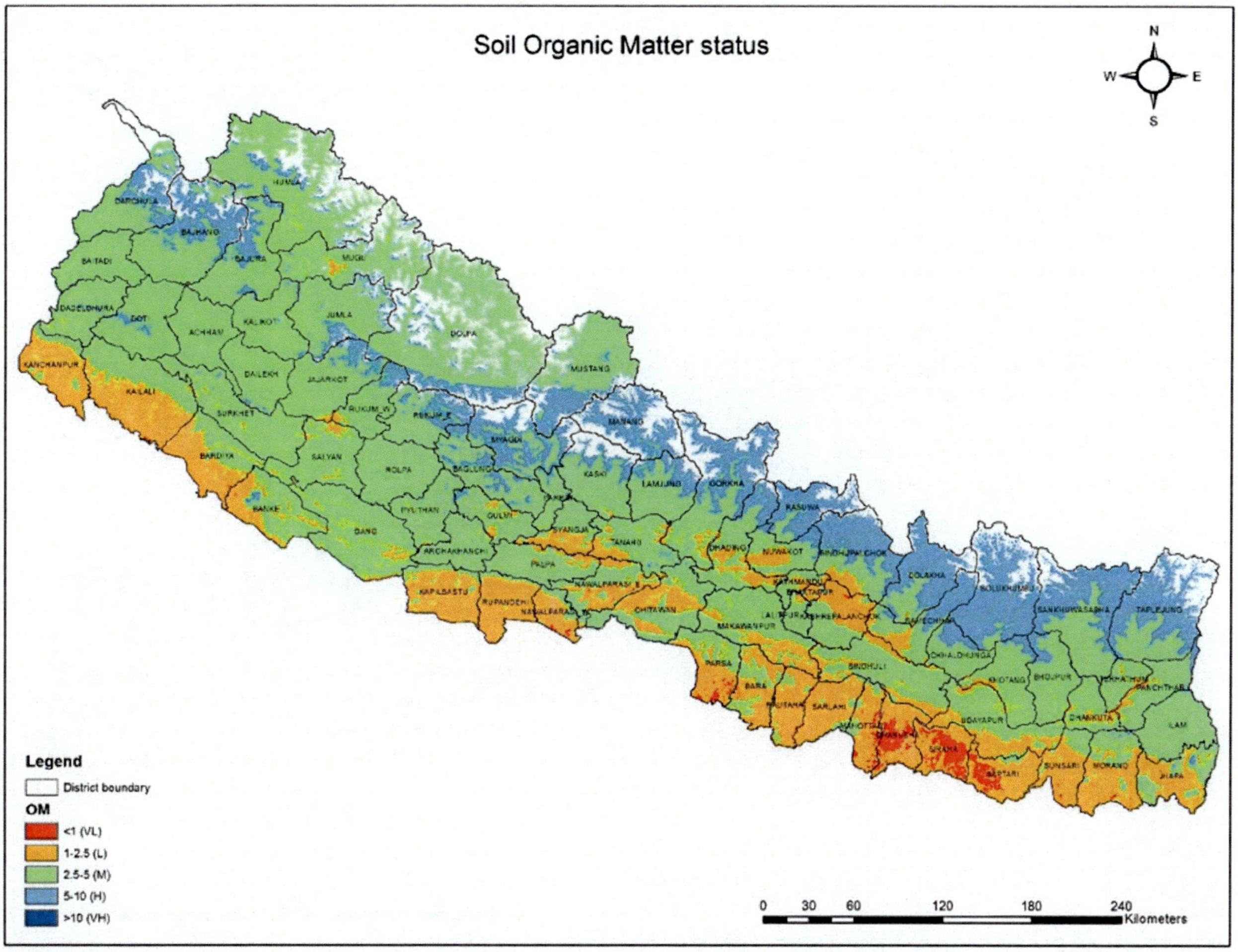

Fig. 8.3 Organic matter content in soils across different agro-ecological zones of Nepal (*Source* Digital soil map 2020, developed by feed the future Nepal seed and fertilizer project and NARC)

Status of micronutrients

Information on the status of soil micronutrients is limited in Nepal. Soil analyses for micronutrients were conducted mostly from NARC's research stations (Table 8.3). The reported analysis was not consistent with the analysis methods; thus, they could show wide variability. A review made by Bajracharya et al. (2007) shows that as with other primary and secondary nutrients, the content of micronutrients in soils was depleted as the addition of organic inputs in soil decreased with crop intensification. Moreover, there are no official recommendations for micronutrients; thus, farmers have limited awareness regarding their use. Some use of micronutrients such as Zn for rice and maize, B for vegetables, and Mo for cauliflower, has been acknowledged. Among the micronutrients, B, Zn, and Mo are deficient in Nepalese soils. Anderson (2007) reported that out of the total samples analyzed across the country based on different reports, proceedings, and articles, 80–90% of soil samples were deficient in B, 20–50% were deficient in Zn, and 10 to 15% were deficient in Mo, while the status of Cu, Fe, and Mn was not a problem.

8.4 Soil Biological Properties

Status of microbes in Nepalese soil

Soil microorganisms differ in type and number based on soil temperature, moisture, management practice, soil type, ecology, and other factors. Nepal has a diverse ecology and its altitude ranges from 60 m asl to the highest peak of the world, Mt. Everest, at 8848.86 m asl. Agriculture is practiced up to about 3500 m asl. The land area above 3500 m asl is used for grazing and pasture. Since microorganisms have a pivotal role in mineralizing soil nutrients and making

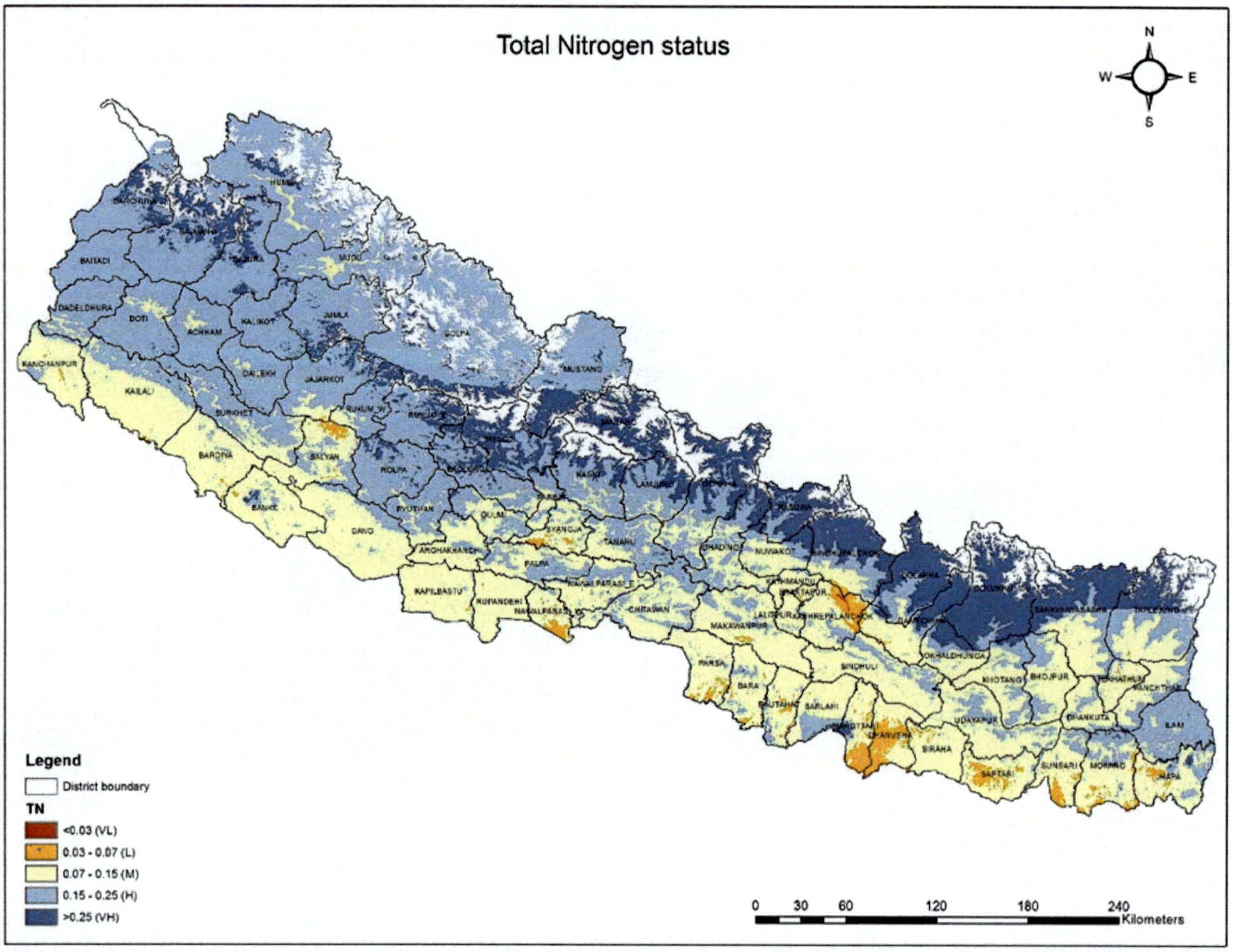

Fig. 8.4 Total nitrogen content (%) of soils across different agro-ecological zones of Nepal (*Source* Digital soil map 2020, developed by feed the future Nepal seed and fertilizer project and NARC)

them available to agricultural crops, it is necessary to study the diversity and population of soil microorganisms at varying altitudes so that the amount of nutrients to be applied can be substantially predicted.

Distribution of microbial diversity in different ecological zones of Nepal

Limited research has been carried out in Nepal with respect to microbial diversity in varying ecological zones. Studies carried out by NARC's Soil Science Division (SSD) in recent years showed that location and treatment have little influence on population and diversity, though a higher degree of fungal and bacterial population was observed in the Terai plains. The Mid hills and High Hills showed a similar level of the fungal and bacterial population. Different treatments of soil under long-term soil fertility trials have not shown a definite trend of the microbial population (SSD 2019).

Altitude is the major factor that has confounded effects on both biodiversity and soil physicochemical properties. High-altitude ecosystems are generally characterized by low temperature, variable precipitation, decreased atmospheric pressure, and soil nutrient stress, all of which have major impacts on biodiversity (Morán et al. 2013). The microbial population varies with varying soil ecology. The *Azotobacter* population was observed to be in higher order in Mid Hill soils compared to Mountain and Terai soils. However, the fungal population was higher in Tearai soil compared to the soil from the Mid and High Hills. This means that low altitudes have a larger fungal population. Different treatments of soil under long-term soil fertility trials have not shown definite trends related to microbial populations. Future studies should focus on microbial population

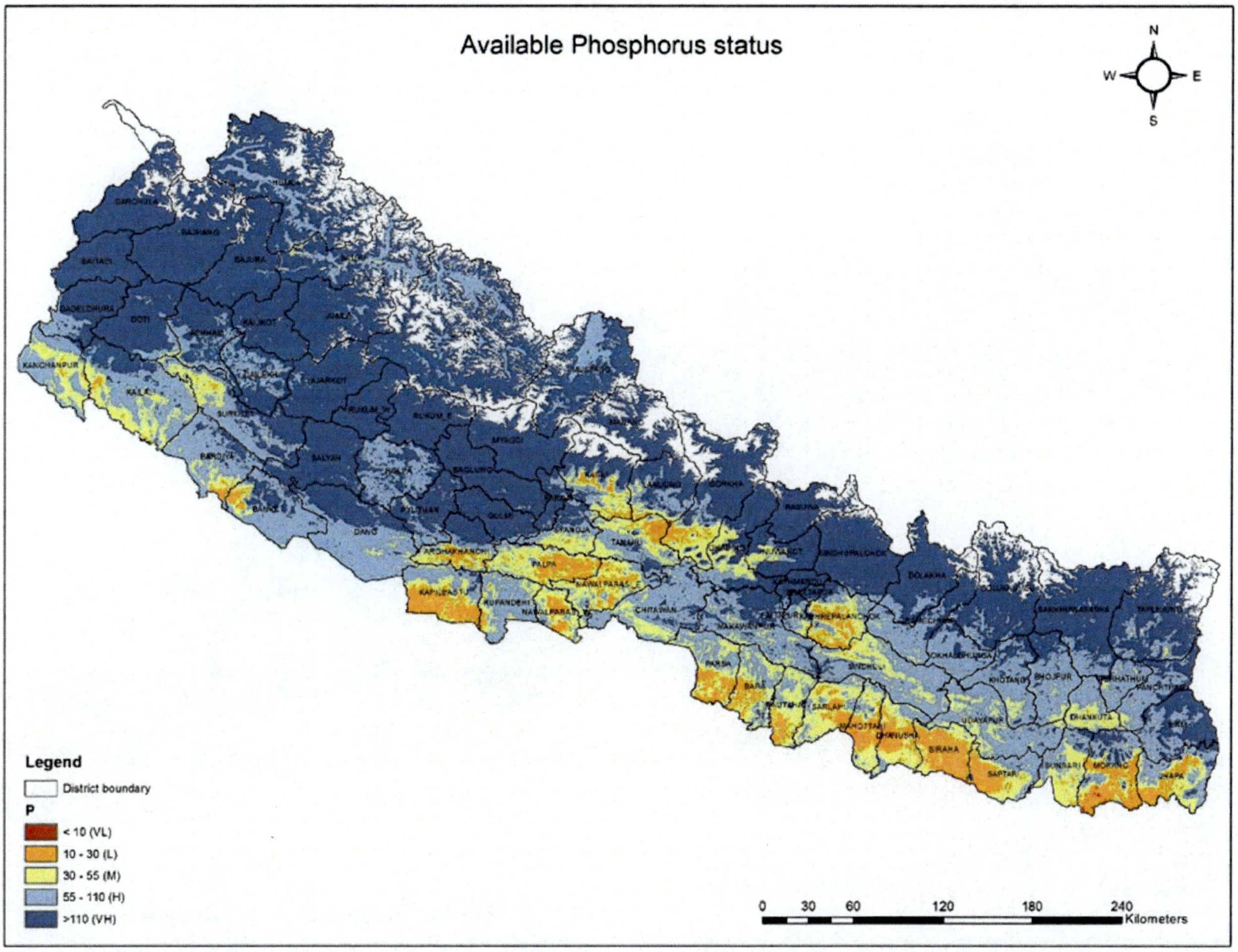

Fig. 8.5 Available phosphorus (kg P_2O_5 ha^{-1}) of soils across different agro-ecological zones of Nepal (*Source* Digital soil map 2020, developed by feed the future Nepal seed and fertilizer project and NARC)

dynamics under different soil management practices across different altitudes.

Agricultural Practices and Soil Biological Properties

Soil is a complex mixture of organic matter, water, air, minerals, and living things formed after the chemical disintegration of rock fragments. Soil quality is the interaction of physical, chemical, and biological properties for agricultural practices and other activities performed in the soil. The biological quality of soil involves a variety of factors occurring within the soil profile, at the surface, and above the ground that is associated with of derive from the living component of the soil ecosystem. This includes the diversity and species composition of soil organisms, namely meso- and macro-fauna, microorganisms, and flora (the types of plant, their root systems, and the vegetative litter produced at the soil surface). We discuss here how agricultural practices influence soil biological properties.

Effect on soil organic carbon stabilization and mineralization

The amount and quality of soil organic matter reflects the nature and abundance of soil flora and fauna and can be considered an indicator of biological soil property. Several studies on the status and dynamics of soil organic carbon (SOC) have been conducted; however, very limited research related to soil fauna and microbial activity has been carried out in Nepal. Bajracharya et al. (2004a) analyzed existing SOC data from the literature and concluded that as expected forest and shrubland soils had higher SOC contents in the top 30 cm; however, due to shallow soils and low density, forest soils had overall lower total OC stock than soils from

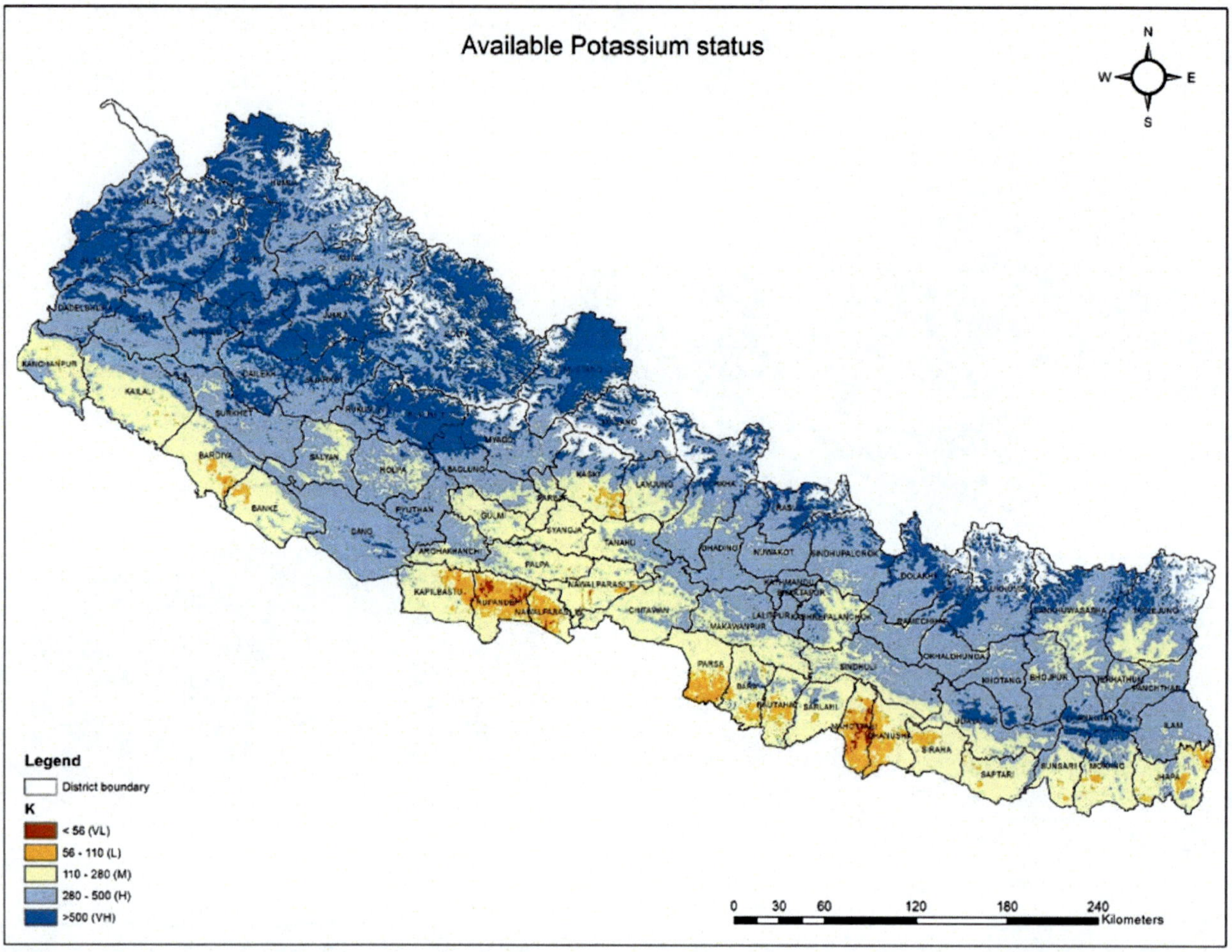

Fig. 8.6 Available potassium (kg K_2O ha^{-1}) of soils across different agro-ecological zones of Nepal (*Source* Digital soil map 2020, developed by feed the future Nepal seed and fertilizer project and NARC)

Table 8.3 Status of micronutrients in soils across different NARC research stations in Terai, Nepal

Micronutrients	Units	Research stations (sites)						
		Tarahara, Sunsari	Hardinath, Dhanusha	Belachapi, Dhanusha	Parwanipur, Bara	Chitwan	Bhairahawa, Rupandehi	Nepalganj, Banke
Boron (B)	mg kg^{-1}	0.08	0.37	0.56	0.59	0.46	0.23	1.13
Zinc (Zn)	mg kg^{-1}	0.35	0.83	0.54	0.51	0.55	2.94	1.82
Iron (Fe)	mg kg^{-1}	244.7	57.79	55.80	85.88	15.89	75.05	21.33
Molybdenum (Mo)	mg kg^{-1}	–	–	–	–	–	–	0.20
Cupper (Cu)	mg kg^{-1}	1.15	0.89	0.30	1.36	0.39	1.87	0.10
Manganese (Mn)	mg kg^{-1}	18.15	6.75	20.50	16.52	3.86	6.87	8.27
Source		Khadka et al. (2017a)	Khadka et al. (2017b)	Khadka et al. (2016)	Khadka et al. (2018)	Khadka et al. (2016)	Khadka et al. (2015)	Baral et al. (2020)

other land use types. While cultivation tends to reduce SOC contents of soils, practices such as mulching, minimum tillage, retention/addition of organic residue, and cover crops can all enhance productivity and reduce erosion losses. Concurrently, degraded forest and grazing lands were noted to be severely depleted in SOC (on the order of 0.1%). For agricultural soils, the SOC contents fell mostly between 1 and 3% in topsoil. Shrestha et al. (2007a,2007b) investigated SOC stocks and C sequestration in a Mid-Hill watershed in western Nepal and reported a significant influence of land use on SOC contents. Among the cultivated land use types, upland soil had higher SOC, while natural forests had the highest overall SOC stocks. A net loss of 29% of SOC stock in the uppermost 40 cm layer of soil was calculated due to changes in land use over the period from 1978 to 1996 (Shrestha et al. 2007).

Effect on microorganism population and diversity

Soil contains many micro- and macroflora and fauna. Soils contain about 8–15 t ha^{-1} of bacteria, fungi, protozoa, nematodes, earthworms, and arthropods (Brady and Weil 2012). A large number of bacteria exist in the soil, but due to their small size, they have smaller biomass. Actinomycetes are 10 times smaller in number than bacteria but larger in size so they have similar biomass in the soil (Hoorman and Islam 2010). Fungal population numbers are smaller but they dominate the soil biomass. Bacteria, actinomycetes, and protozoa can tolerate more soil disturbance than fungal populations so they dominate in tilled soils while fungal and nematode populations tend to dominate in untilled (Silva et. al. 2013). A long-term study carried out across different regions of Nepal by NARC's SSD shows a greater number of fungal populations in the Terai region followed by the High Hill region (SSD 2019). Similarly, a greater number of bacterial populations were found in the Terai region followed by the High Hill and Mid-Hill regions. A greater number of the Azotobacter population was found in the Mid Hill regions. However, there was no distinct trend of the microbial population across fertilizer treatments.

Soil aeration and microorganism population

Microorganism populations in soil could be controlled by soil porosity, as greater amounts of pore space meaning higher counts of microbes (Collins 2010). Well-tilled soil is well aerated and favors microorganism growth. The microbial population is found to be more in aerobic (O_2-rich) soil compared to anaerobic (CO_2–rich) soil (McNabb and Startsev 2009). A major contributor to poor aeration is soil compaction. A count of Azotobacter, Azospirillum, Rhizobium, cyanobacteria, and phosphorus- and potassium-solubilizing microorganisms and mycorrhizae is found to be higher under long-term no-till or minimum-tillage soil (Bhardwaj et al. 2014).

Organic matter and aeration have a positive correlation in building the microorganism population where depth has a negative correlation. Further, soil aeration and organic matter decreases with increasing soil depth, thus decreasing the microorganism population. This indicates that surface soil is rich in microorganisms. In the present context, the soil is being compacted with heavy agricultural equipment, which has created a soil horizon devoid of air space. The soil density has increased resulting in the decrease in the porosity of the soil and limiting microorganism growth. Similar is the case with organic matter, where present agricultural practice uses chemical fertilizer, limiting the use of organic matter. Consequently, microorganisms are deprived of food and their growth has been checked. This has created an imbalance in the soil ecosystem that has resulted in poor structure and less fertile soil.

Effect on greenhouse gas (GHG) emissions from soil

After carbon dioxide (CO_2), methane (CH_4), and nitrous oxide (N_2O) are two important greenhouse gases that are emitted from agriculture practices. Agricultural intensification contributes directly to emissions through a variety of processes. Most of the N_2O from soils is produced mainly by two biological processes, namely, nitrification and denitrification. There are very few studies in Nepal observing the effects on GHG emissions from soil/land use change, though Awasthi (2004) found that land-use changes in the Mid-Hills from the forest or grassland to flooded rice fields were significant sources for CH_4 emission into the atmosphere.

Effect of agrochemicals on soil microbial activities

Several herbicides can alter the symbiotic association between legume plants and rhizobacteria and hinder the vital processes of N-fixation (Singh and Wright 2000; Meena et al. 2015). Some of the examples are paraquat, glyphosate, pendimethalin, which reduce N-fixation in legumes (Dos Santos et al. 2005; Strandberg et al. 2004).

Similarly, fungicides, especially copper (Cu)-based fungicides, have a deleterious effect on the population of N-fixing bacteria (Van Zwieten et al. 2003). Both mancozeb and chlorothalonil can decrease the process of nitrification and denitrification (Kinney et al. 2005). Similarly, insecticides have a negative effect on soil microbes. The growth and population of Azotobacter are significantly inhibited by phosphamidon, malathion, fenthion, methyl phosphorothioate, and parathion (Panday and Singh 2004).

Therefore, to create healthy and fertile soil, we must provide an environment that favors the growth of microorganisms. For this addition of organic matter, loosening the soil mass, providing the optimum moisture in soil, reducing heavy agricultural equipment, and replacing chemicals with alternative sources of manure are steps toward creating eco-friendly soil with high microbial populations.

Microbial Uses in Nepalese Agriculture

Microbes are used for various purposes, including making biofertilizers, biopesticides, west decomposers, food and beverage manufacture, and fermentation of industrial products. Special reference is made here for managing nutrients as a fertilizer, growth-promoting substance, or as a decomposer and symbiotic nutrient supplier.

In Nepal, the exploitation of microbes as biofertilizers is carried out through research and production activities such as isolation, identification, and characterization of Rhizobium and Azotobacter, and symbiotic and asymbiotic N fixers by NARC's SSD in the government sector. In addition, some private laboratories have also initiated the production of biofertilizers and biopesticides.

Microbes used as Biofertilizers in Nepal

Biofertilizers are substances produced through biological processes with a significant nutrient value that can be used as effective sources of nutrients. These contain living microorganisms that, when applied to soil, seed, or plant surfaces, colonize the rhizosphere or the interior of the plant and promote growth by increasing the availability of nutrients to the host plant. Unlike manures and fertilizers, biofertilizers add nutrients to the system through the natural process of atmospheric fixation (e.g., of N) or by making nutrients available through the process of mobilization and solubilization (e.g., for phosphorus, P). Some of these species also stimulate plant growth through the synthesis of growth-promoting substances. The microorganisms in biofertilizers also restore the natural nutrient cycle and build SOM that supports healthy plant growth and soil health, which are fundamental for sustainable soil fertility. Therefore, biofertilizers can contribute significantly toward reducing the use of chemical fertilizers and can play a significant role in Nepal for enriching soil fertility and fulfilling plant nutrient requirements in a sustainable way. Biofertilizers such as Rhizobium, Azotobacter, Azospirillum, and blue-green algae (BGA) have been in use for a long time in Nepal. Based on their mode of nutrient synthesis or release in soil or plants, biofertilizers can be categorized into several groups, which are as follows:

Rhizobium species

Rhizobium sp., the symbiotic N-fixing bacteria, is an important group among the micro-organisms used as biofertilizers. Rhizobia fix atmospheric N (40–250 kg N ha^{-1} yr^{-1}) symbiotically into the root's nodules of host legume plants. Rhizobia can also be utilized through the growing of leguminous plants as a green manure that adds N through fixation as well as through the decomposition of N-rich legume biomass in soil. During the fiscal year 2018/19, NARC's SSD produced 4507 packets of biofertilizers for different crops including lentil, soybean, cowpea, pea, black gram, and some grasses, and 3510 packets were distributed to farmers and Agricultural Agencies for practical use in their fields (SSD 2018/19).

Both inoculation successes and failures at the field level have been reported in the literature. Responses to Rhizobium inoculation have mostly been demonstrated with grain legume crops, particularly lentil, chickpea, soybean, groundnut, black gram, and faba bean. Inoculation of Rhizobium increased the grain yield of different legumes from 17 to 60% and are known to leave behind some residual nitrogen in the soil (Maskey et al. 2001). In some experiments, a significant residual effect was reported in the yield of subsequent crops such as wheat, rice, and maize. The maximum residual effect was seen in soybean, which increased the yield of subsequent crops of wheat by 65.9% over un-inoculated crops (SSD 2015).

Azolla

Azolla is a free-floating water fern that fixes atmospheric N in association with BGA (Anabaena azollae). Anabaena in association with Azolla contributes up to 60 kg N ha^{-1} $season^{-1}$ and also enriches soils with OM (Stewart et al. 2005). Azolla as a biofertilizer is commonly used in rice farming systems. Azolla is most found naturally in stagnant water and is also used as a biofertilizer in rice in Nepal. There remains the need for wider application of Azolla because farmers in Nepal still lack knowledge about its use.

In Nepal, A. Pinnate and A. filiculoides are commonly found in natural water and swampy lands. Azolla requires phosphorus (15–20 kg ha^{-1}) and in very deficient soil deficiency symptoms such as purple color occur (Maskey and Bhattarai 1984). Molybdenum is beneficial for the adequate growth of Azolla (Adhikary and Bhattrai 2000). Azolla alone can increase rice yield by at least 12–14% without any additional N fertilizer, though recommended doses of P and K fertilizers are needed to meet the nutrient requirement of the crop (Adhikary et al. 2015).

Blue Green Algae (BGA)

Blue-Green Algae (BGA) are also known as Cynobacteria. They are either single-celled or filamentous multicellular. The N-fixing BGA possesses a special structure called Heterocyst. The standing water of the rice field encourages the growth of Blue-Green Algae (BGA), and these BGA possess photosynthesis abilities as well as biological nitrogen fixation abilities. BGA belonging to the general genus Nostoc, Anabaena, Tolypothrix, or Aulosira fix atmospheric N and are used as inoculants for upland and lowland paddy rice.

Blue-green algae are the dominant N-fixer. In addition to N-fixing, they excrete vitamin B12, auxin, and ascorbic acid, which contribute to rice growth. They fix atmospheric N equivalent to 20–30 kg h^{-1} yr^{-1}. Azollae and BGA (Anabaena azollae) are actively involved in the symbiotic association and fix atmospheric N, after which the host plant (Azolla) provides a carbon source while Anabana azollae fix atmospheric N and transfer it to the azolla, which then multiplies rapidly. In Nepal very limited research, demonstration, and promotion activities are carried out regarding BGA.

Free-living N-fixing bacteria

Free-living N-fixing bacteria such as *Azotobacter, Azospirillum*, and *Clostridium* sp. also fix N in nonlegume crops such as rice, wheat, barley, millet, and cotton. These are not as common as Rhizobia, but they have a potential for N-fixing in non-legume crops. Azotobacter have been used in cereals (e.g., wheat and barley), potatoes, and vegetables, while Azospirillum inoculations are recommended mainly for use in sorghum, millets, maize, sugarcane, and wheat. The population of Azotobacter in Nepalese soils is very low, that is, not more than 10,000–100,000 g^{-1} soil. The population of Azotobacter is mostly influenced by other microorganisms present in the soil; for instance, Cephallosporium is mostly found in soil that restricts the growth of Azotobacter. Several field experiments have been conducted in Nepal on the Azotobacter inoculation of seeds and seedlings in rice, wheat, maize, tomato, potato, cabbage, and other crops under different agroclimatic conditions. The lack of organic matter in soil is a limiting factor for its multiplication, though the effect of Azotobacter was increased by 154% when compost was used. Maskey and Bhattrai (1984) reported yield increases of 6–12% in rice and 8–12% in wheat after inoculating Azotobacter. Baral and Adhikary (2013) reported that inoculation of only Azotobacter increased grain yield in maize 15–35% and that the benefit is higher in the absence of chemical fertilizer application. Inoculation of Azotobacter also increased yield 5% in tomatoes and 3.5–55% in cauliflower (SSD2000).

In Nepal, Azotobacter is mainly isolated, cultured, produced, and supplied by the Soil Science Division of NARC. Besides the Soil Science Division, private agro supply companies import biofertilizers with different trade names, though their quality is not assured.

Phosphate-solubilizing bacteria (PSB)

Phosphate-solubilizing bacteria (PSB), such as *Pantoea agglomerans* or *Pseudomonas putida* can solubilize the insoluble phosphate from organic and inorganic phosphate sources. Due to the immobilization of phosphate by mineral ions such as Fe^{2+}, Fe^{3+}, Al^{3+}, and Ca^{2+} or organic acids, the available phosphate (H_2PO_4) in soil absorbed by plants can be as low as 20% of added P fertilizer. In Nepal, very limited research work has been carried out on PSB, though PSB as a bio-fertilizer in various trade names is now available through importation from other countries.

Vesicular Arbuscular mycorrhizal (VAM)

A mutually beneficial (symbiotic) association between numerous fungi and the roots of higher plants is called mycorrhiza. The available technique of inoculation is to collect infested roots and use them for subsequent infection, a method that limits widespread application. These inoculations are to prepare per-inoculated transplants for coffee, tea, cocoa, papaya, and oil palm.

Mycorrhizas increase the longevity of feeder roots and root surface area by forming a mantle and spreading mycelia into the soil, which in turn enhances the rate of absorption of macro- and micronutrients and water. Mycorrhizas also play a key role in the selective absorption of immobile (P, Zn, and Cu) and mobile (S, Ca, K, Fe, Mn, Cl, and N) elements to plants (Tinker 1984). Vesicular Arbuscular mycorrhizal (VAM) fungus reduces the plant response to soil stresses caused by high salt, drought, and toxicity associated with heavy metals, mine spoils, and minor element (e.g., Mn) deficiencies. Among the different types of endomycorrhizae Glomus and Acaulosporus are predominantly found in upland soil (Bajracharya et al. 2004b). The organic amendment significantly increases the biomass of VAM and enhances the rehabilitation of eroded soil (Vaidya et al. 2007).

Limitations of Biofertilizer Adoption in Nepal

Although the use of biofertilizers has more benefits compared to the use of inorganic fertilizers, there are certain limitations to its widescale adoption in Nepal. Some of these limitations include (i) unavailability of appropriate inoculum, (ii) preservation and transport of inoculum, (iii) poor farmers' awareness of the use of biofertilizers, (iv) high cost of production, (v) lack of commercial operation, and (vi) slow effects on crops. These issues should be addressed to utilize the potential of biofertilizers for developing commercial products available to Nepalese farmers.

8.5 Summary and Conclusion

To keep our soil healthy, all of the physical, chemical, and biological properties have to be properly managed. Nepalese soils are developed from micaceous parent materials such as phyllites, schists, gneisses, granites, and others, and therefore most of the soil contains higher proportions of mica. Loam and sandy loam are the most dominant soil textures in both hills and alluvial terraces. However, this varies with agroecology and land use types. The sub-angular blocky structure is most common in Nepal. In Terai, angular blocky is a common structure in clayey-type soils. Because the use of organic inputs is common in Nepalese agriculture, particularly in the Hills and Mid-Hills, most of the soils in the hills and mountainous are friable. However, hard consistency predominates in areas where organic manure is limited, particularly in the Terai region.

Nepalese farmers in the hills and river valleys apply a sufficient amount of organic inputs, including farmyard manure and compost, to their field crops. Soils are more acidic at higher altitudes (hills and mountains) and in the eastern Terai compared to the western part of the country. The total N content in Nepalese soil lies mostly in the medium range, while available phosphorus and potassium content are in the higher range. Among the micronutrients, B, Zn, and Mo are deficient in the soil (though in patches), while (Cu), iron (Fe), and manganese (Mn) are at adequate levels.

The soils of Nepal are heterogeneous, therefore requiring multiple technologies for their sustainable management. Most farmers manage soils using indigenous practices that are organic based in the Hill and Mountain regions, where chemical fertilizers are used as a supplement. Farmers in the Terai region use more chemical fertilizers compared to farmers in the Hill and Mountain regions, and organic inputs are used as a supplement. Soil erosion and acidification are the most pertinent soil problems in Nepal. Careful management of soil with adequate and sustainable management practices is the need of the day to improve soil fertility in Nepal.

References

Adhikari BH, Karki KB (2007) Use of fertilizers and lime for enhancing productivity of maize genotypes in western hill of nepal. Nepal Agric Res J 8:42–49

Agvise (2020) Soil testing and laboratory service (Online service). In: Water holding capacity. Benson, MN 56215. https://www.agvise.com/educational-articles/water-holding-capacity/

AICC 2076. Krishi Diary 2076. Agriculture Information and Communication Centre, Ministry of Agriculture and Livestock

Amirkhanov DV, Nikolenko AG, Bagautdinov FY, Kirillova SS (1994) Effect of production dosage of gamma-HCCH, foxim, cypermethrin and chlorfluazuron on soil microorganisms [grey forest soils] [Russian]. Agrokhimiya 2:83–88

Andersen P (2007) A review of micronutrient problems in the cultivated soil of Nepal. Mt Res Dev 27:331–335

Angst G, Messinger J, Greiner M, Häusler W, Hertel D, Kirfel K, Kögel-Knabner I, Leuschner C, Rethemeyer J, Mueller CW (2018) Soil organic carbon stocks in topsoil and subsoil controlled by parent material, carbon input in the rhizosphere, and microbial-derived compounds. Soil Biol Biochem 122:19–30

Awasthi KD (2004) Land-use change effects on soil degradation, carbon and nutrient stocks and greenhouse gas emission in mountain watersheds. PhD dissertation submitted to Department of Plant and Environmental Science, Agricultural University of Norway

Ba¨umler R, Zech W (1994) Soils of the high mountain region of Eastern Nepal: classification, distribution and soil forming processes. CATENA 22:85–103

Bajracharya R, Lal R (1999) Land use effects on soil crusting and hydraulic response of surface crusts on a tropical Alfisol . Hydrol Process 13:59–72

Bajracharya SK, Maskey SL, Bhattarai S (2004) Survey, Isolation and identification of endomycorrhizae. In: Proceedings of 4th national conference on science and technology organized by RONAST held on March 23–26, 2004, Kathmandu, Nepal

Bajracharya RM, Sitaula BK, Shrestha BM, Awasthi KD, Balla MK, Singh BR (2004) Soil organic carbon status and dynamics in the central Nepal Middle Mountains. J Inst Forest 12:28–44. Institute of Forestry, Pokhara, Nepal

Bajracharya RM, Sitaula BK, Sharma S, Jeng A (2007) Soil quality in the Nepalese context—an analytical review. Int J Ecol Environ Sci 33(2–3):143–158

Baral BR, Adhikari P (2013) Effect of Azotobacter on growth and yield of maize. SAARC J Agri 11(2):141–147

Baral BR, Pande KR, Gaihre YK, Baral KR, Sah SK, Thapa YB, Singh U (2020) Increasing nitrogen use efficiency in rice through fertilizer application method under rainfed drought conditions in Nepal. Nutr Cycl Agroecosyst 118:103–114

Begum F, Bajracharya RM, Sitaula BK, Sharma S (2013) Seasonal dynamics, slope aspect and land use effects on soil mesofauna density in the mid-hills of Nepal. Int J Biodiv Sci Ecosys Serv Manage 9:290–297

Bhardwaj D, Ansari MW, Sahoo RK, Tuteja N (2014) Biofertilizers function as key player in sustainable agriculture by improving soil fertility, plant tolerance and crop productivity. Microb Cell Fact 13:66

Brady NC, Weil RC (2012) The nature and properties of soils, 14th edn. Dorling Kindersley India Pvt. Ltd, Noida, India

Brady NC, Weil RR, Weil RR (2008) The nature and properties of soils. Prentice Hall Upper Saddle River, NJ

Bremner JM (1997) Sources of nitrous oxide in soils. Nutr Cycl Agroecosyst 49:7–16

Brown S (1997) Soil fertility, nutrient dynamics, and socio-economic interactions in the Middle Mountains of Nepal. PhD. Thesis, Interdisciplinary Studies in Resource Management Science, University of British Columbia, Vancouver, BC

Carson B (1992) The land, the farmer and the future: a soil management strategy for Nepal. ICIMOD occassional paper no. 21, International Centre for Integrated Mountain Development, Kathmandu, Nepal

Cass A, Gusli S, MacLeod DA (1994) Sustainability of soil structure quality in rice paddy— soya-bean cropping systems in South Sulawesi, Indonesia. Soil Tillage Res 31:339–352

Collins H (2010) Impacts of fumigation and crop rotation on soil microbial populations. USDA-ARS Irrigated Research Center, USA

DOA/STSS (Department of Agriculture/Soil Testing & Service Section) (1999). In: Workshop on soil fertility management activities in Nepal: past, present and future, 11 June 1999, Department of Agriculture. Soil Testing & Service Section, and Japan International Cooperation Agency. Kathmandu, Nepal, 18 p

dos Santos JB, Ferreira EA, Kasuya MCM, da Silva AA, de Oliveira Procópio S (2005) Tolerance of Bradyrhizobium strains to glyphosate formulations. Crop Prot 24:543–547

Dawadi DP, Thapa M (2015) Soil fertility status of Nepal: report from laboratory analysis of soil samples of five development regions. In: Proceedings of the second national soil fertility research workshop, soil science division, Khumaltar, pp 42–52

Ela S, Gupta SC, Rawls W (1992) Macropore and surface seal interactions affecting water infiltration into soil. Soil Sci Soc Am J 56:714–721

FAO (2006) World reference base for soil resources 2006, A framework for international classification. Correlation and communication World Soil Resources Reports No 103. Food and Agriculture Organization, Rome, Italy

Gabriels D, Horn R, Villagra M, Hartman R (eds) (1997) Assessment, prevention and rehabilitation of soil structure caused by soil surface sealing, crusting and compaction, pp 1–129–165

Gami S, Ladha J, Pathak H, Shah M, Pasuquin E, Pandey S, Hobbs P, Joshy D, Mishra R (2001) Long-term changes in yield and soil fertility in a twenty-year rice-wheat experiment in Nepal. Biol Fert Soils 34:73–78

Gangwar K, Singh K, Sharma S, Tomar O (2006) Alternative tillage and crop residue management in wheat after rice in sandy loam soils of Indo-Gangetic plains. Soil Tillage Res 88:242–252

Gardner R, Gerrard A (2003) Runoff and soil erosion on cultivated rainfed terraces in the Middle Hills of Nepal. Appl Geogr 23:23–45

Ghimire R, Lamichhane S, Acharya BS, Bista P, Sainju UM (2016) Tillage, crop residue, and nutrient management effects on soil organic carbon sequestration in rice-based cropping systems in South Asia: a review. J Integr Agric 15:60345–60347. https://doi.org/10.1016/S2095-3119(16)61337-0

Gilmour DA, Bonell M, Cassells D (1987) The effects of forestation on soil hydraulic properties in the Middle Hills of Nepal: a preliminary assessment. Mount Res Dev 239–249

Grable AR (1966) Soil aeration and plant growth. In: Norman AG (ed) Advances in agronomy, vol 18, pp 57–106. Academic Press

Hoorman JJ, Islam R (2010) Understanding soil microbes and nutrient recycling. Agric Nat Resour 16(10)

Horn M, Rutledge E, Dean H, Lawson M (1964) Classification and genesis of some solonetz (Sodic) soils in Eastern Arkansas 1. Soil Sci Soc Am J 28:688–692

Ivarson K, Heringa P (1972) Oxidation of manganese by microorganisms in manganese deposits of Newfoundland soil. Can J Soil Sci 52:401–416

Jat ML, Gathala MK, Saharawat YS, Tetarwal JP, Gupta R, Yadvinder S (2013) Double no-till and permanent raised beds in maize–wheat rotation of north-western Indo-Gangetic plains of India: Effects on crop yields, water productivity, profitability and soil physical properties. Field Crop Res 149:291–299

Kaiser K, Guggenberger G (2005) Storm flow flushing in a structured soil changes the composition of dissolved organic matter leached into the subsoil. Geoderma 127:177–187

Karki KB (1986) Assessment of land degradation and formulation of productivity improvement strategies in Southern Lalitpur District of Nepal. Partial Fulfillment for the degree Central Luzon State University, The Philippines

Karki KB, Dacayo JB (1990) Assessment of land degradation in Southern Lalitpur of Nepal. In: Proceedings 14th congress of international society of soil science, Kyoto, Japan, August 12–19, 1991. Karki KB, Joshy D (1993) Soil fertility and fertilizers use in Nepal. In: Hls T (ed) Soil fertility and fertilizer use in South Asia. Development and Consultation Organization, New Delhi, India

Khadka D, Lamichhane S, Bhurer KP, Chaudhary JN, Ali MF, Lakhe L (2018) Soil fertility assessment and mapping of regional agricultural research station, Parwanipur, Bara Nepal. J Nepal Agric Res Council 4:33–47

Khadka D, Lamichhane S, Thapa B, Baral BR, Adhikari P (2016) An assessment of soil fertility status of National Maize Research Program, Rampur, Chitwan, Nepal. Int J Interdisc Res 2:1798–1807

Khadka D, Lamichhane S, Thapa B, Rawal N, Chalise DR, Vista SP, Lakhe L (2015) Assessment of soil fertility status and preparation of their maps of National Wheat Research Program (NWRP), Bhairahawa, Nepal. In: Proceeding of the second national soil fertility research workshop, 24–25 March 2015. Soil Science Division, NARC, Khamulatar, Lalitpur, Nepal

Khadka D, Lamichhane S, Tiwari D, Mishra K (2017) Assessment of soil fertility status of national rice research program, hardinath, dhanusha Nepal. Int J Agric Environ Res 3:86–105

Khadka D, Lamichhane S, Shrestha SR, Pant BB (2017) Evaluation of soil fertility status of regional agricultural research station, Tarahara, Sunsari Nepal. Eurasian J Soil Sci 6:295

Khaleel R, Reddy K, Overcash M (1981) Changes in soil physical properties due to organic waste applications: a review. J Environ Qual 10:133–141

Kinney CA, Mandernack KW, Mosier AR (2005) Laboratory investigations into the effects of the pesticides mancozeb, chlorothalonil, and prosulfuron on nitrous oxide and nitric oxide production in fertilized soil. Soil Biol Biochem 37:837–850

Lucas RW, Klaminder J, Futter MN, Bishop KH, Egnell G, Laudon H, Högberg P (2011) A meta-analysis of the effects of nitrogen additions on base cations: Implications for plants, soils, and streams. For Ecol Manage 262:95–104. https://doi.org/10.1016/j.foreco.2011.03.018

Lui W, Upadhyaya SK, Kataoka T, Shibusawa S (1996) Development of a texture/soil compaction sensor. Precis Agric 617–630

Lund ZF (1959) Available water-holding capacity of alluvial soils in Louisiana. Soil Sci Soc Am J 23:1–3

Maskey SL, Bhattarai S (1984) Effect of Azolla on rice. IAAS J 3(1)

Maskey SL, Bhattarai S, Peoples MB, Herridge. (2001) On farm measurement of nitrogen fixation by winter and summer legumes in the hill and Tarai region of Nepal. Field Crop Res 70:209–221

McLaren RG, Cameron KC (1996) Soil sciences, sustainable production and environmental protection. Oxford University Press Auckland

McNabb DH, Startsev AD (2009) Effects of compaction on aeration and morphology of boreal forest soils in Alberta Canada. Can J Soil Sci 89(1):45–56

Meena RS, Meena VS, Meena SK, Verma JP (2015) The needs of healthy soils for a healthy world. J Clean Prod 102:560–561

Meki MN, Snider JL, Kiniry JR, Raper RL, Rocateli AC (2013) Energy sorghum biomass harvest thresholds and tillage effects on soil organic carbon and bulk density. Ind Crops Prod 43:172–182

MoALD and Practical Action (2020) Multistakeholder action plan for improving soil organic matter in Nepal (Draft report). 28 February 2020

Mohammed AK, Hirmas DR, Nemes A, Giménez D (2020) Exogenous and endogenous controls on the development of soil structure. Geoderma 357:113945

Morán-Tejeda EL, Moreno JI, Beniston M (2013) The changing roles of temperature and precipitation on snowpack variability in Switzerland as a function of altitude. Geophys Res Lett 40 (10):2131–2136

Morin J, Van Winkel J (1996) The effect of raindrop impact and sheet erosion on infiltration rate and crust formation. Soil Sci Soc Am J 60:1223–1227

Motomura S (1969) Dynamic behavior of ferrous iron in paddy soils. Bull Natl Inst Agric Sci Ser B 21:1–114

Mücher H, van Steijn H, Kwaad F (2018) Chapter 2—Colluvial and mass wasting deposits. In: Stoops G, Marcelino V, Mees F (eds) Interpretation of micromorphological features of soils and regoliths, 2nd edn, pp 21–36, pp. 21–36. Elsevier

Ogban PI, Babalola O (2003) Soil characteristics and constraints to crop production in inland valley bottoms in southwestern Nigeria. Agric Water Manag 61:13–28

Ojha RB, Shrestha S, Khadka YS, Panday D (2021) Potassium nutrient response in the rice-wheat cropping system in different agro-ecozones of Nepal. Plos One 16(3):e0248837. https://doi.org/10.1371/journal.pone.0248837

Pal D, Srivastava P, Durge S, Bhattacharyya T (2001) Role of weathering of fine-grained micas in potassium management of Indian soils. Appl Clay Sci 20:39–52

Pandey S, Singh DK (2004) Total bacterial and fungal population after chlorpyrifos and quinalphos treatments in groundnut (Arachis hypogaea L.) soil. Chemosphere 55:197–205

Paudel S, Sah JP (2003) Physiochemical characteristics of soil in tropical sal (Shorea robusta Gaertn.) forests in eastern Nepal. Himal J Sci 1:107–110

Rabenhorst MC (1990) Micromorphology of Induced Iron Sulfide Formation in a Chesapeake Bay (Usa) Tidal Marsh. In: Douglas LA (ed), Developments in soil Science, vol 19, pp 303–310. Elsevier

Regmi BD, Zoebisch MA (2004) Soil fertility status of Bari and Khet land in a small watershed of middle hill region of Nepal. Nepal Agric Res J 5:124–129

Rigueiro-Rodriguez A, Mosquera-Losada MR, Fernandez-Nunez E (2012) Afforestation of agricultural land with Pinus radiata D. don and Betula alba L. in NW Spain: effects on soil pH, understorey production and floristic diversity eleven years after establishment. Land Degrad Dev 23:227–241. https://doi.org/10.1002/ldr.v23.3

Schreier H, Shah PB, Lavkulich L, Brown S (1994) Maintaining soil fertility under increasing land use pressure in the Middle Mountains of Nepal. Soil Use Manage 10:137–142

Schreier H, Shah P, Nakarmi G, Shrestha B (2000) Prevention, degradation and rehabilitation: a new dilemma in Himalayan watersheds. In: Environmental reconstruction in headwater areas, pp 147–159. Springer

Sen H (2003) The 10th Dr. DP Motiramani Memorial Lecture. J Indian Soc Soil Sci 51:388–408

Shrestha B, Singh B, Sitaula B, Lal R, Bajracharya R (2007) Soil aggregate-and particle-associated organic carbon under different land uses in Nepal. Soil Sci Soc Am J 71:1194–1203

Shrestha BM, Singh BR, Sitaula BK, Lal R, Bajracharya RM (2007) Soil aggregate- and particle associated organic carbon under different land uses in Nepal. Soil Sci Soc Am J 71(4):1194–1203

Shrestha CL (1985) River sedimentation and remedial measures in Nepal. Water Int 10:14–17

Shrestha RM, Malla S (1996) Air pollution from energy use in a developing country city: the case of Kathmandu Valley. Nepal Energy 21:785–794

Silva AP, Babujia LC, Matsumoto LS, Guimaraes MF, Hungaria M (2013) Bacterial diversity under different tillage and crop rotation systems in an oxisol of Southern Brazil. Open Agric J 7(Supp 11-M6):40–47

Singh N, Dhillon BS, Sandhu PS (2014) Alleviation of soil physical constraints for improved crop production: a review. Ecol Environ Conserv 20:29–36

Sitaula BK, Bajracharya RM, Singh BR (2004) Factors affecting organic carbon dynamics in soils of Nepal/Himalaya region—a review and analysis. Nutr Cycl Agroecosyst 70(2):215–229

SMD/DoA 2013. Annual report. Soil Management Directorate, Department of Agriculture, Harihar Bhawan, Lalitpur, Nepal

Srivastava P, Parkash B, Pal DK (1998) Clay minerals in soils as evidence of Holocene climatic change, central Indo-Gangetic Plains, north-central India. Quatern Res 50:230–239

SSD (2000) Annual Report, Soil Science Division, National Agriculture Research Institute, Nepal Agricultural Research Council (NARC), Nepal

SSD (2015) Annual Report, Soil Science Division, National Agriculture Research Institute, Nepal Agricultural Research Council (NARC), Nepal

SSD (2016) Annual Report, Soil Science Division, National Agriculture Research Institute, Nepal Agricultural Research Council (NARC), Nepal

SSD (2017) Annual Report, Soil Science Division, National Agriculture Research Institute, Nepal Agricultural Research Council (NARC), Nepal

SSD (2019) Annual Report, Soil Science Division, National Agriculture Research Institute, Nepal Agricultural Research Council (NARC), Nepal

Stewart WM, Dibb DW, Johnston AE, Smyth TJ (2005) The contribution of commercial fertilizer nutrients to food production. Agron J 97:1–6

Strandberg M, Scott-Fordsmand JJ (2004) Effects of pendimethalin at lower trophic levels—A review. Ecotoxicol Environ Saf 57:190–201

Talchabhadel R, Karki R, Yadav M, Maharjan M, Aryal A, Thapa BR (2019) Spatial distribution of soil moisture index across Nepal: a step towards sharing climatic information for agricultural sector. Theoret Appl Climatol 137(3):3089–3102

Umiker KJ, Johnson-Maynard JL, Hatten TD, Eigenbrode SD, Bosque-Perez NA (2009) Soil carbon, nitrogen, pH, and earthworm density as influenced by cropping practices in the Inland Pacific Northwest. Soil Tillage Res 105:184–191. https://doi.org/10.1016/j.still.2009.09.001

Vaidya GS, Shrestha K, Khadge BR, Johnson NC, Wallender H (2007) Organic matter stimulates bacteria and AM Fungi in purpurea and Luecaena diversifolia plantation on eroded slopes in Nepal

Van Zwieten L, Ayres MR, Morris SG (2003) Influence of arsenic co-contamination on DDT breakdown and microbial activity. Environ Pollut 124:331–339

Vengadaramana A, Jashothan PJ (2012) Effect of organic fertilizers on the water holding capacity of soil in different terrains of Jaffna peninsula in Sri Lanka. 2:500–503

Walker C, Lin H (2008) Soil property changes after four decades of wastewater irrigation: a landscape perspective. CATENA 73:63–74

Zhang M, Karathanasis AD (1997) Characterization of iron-manganese concretions in kentucky alfisols with perched water tables. Clays Clay Miner 45:428–439

Soil Fertility and Nutrient Management

9

Krishna Bahadur Karki, Dil Prasad Sherchan, Dinesh Panday, and Rajan Ghimire

Abstract

Soil fertility is the key to food security and a sustainable livelihood. Soil fertility depletion has become a significant challenge for crop production in Nepal. The main causes of low soil fertility are soil erosion, soil acidity, and organic matter depletion. This chapter discusses an overview of the climate and topography, inherent factors determining soil fertility, historical perspective on soil fertility management, status, constraints, and policy recommendations for improved soil fertility management in Nepal. The authors demonstrate research gaps in soil fertility-related issues, such as fertilizer use efficiency, recycling of residual nutrients, secondary and micronutrient availability and management, and the policy gap in the successful implementation of soil fertility management programs. More research on region-specific soil fertility issues and the formulation of agriculture policies addressing marginal farmers' needs will help sustain crop production and improve food security in Nepal.

Keywords

In-situ manuring • Mountain agriculture • Nepal • Soil fertility • Traditional farming

9.1 Introduction

Soil fertility—the capacity of soil to supply essential nutrients to crops—is an inherent property of soil that supports food production. There are 17 essential elements required by plants for their growth and production. Plants take up Carbon (C), Hydrogen (H), and Oxygen (O) from air and water, and farmers do not need to apply these three elements. The Nitrogen (N), Phosphorous (P), and Potassium (K) are required in large amounts, so they are considered the primary or major nutrients, followed by the secondary nutrients Ca, Mg, and S needed by plants in lesser amounts. Boron (B), Chlorine (Cl), Copper (Cu), Iron (Fe), Molybdenum (Mo), Manganese (Mn), Zinc (Zn), and Nickel (Ni) are required in minute quantities and are referred to as micronutrients or trace nutrient elements (Havlin et al. 2016; Mengel and Kirkby 1987). Brown et al. (1987) exhibited Ni as an essential plant nutrient for higher plants. Later, several studies confirmed the critical role of Ni in crops and considered as essential plant nutrient (de Queiroz Barcelos et al. 2017; Hänsch and Mendel 2009; Rahman et al. 2005).

Nepal lies between the two giants of Asia, China in the north and India in the east, south, and west. On average, it is 885 km long east to west and 185 km north–south. The flat plain of Tarai in the south lies as low as 66 m asl and the towering Himalayan Mountains in the north peak at Mt. Everest (8848.86 m asl). With rapid topography changes in such a narrow strip of land, this country enjoys a varied climate. The Tarai region enjoys a warm summer and mild winter climate, but the Himalayan region has an alpine to tundra climate. The middle and high mountains have chill winters and mild summers. The country's soils are mainly developed from igneous and metasedimentary rocks consisting of quartzite, phyllites, and slates of Precambrian to Miocene age. The soils are medium to light textured, acidic in pH, low to medium in organic matter, and low in primary macro-nutrients (NPK) (Dawadi and Thapa 2015).

K. B. Karki (✉) · D. P. Sherchan
Nepal Agricultural Research Council, Kathmandu, Nepal

D. Panday
Department of Biosystems and Engineering and Soil Science, The University ofTennessee-Knoxville, Knoxville, TN, USA

R. Ghimire
Agricultural Science Center, New Mexico State University, Las Cruces, NM, USA

R. B. Ojha and D. Panday (eds.), *The Soils of Nepal*, World Soils Book Series,
https://doi.org/10.1007/978-3-030-80999-7_9

Crop production in Nepal was a chemical-free production practice until 1954 when mineral fertilizer was first imported (Karki 2015). Specifically, farmers in the hills and mountains applied mostly farmyard manure (FYM) to upland crops where maize and millets were grown, but rice crops did not receive any mineral or organic fertilizer; instead, they depended on flood sediments. Farmers applied a large amount of manure mainly to maize crops, though it was insufficient to supplement the required amount of nutrients (Karki 2008; Maskey et al. 2004). Finger millet, tori (mustard), and other legumes relied on residual nutrients from manure applied to maize crops.

Nepalese farmers have explored ways to improve soil fertility and supply nutrients to plants. Farmers have used green leaves of Ashuro (*Adhatoda vasica*), Tite Pati (*Artemesia Vulgaris*), Khirro (*kapium insigne*), Ankhitare (*Walsura trijuga*), Sajion (*Jatropha curcas*), Siris (*Albizia lebbeck*), Padke (*Albizia odoretissima*), taramandal (*Tithonia diversifolia*), and many other succulent twigs and leaves of wild plants as fertilizer to rice nurseries in the hills and mountains (Joshi 1997). In situ manuring by halting flocks of animals on different fallow lands and shifting animal sheds during winter was practiced in the early Nepalese farming system.

In 1954 (2011 BS), a massive flood occurred in the hills and mountains that took many lives and destroyed large amounts of property, leading to widespread famine, specifically in the western hills (Skerry et al. 1991). The governments of India and the United States supported Nepal with food aid. In later years, instead of food, these donors assisted Nepal with fertilizers. Ammonium sulfate, single superphosphate, and muriate of potash were imported through Tribhuvan Gram Vikash Samiti, an integrated rural development program that ran for several years starting in 1952 (Skerry et al. 1991). This fertilizer aid coincided with the Malaria Eradication Program in the Tarai and river valleys. Soil surveying allowed the soil fertility of the area to be understood and for flood and landslide victims from the hills and mountains to be resettled. The Rapti Valley (now Chitwan Valley) Development Program was one such area where these flood victims were resettled, and modern agriculture was practiced using improved crop varieties and fertilizers. However, most farmers harvested single crops until the 1970s. Soil fertility was maintained to harvest strong crops, and improved fertility management was not a major issue. Later in the mid-1980s, Nepal experienced a sudden population rise and the country felt acute food insecurity. Consequently, farmers themselves felt the need to harvest two to three crops on a single piece of land per year. Double- or even multiple-crop harvesting with the imbalanced use of mineral fertilizer depleted soil fertility status (Karki and Joshi 1993). Although some farmers apply a large amount of mineral and organic fertilizer to their upland crops, in most cases this is specifically only N fertilizer, and soil fertility status has not improved because of this imbalanced fertility management (Shah and Schreier 1995).

9.2 Climate and Topography

Nepal has a monsoon type of climate in which the monsoon rains start from the middle of June and end in the last week of September. Monsoon rain originates from the Bay of Bengal, with precipitation starting from the eastern part of the country and moving west, reaching western Nepal after a week or sometimes 2 weeks later. Most of the winter is dry and cold. The country receives one or two showers during winter because of the monsoons coming from the Arabian Sea. The winter monsoons provide enough moisture for winter rainfed crops such as wheat, barley, and winter legumes. The country's precipitation rate is higher in the mountains over 3000 m asl in central Nepal, which typically receive more than 2000 mm rainfall per year (Karki et al. 2017). The western hills receive precipitation only slightly higher than 1000 mm. The winter rain starts from the west and drops most of the rainfall in that region. The precipitation declines when it reaches to the eastern part of the country (Ichiyanagi et al. 2007). Therefore, the east–west gradient of decreasing precipitation exists during the summer monsoon, and the opposite trend is common during the winter. Temperature steadily decreases as one moves north. The climate changes drastically from subtropical monsoons to alpine tundra as the elevation gradually increases from the Tarai to the High Himalayas. An array of land use and related soil fertility challenges are observed as one moves from Tarai to the High Himalayas. The contrast in physiographic and climatic conditions make agricultural decision-making complex and soil fertility management challenging.

Nepal has five physiographic regions that run almost parallel to one another, extending from east to west. The High Himalayan region (5000–8848.86 m asl) represents 23.7% of the total landmass of the country and is not suitable for agriculture. The High Mountain region (3000–5000 m asl) covers almost 19% of the country's land area and is used for rearing of livestock such as sheep, mountain goats, and yak. There are lower valleys where people live and grow potatoes, barley, buckwheat, and temperate fruits and vegetables. This region is home to valuable medicinal herbs. The Middle Mountain region (2000–3000 m asl) occupies nearly 29.5% of the country and is home to about 50% of the population. This region includes low fertile valleys and tars that are intensively cultivated. All types of livestock are raised in this region. Though farmers harvest good crops, they are not sufficient for this region's population, and food needs to be imported from surplus districts in other areas within the country and sometimes from outside. The Siwalik Region (300–2000 m asl) lies in the southern part of the country and covers 12.7% of its area. Large valleys where farmers grow all types of cereals, vegetables, and subtropical and tropical fruits are part of the Siwalik region. It is home to valuable non-timber forest products (NTFP). The fifth region is the Tarai, which starts from 60 to 300 m asl and occupies about 14% of the country. It is a flat plane with a slope of about 0.2–1% gradient except for some uplifting with the action of rivers. It is known as the breadbasket of Nepal. All types of food, vegetables, and subtropical and tropical fruits are produced in surplus quantities and supplied to deficit areas. Because of the ample opportunity to grow more food, population pressure has increased in this region due to migration from the hills and mountains.

There are about 6000 rivers and streams in Nepal, 1000 of which are more than 10 km long, and about 100 of these are longer than 160 km. With all these rivers and rivulets, four river systems are formed, the Koshi in the east, the Gandaki (Narayani) in the middle, the Karnali in the Mid-west, and the Mahakali in the far west, which drains the country. All of these rivers originate from the Tibetan plateau and enter the Nepalese border at different points, where they flow toward India and ultimately to the Bay of Bengal (Paudel et al. 2015; Yamada and Sharma 1993). While in Nepal, these rivers pass through varied topographies, erode them, and carry enormous debris and sediments. Mountains and hills constituting three-fourths of Nepal's area comprise four macro-, 24 meso-, and 6,000 microwatersheds, with diverse biophysical conditions and biological deteriorations under the influence of ongoing human activities (Wagley 1995).

9.3 Components of Soil Fertility

Soil Texture

Soil texture constitutes the relative proportion of sand, silt, and clay. Materials larger than 2 mm are considered gravel and not included in soil texture. Based on the international system, particle sizes between 2.00 and 0.2 mm are considered sand, 0.2–0.02 are considered silt, and those smaller than 0.02 mm are clay. The proportion of these three mineral particles is important for aeration and the holding of water and nutrients. If the soil contains a higher amount of clay, it can hold more water and nutrients, but it becomes muddy when wet and hard when dry, making it difficult for intercultural operation. Soils with more silt lack cohesiveness and are slippery, whereas soils with more sand have low water and nutrient-holding capacity and are loose and workable. Therefore, a near-equal ratio of sand, silt, and clay is ideal for a good crop harvest.

The Mountain soils of Nepal contain a high amount of silt and sand, followed by clay. This can be due to the slow weathering process as well as to increased soil erosion. The hills are sloppy, and the soil cannot remain in situ due to soil

Table 9.1 Soil texture from surface soil of the 18 different districts of Nepal

Districts	Sand (%)	Silt (%)	Clay (%)	Soil texture
Kailali (128)	37.48	47.56	14.96	Loam
Kathmandu (210)	63.29	31.24	5.47	Sandy loam
Lalitpur (148)	30.58	11.53	57.89	Clay
Banke (131)	34.37	47.11	18.52	Silt loam
Bara (124)	37.53	49.73	12.74	Loam/silt loam
Nawalparasi (190)	25.24	57.48	17.28	Loam
Surkhet (52)	31.67	51.78	16.55	Sandy loam
Morang (187)	46.46	45.42	8.12	Loam
Taplejung (350)	65.78	24.76	9.46	Sandy loam
Teherathum (350)	56.37	29.06	14.56	Sandy loam
Sallyan (350)	51.53	31.59	16.88	Silt loam
Rolpa (350)	41.67	32.49	25.84	Loam
Dhanusha (219)	33.01	8.98	58.00	Clay
Saptari (227)	35.11	8.91	55.98	Clay
Sirha (192)	27.88	34.98	37.14	Clay loam
Sunsary (176)	43.20	52.75	4.05	Silt
Kapilavastu (211)	12.66	15.25	72.09	Silt loam
Jhapa (201)	8.55	54.28	37.18	Silt clay loam

Source Collected and calculated from the authors' various works, the figures in parenthesis indicate the number of samples analyzed

Table 9.2 Particle size distribution in Rhodic Ustochrept soil profile opened at Kallaritar of Dhading District

Depth (cm)	Sand (%)	Silt (%)	Clay (%)	Soil Texture (USDA)
0–10	20.3	61.6	18.0	Silt loam
10–31	19.0	61.0	20.0	Silt loam
31–54	29.0	44.5	26.0	Loam
54–80	30.0	32.0	38.0	Clay loam
80–140	27.0	28.5	44.0	Clay

Source Karki (2006)

erosion. Results of 1300 samples analyzed from Tehrathum, Taplejung, Sallyan, and Rolpa districts show sandy loam and loamy sand texture as predominant in Tehrathum and Taplejung districts, whereas loam and sandy clay loam soils are common in Salyan and Rolpa (Table 9.1). A similar result was reported by Baumler and Zech (1994). Soils in the Tarai region are mostly loam and silt clay loam (Tandan et al. 2015); however, Dhanusha, Saptari, Siraha, and Jhapa districts have heavy-textured soils. Although it is made up of predominantly hills, mountains, and valleys, Lalitpur District has heavy-textured soils (Table 9.1).

Clay content is typically less in the surface soil than in subsequent depths. A soil profile from an irrigated rice field shows the depth distribution of various particles (Table 9.2). Due to puddling, the finer soil particles eluviate to the second layer and illuviate. The horizon where the finer soil particles are illuviated and form hardpan helps water ponding on the surface.

Soil pH

One of the most important properties of soil is its reaction. Soil microorganisms and higher plants respond remarkably well to soil reactions and tend to control much of their chemical environments. Soil acidity is common in high-rainfall areas where basic cations are lost and acidic cations accumulate (Sumner and Noble 2003). Soils are generally acidic in Nepal because of the parent material, vegetation in the natural landscape, and high precipitation favoring nutrient leaching and runoff of base cations (Ca^{2+}, Mg^{2+}, K^{+}, etc.). Increased acidity is attributed to acidic parent materials such as sandstone, granite schist, and phyllites (Karki 1986; Schreier et al. 1994). Another cause of increasing soil acidity is farmers applying nitrogenous fertilizers every season to every crop who deviate from the judicial recommended dose of fertilizer (Karki 1986; Karki and Joshi 1993). In acidic soils, micronutrients such as iron, copper, manganese, and zinc will be active and bind with phosphorus, making it unavailable to crops. Most of our upland soils are the victims of low pH. In contrast, calcium, magnesium, and sodium become active in higher pH soils while minimizing micronutrient availability. Soil test results show that 53% of the soil tested in Nepal is acidic (Dawadi and Thapa 2015), and the majority of these acidic soils are found in the middle mountain region.

Water Holding Capacity

Water holding capacity is the amount of water that the soil can retain in its pore spaces. When there is irrigation or rain, the surface soil gets wet and water slowly fills the pores. If water from the surface saturates the pores, it moves downward, recharging groundwater. This water-moving process is known as infiltration, which is generally slow in irrigated fields and rapid in upland soils. As soon as the water added from the external source is stopped, the downward movement of water also slows down and ultimately ceases, though the water fills the porous space and the soils hold the water. The amount of water held by the soil at this point is known as the water holding capacity of the soil.

Water holding capacity is related to the total clay and organic matter content of the soil (Paudel and Sah 2003). Organic matter content in cultivated land is mostly lower than in forestland, but bulk density in cultivated land is still favorable for crop growth (Dahal and Bajracharya 2012). Cultivation increases bulk density and other soil properties such as pore space and soil structure (Celik 2005). Research results related to the water holding capacity of hill soils are scarce, and studies in Tarai soils show less than 50% water holding capacity (Gautam and Mandal 2013). Hills and mountains cover more than 70% of Nepal, and despite the higher amount of organic matter in hill soils, the clay content is very low, leading to low water holding capacity (Paudel and Sah 2003; Gautam and Mandal 2013). The hill soils, specifically the bari lands, are low in clay content and highly prone to erosion and nutrient leaching (Acharya et al. 2007; Atreya et al. 2006).

Soil Organic Matter

Soil organic matter is a critical component of soil fertility. The addition of compost and farmyard manure (FYM) and the decomposition of crop roots adds to soil organic matter. In modern agriculture, crop residue is left in the field and buried in the soil, leading to an increase in soil organic matter (Ghimire et al. 2012). Nepalese farmers keep many low productive animals on their farms that depend on forest grazing or on fodder collected from the forest. Additional FYM from these animals adds to the fertility status of farms. However, most crop residues such as rice, maize, wheat, and barley are good sources of animal feed during the dry period. Crops such as mustard, black grams, green grams, and chickpeas are harvested by uprooting the plants, leaving a minimum amount of organic residue in the ground. Some farmers even uproot stubbles left in the field after the rice and maize harvests and use them as fuel for their kitchens. Recent changes in community forestry practices have prohibited farmers from collecting fodder from forests, hindering them from maintaining livestock herds so that they ultimately produce less FYM. About 8 M t of dung are burnt in Nepalese kitchens each year, which would produce food grain sufficient to feed 1 million people (Karki and Joshi 1993). Further, interest in mineral fertilizer has been increasing in recent years. Mineral fertilizer is easy to handle and quickly responds after application, which has reduced farmers' interest in organic manure production (Dahal and Bajracharya 2011). Nevertheless, imbalanced application rates of chemical fertilizer can decrease organic matter storage and deplete soil fertility in the long term.

There are several ways to increase organic matter in the soil, which improves soil physical properties and increases soil's nutrient-holding capacity and crop yield over the long term (Sah et al. 2014). In this study, the authors manifest that soil organic matter is substantially increased when rice and wheat stubble mulching is incorporated into the soil, but that soil organic matter is decreased when only chemical fertilizer is added to the soil. Various green manuring plants can supply nutrients to crops (Table 9.3), and the use of green manuring plants and crop stubbles increased soil organic matter in the Long-Term Soil Fertility Trial (LTFT). In situ incorporation of Sesbania had significantly increased residual soil nutrients and crop yield (Regmi et al. 2002; Sah et al. 2014). However, an experiment conducted to compare *Sesbania canabina* and *Sebania rostrata* with other organic manure sources at Soil Science Division, Khumaltar, manifests that *S. canabina* produced a substantially higher yield than other sources of organic manure (Maskkey et al. 2004). The compost amount is lower in quantity but can produce a yield comparable with other green manure (Table 9.4). The use of other legumes such as mung bean, fava beans, and lupins, as well as wild plants, for example, ashuro (*Adhatoda vasica*) as green manure also add to soil organic matter. Despite the tremendous efforts made from various sectors, improvement of organic matter content in farms is not encouraging.

Sewage sludge and solid waste compost can be excellent sources of soil organic matter. Though it contains large numbers of heavy metals, biosolid contains many micronutrients that other sources of organic fertilizers lack. City waste compost collected from Bhaktapur municipality had a comparable amount of nutrients with farmyard manure (Table 9.4). Most of the elements were in higher concentration in FYM, though the calcium content was higher in city compost.

There are over 330,000 biogas plants in Nepal (Subedi 2015), and over the past several years, the installation of these biogas plants was expected to decrease dung cake burning and increase farm use of biogas effluent in crop production. However, farmers have not taken the maximum benefit from biogas slurry due to a lack of proper training on slurry handling. Farmers understood that after the installation of biogas, the amount of farmyard manure and compost is reduced. The manure from biogas effluent contains readily

Table 9.3 Nutrient content in some indigenous green manure plants

Common name	Scientific name	N (%)	P (%)	K (%)
Tite Pati	*Artemisia vulgaris*	2.4	0.42	4.9
Ashuro	*Adhatoda vasica*	4.3	0.88	4.49
Taramandal	*Tithonia diversifolia*	4.9	0.87	5.23
Banmara/Banmasa	*Eupatorium Adenoforum*	2.35	0.71	5.43
Khirro	*Kapium Insigne*	2.79	0.79	0.89

Table 9.4 Plant nutrient contents of organic manures as extracted by Aqua Regia

Element	City compost	Farmyard manure
Total N (%)	0.44	0.96
Organic C (%)	14.74	14.50
Total P (mg kg^{-1})	3094.74	4235.89
Total K (mg kg^{-1})	7866.02	17,917.91
Magnesium (mg kg^{-1})	4426.51	9784.70
Calcium (mg kg^{-1})	15,552.40	11,370.22

Source Karki (2006)

Table 9.5 Plant nutrient contents in dung, biogas slurry, and biogas compost content

Nutrient	Biogas slurry	Fresh dung	Biogas compost
Moisture	98.84	88.37	80.62
PH	7.5	7.9	7.7
OC (%)	43.5	32.5	27.02
Total N (%)	0.28	1.4	1.5
Total P (%)	0.02	1.34	0.8
Total K (%)	0.017	1.25	1.61

Source Karki (2004)

available nutrients, especially nitrogen, which increased yields of maize, rice, wheat, and cabbage by 30, 23, 16, and 25%, respectively, over a control plot (Karki 2004). Biogas reduced forest cutting for firewood and improved the health of rural women who frequently cooked using firewood and dung cakes (Sapkota et al. 2013). The installation of biogas using animal dung and human excreta has also improved the environment. However, effluent use in the field has not been successful. This is mainly due to the effluent being in liquid form, and as many Nepalese farmers have scattered parcels of land, transportation is often difficult. An inbuilt program to train farmers who construct biogas plants in composting the biogas slurry, which is easy to transport, has been initiated by the Alternative Energy Promotion Centre (AEPC), an apex body of the Government of Nepal. The slurry composted shows higher N and Potassium content (Table 9.5), and the total P was higher in the fresh dung. After being composted with available dry matter applied in combination with a 50% recommended dose of fertilizer, biogas slurry demonstrated 36% higher crop yield over a control plot. Sole application of composted slurry resulted in a 28% yield increment. These results showed that maize and cabbage yielded higher with biogas slurry than with the recommended mineral fertilizer doses (Karki and Karki 2001).

9.4 Soil Nutrients

Primary nutrients

Nitrogen: Nitrogen is an essential element for crop production. Air contains 78% nitrogen by volume in the form of N_2, an inert gas. It combines with oxygen and forms nitrate, nitrite, and nitrous oxide, whereas it combines with hydrogen to form ammonia. In soil, when nitrogen from organic matter is mineralized, NO_3 leaches if there is enough moisture to leach down. In alkaline conditions, the ammoniacal forms of nitrogen volatilize as NH_3. If the plants do not take up NO_3 soon after mineralization, it is either leached or converted to other forms of nitrogen, including N_2O, a greenhouse gas that causes global warming once it is released into the atmosphere (Ghimire et al. 2017). Plants need N for development and reproduction, and in the presence of sunlight, through the process of photosynthesis, plants manufacture amino acids and nucleic acid. The amino acid produces the protein building blocks needed for all physiological and biochemical reactions in plants (Uchida 2000). Nitrogen is also a component of several vitamins. The N improves the quality and quantity of leafy vegetables and increases grain production and is hence required in a higher amount, which is lost with the crop harvest.

Even with the enormous amount of nitrogen in the atmosphere, plants are always starving for it (Dawadi and Thapa 2015; Dawadi et al. 2015). Nepalese farmers depend mostly on nitrogen from organic manure, but the plant nutrients they supply through manure are not sufficient for sustainable crop production. Moreover, farmers do not apply fertilizer at the right time and the right amount (Joshy 1997; Karki 2015; Karki and Joshi 1993). Adhikari et al. (1999) published the results of surface soil samples (0–15 cm) collected from 42 farmers from Rupandehi District, which indicate that most of the soils are low in N content and nitrogen recovery was also low. Similar results were also reported by Cassman et al. (1996), with both cultivated and forest soils having low nitrogen contents. The total N in Nepalese soil also decreases by 0.07% with every increase of 1000 m in altitude (Sah and Brumme 2003).

Green manure is mostly used in rice cultivation because the water content in the soil is higher and decomposes easily. However, green manure plants have sufficient moisture to activate microorganisms even in upland conditions. Results of experiments conducted by the Soil Science Division of NARC show rice response to some of the green manure in the Khumaltar condition (Table 9.6). This study was conducted because *Sesbania rostrata* developed nodules in its stem and is believed to contribute more nitrogen than traditionally used species of *S. canabina* and other green manuring plants. The plant nitrogen content is less than *S. canabina*, which is even less than in mung bean. The rice grain yield results show that all the green manures, including compost, produced statistically comparable grain yield with numerically higher yield under *S. canabina*. Green manure of Ashuro (*Adhatoda vasica*) proved better than other

Table 9.6 Rice crop response to some green manuring plants

Treatments	Biomass incorporated		Dry weight basis		Grain yield (kg ha^{-1})	
	1989/90	1990/91	N (%)	1989/90	1990/91	Mean
Control	–	–		4195	1927	3061
S. canabina	23.6	23.6	5.6	6190	3606	4898
S. rostrata	21.6	21.6	4.1	6175	2246	4210
Mungbean	11.8	11.8	4.8	6511	2518	4514
Azola	10.0	10.0	4.0	5156	2259	3768
Water hyacinth	10	10	0.6	4457	1831	3144
Compost	10	10	2.6	5869	2586	4228

Source Soil Science Division, Khumaltar

indigenous green manure plants, producing 4.8 t ha^{-1}, which is more than *E. adinoforum* and compost and mineral fertilizer (Subedi 1993). In rice cropping, mung bean and *Sesbania* are mostly grown in situ and manured, while Azola and Water hyacinth were brought in from outside. The treatment materials in the following experiment are calculated on the N content basis. The rice yield influences different green manuring plants and azola. Though the rice yield is at par, *Sesbania,* mungbean, and compost produced a notably higher yield over others.

Biological N Fixation: Research work on biological nitrogen fixation with legume crops and non-legume crops and its use in crop production in Nepal has been achieved significantly (Bhattarai et al. 1987). Production of effective Rhizobium strains for soybean, black gram, cowpea, lentil, chickpea, peanut, mung bean clover, desmodium, stylosanthus, ipil-ipil, medicago, and vetch has been maintained at Soil Science Division, NARC. The productivity of major summer and winter legume crops such as soybean and lentil has been increased using effective rhizobium inoculums. The rhizobium is widely used in Nepal, and demand for rhizobium inoculum has increased. A survey covering 107 different crops indicated that soybean fixed 59 kg N ha^{-1}, black gram (*Vigna mungo*) 28 kg N ha^{-1}, groundnut (*Arachis hypogea*) 153 kg N h^{-1}, lentil (*Lens culinaris)* 72 kg N ha^{-1}, Chickpea (*Cicer arietinum*) 84 kg N ha^{-1}, pigeon pea (*Cajanus cajan*) 412 N ha^{-1}, and grass pea (*Lathyrus sativus)* and fava bean (*Vicia faba*) 80 N ha^{-1} (Maskey et al. 2001b).

Phosphorus: Phosphorus is the second essential element in crop production. Soil developed from limestone and mudstone generally contains phosphorus as impurities, but can support plant growth and production (Dattilo et al. 2019). Phosphorus is needed in plants for transportation and storage of energy obtained from photosynthesis and carbohydrate metabolism in adenosine triphosphate (ATP). During enzymatic reactions, Adenosine Triphosphate (ATP) is converted to adenosine diphosphate (ADP) and inorganic P, releasing energy useful for several biochemical reactions. Therefore, soil deficient in plant-available P faces serious consequences because P is needed in cell division and stimulation of root growth. Phosphorus is also a structural component of ribonucleic acid (RNA) and deoxyribonucleic acid (DNA). It is needed for flowering, fruiting, seed formation, and tillering in cereals (Havlin et al. 2016; McLaren and Cameron 1996).

Phosphorus-containing rocks are not widely available in Nepal. Therefore, the phosphorus content of the soil is mostly low. Only cultivated soils where farmers apply phosphatic fertilizer to every crop in succession contain a higher amount of plant-available phosphorus despite low soil pH (strongly acidic condition). The data presented by Dawadi and Thapa (2015) shows a low level of plant-available P even with low pH. Similar results are presented by Karki (2015) and Karki and Joshi (1993). Most P fertilizers sold in Nepalese markets are Diammonium Phosphate (DAP 18% N and 46% P_2O_5), and Triple superphosphate. However, most Nepalese farmers do not balance the P nutrients required for crops properly. Hence, farmers either do not apply P fertilizer or apply it at a higher dose, leading to nutrient imbalance in crops. Most micronutrients are not available for plant uptake if there is high P content (Karki 2004). Soil P is also lost with erosion and crop harvests (Quinton et al. 2001; Stevens et al. 2009).

Potassium: Potassium (K) is the third essential element for plant production. In soils, 90–98% of potassium is unavailable, with between 1 and 10% in a slowly available form and only 0.1–2% in a readily available form (Mengel 2015). Potash releases in the soil solution after it is decomposed from K-bearing rocks such as potash feldspar, muscovite, or biotite. When released from the parent materials, it is lost to drainage, remaining in an exchangeable form with clay particles, and converted to a slowly available form. Part of the released K is fixed in exchange with ammonium (Marschner 2011; Mengel 2015; Mengel and Kirkby 1987). The opening and closing of stomata are one of the most critical functions of K in plants, and this function assists with the entry of CO_2. Potassium also regulates water uptake and the transpiration of water, which is essential during drought. Soil K, when it enters plants, activates enzymes and takes part in biochemical activities (Leigh and Wyn Jones 1984). Potassium in plants develops resistance to diseases and pests, transporting nitrogen and starch produced from photosynthesis to other parts of the plant where it is needed. It also transports starch to the roots and stems of bulb crops such as potatoes and onions. Thus, a higher amount of K fertilizer is required for potato production (Adhikari and Karki 2006).

Nepalese soils are rich in silt, more so than Tarai soils because the soils in Tarai are flooded every year with large quantities of silt carried by the massive tributaries of the Ganga Rivers. As a result, some soils do not respond to K fertilizer application (Pal et al. 2001; Regmi et al. 2002) because of the K release from the silt developed on fine-grained mica. Farmers stopped applying K fertilizer for several years due to no or low response of applied K fertilizers. A study comparing five intensively cultivated soils (*Typic Ustochrepts, Dystric Ustochrepst, Fluvaentic Ustochrepts, Typic Fluvaquents,* and *Aquic Ustochrepts*), however, revealed that K reserve in the soil is exhausted (Karki 2003). A long-term soil fertility experiment at Bhairahawa showed that rice yield significantly increased with 84 kg ha^{-1} of K fertilizer, a yield response comparable with FYM application in rice and wheat (Rawal et al. 2017; Regmi et al. 2002). The FYM also contained a considerable amount of K (Karki 2006), which became available for crop uptake during decomposition. Westarp et al. (2004) noted the need for addressing exchangeable K in mountain valley soils; otherwise, the yield of vegetables and cereals will be limited.

In a study comparing nutrient uptake by the crop under compost, combinations of N, P, and K with or without one of the nutrient elements and a control (without any fertilizer) showed that the highest NPK uptake by wheat occurred under 15 t compost + 30 kg N ha^{-1} top-dressed (Karki 2006) (Table 9.7). The same treatment shows the highest NPK uptake by rice crops as well. The zero-phosphate treatment shows better uptake of all three elements, though the NPK uptake was very poor without K application.

Secondary nutrients

Nutrients like, calcium (Ca), magnesium (Mg), and sulfur (S) are required by plants in smaller amounts than nitrogen, phosphorus, and potassium, and are hence known as secondary elements (macroelements). Calcium is mostly needed for tissue building in plant systems (Upadhyay 2017), whereas Mg is used for chlorophyll (Yong et al. 2005) and S for the formation of different enzymes (Lang et al. 2007). Both Ca and Mg are strong reducing agents, and deficiency of Ca and Mg is observed in strongly acidic soils, mostly in the hills of Nepal, whereas deficiency of sulfur is observed in intensively rice/paddy cultivated soils in the Tarai region, especially, where sulfur-free fertilizers and low amounts of organic manure are used (Das et al. 2019). Khadka et al. (2017, 2018) evaluated soil fertility of research farms and stations in Nepal and found that while lack of Ca is not a problem in the Tarai, its content is low in hill soils. The low level of Ca in hill and mountain soils is mainly due to the loss of Ca during soil erosion, though this does not occur in plain areas where soil erosion is minimal. The status of Mg and S is low in both the hill and Tarai stations.

In Nepal, secondary nutrients are not studied to determine their fertilizing value. Soil is mostly acid because of the loss of bases such as Ca and Mg. On average, 159 kg of Ca, 12 kg of Mg, and 30 kg of S ha^{-1} are lost each year (Eltun et al. 1997). The loss of these elements occurs mainly through soil erosion, leaching, and crop removal (Tripathi 1999). A study conducted in 1996/97 with three crops

Table 9.7 Nutrient uptake by rice and wheat crops compared to organic manure in the Eastern Hills of Nepal

Treatments	Nutrients uptake by rice grain (kg ha^{-1})			Nutrients uptake by wheat grain (kg ha^{-1})		
	N	P_2O_5	K_2O	N	P_2O_5	K_2O
Control	21.0	19.52	10.0	24.3	21.2	17.1
Compost at 15 t kg ha^{-1}	24.2	25.3	12.6	32.3	32.6	29.7
Compost at 15 t ha^{-1} + 30 kg N top dressing	34.6	28.3	14.3	61.9	44.8	37.1
60:30:30 N: P_2O_5:K_2O kg ha^{-1} 1/2 N top dressing	25.5	25.6	12.5	48.2	43.4	30.5
60:0:30: N:0P: K_2O kg ha^{-1} ½ N top dressing	23.9	18.8	12.3	43.4	29.0	26.9
60:30:0 N: P_2O_5:0K_20 kg ha^{-1}, ½ N top dressing	18.7	19.0	10.0	34.0	24.5	52.2
60:30:30 N:P_20_5: K_2O kg ha^{-1} and 15 t ha^{-1}, compost	29.0	29.8	14.7	63.8	42.2	48.6
Average uptake	* 3.02	Ns 4.51	* 0.93	** 6.79	Ns 6.12	**

Source Sherchan and Gurung (1998)

showed an S uptake of 5–16 kg ha^{-1} by a single crop (Ghani and Brown 1997). However, S is replenished by applying organic manure, as most upland crops in Nepal are grown with organic manure, whereas rice is fertilized by flood sediments (Abe et al. 2020; Pandey 2020).

Micronutrients

Trace elements are present on earth in minute quantities that are less than 1000 mg kg^{-1}. These elements can be grouped into two categories. Plants need these elements as essential trace elements (TEs), and without them they cannot complete their life cycles. These elements are B, Cl, Cu, Fe, Mo, Mn, and Zn. Nickel (Ni) is an activator of urease in higher plants needed for growth and chlorophyll content, so it is included as an essential trace element. Its concentration of 0.01–0.05 mg g^{-1} soil is recommended (Gheibi et al. 2009). There are some other elements that are not essential but that promote growth and develop biological stress resistance. These beneficial elements are aluminum (Al), cobalt (Co), sodium (Na), selenium (Se), and silicon (Si) (Hänsch and Mendel 2009). Elements such as Al are essential in tea (*Camellia sinensis*) to promote root growth. Likewise, Co is needed by rhizobium bacteria and cyanobacteria in biological nitrogen fixation. Other elements such as Na help plants with drought resistance, while Se promotes oxidative stress, especially from UV resistance, along with supporting the growth of aging seedlings (Xue et al. 2001). Silicon also aids rice plants in developing resistance from fungal disease (Luyckx et al. 2017). Furthermore, animals require Arsenic, Cobalt, Chromium, Fluorine, Iodine, Selenium, and Silicon as trace elements. Humans need more trace elements than plants and animals, including Vanadium, Arsenic, and Tin (Welch and Graham 2004). Tin (Sn) is an essential element for humans and animals (Sroor et al. 2003), but more research needs to be carried out on the special function it plays in the body. Cobalt deficiency leads to a deficiency of B12, reflecting neurological and muscular lesions and causing hepatic damage (Graham et al. 2007; Salwen 2017). When soils are deficient in micronutrients, humans and animals subsequently become deficient and their health is adversely affected (Gupta and Gupta 2005; Steinnes 2009).

Quantities of micronutrients in soils are always greater in surface soils and less in the subsoil. They are trace elements present in the soil in microamounts that are needed by plants and animals in microquantities. Though plants and animals only need them in microamounts, their deficiency causes serious consequences (Gupta et al. 2008). Different plants respond to different micronutrients, as brassica and legumes are deficient in Mo and B, whereas cereals such as rice, wheat, and maize are deficient in Zn and Cu.

Certain trace elements are used in human disease treatments and animal fertility. For example, Se deficiency in goats led to infertility and the birth of stillborn kids. A study conducted in Dhanusha District administering Se to goats increased their fertility rate and improved both the weight of kids and their growth (Shrestha et al. 2005). Inadequate micronutrients result in poor growth and development of newborn children, though it can be improved through supplementation (Friel et al. 1993). Micronutrient deficiencies related to chronic diseases have been gaining serious attention, such as anemia caused by Fe deficiencies, and Zn deficiencies causing shooting diarrhea and stunted growth of children under-five, as these diseases affect millions of people around the world. Micronutrient deficiency in soil is related to human deficiency and is widely studied, especially Zn deficiency in crops and its effect on human health (Imtiaz et al. 2010). Andersen (2007) found that over 95% of South Asians are at risk of Zn deficiency. Specifically, symptoms of micronutrient deficiency are observed in Nepalese children aged 6 and below. Over 56% of children under 5 years old in the eastern hills have stunted growth due to Zn deficiency, and this figure is 72% in the western mountain region. Surveys conducted in the eastern districts of Dhankuta and Sunsari analyzing blood plasma from school-going children revealed that 83.9% in Dhankuta and 87.3% in Sunsari were low in Zn. Children aged 6 to 8 years were more vulnerable to Zn deficiency than those aged 9–12 years (Nepal et al. 2014). To tackle the undernourishment problem, the Government of Nepal created a program to support children with I, Fe, and Zn supplementation, as well as the National Vitamin A Program (Khor and Misra 2012).

Legume crops are richer in TEM than cereals. Certain genes of legumes are responsible for the accumulation of TEM, and if these genes are transferred to cereals, their uptake of TEM could be enhanced and lead to successful biofortification. The iron-storing gene ferritin, which is generally found in soybean and *Aspergillus fumigatus*, has been successfully transferred to maize and rice crops and enriched with I, Ca, Fe, Se, and Zn (Gómez-Galera et al. 2010).

Nepalese soils are deficient in almost all micronutrients. Sillanpää (1982) published a global survey that showed Nepalese soils and plants in short supply of micronutrients. This is the first report about micronutrients in Nepalese soils. A separate report on cadmium, lead, cobalt, and selenium status was later created using the results of the same soil and plant samples collected and reported in 1982, and pointed out that Co and Se are low in Nepalese soils (Sillanpää and Jansson 1992). Later, Karki et al. (2005) reported that with the exception of Cu and Mn, most of the micronutrient elements are low in Nepalese soils. Boron (B) is widely studied because its deficiency is related to the sterility of wheat and some legumes in Nepal, though B applied at the rate of 2–2.5 kg ha^{-1} has improved fertility and increased wheat yield (Karki 1995a; Rerkasem et al. 1996). Boron deficiency is usually corrected by soil application of borax (2 kg B ha^{-1}) (Ghasemi et al. 2016). Introducing modern high-yielding rice and wheat varieties in soils of the Indo-Gangetic Plain (IGP) that were previously long-duration monocultural rice-cultivated increased their micronutrient deficiency.

9.5 Soil Fertility Management

The demand for mineral fertilizer is always increasing around the world. In Nepal, nitrogenous fertilizer demand increased from 2014 to 2018, though the rate of increase in demand is not consistent among years (Table 9.8). The same trend can be observed with Phosphate fertilizer demand, while potash fertilizer demand is continuously increasing.

Prior to 1954/55, Nepalese agriculture was purely organic. Farmyard manure was the primary source of plant nutrients, and animal shed sifting and herd resting during winter were common for manuring the fields. In 1953/54, the Government of Nepal decided to import mineral fertilizers rather than importing food grain. India and the US supported this agriculture extension and taught farmers about fertilizer use and management.

After the opening of National Trading Ltd. (NTL) in 1962 and Salt Trading Corporation in 1963, some fertilizers were imported from China and USSR. The types imported were mainly ammonium sulfate (21% N), single superphosphate (16% P_2O_5), and muriate of potash (60% KCl), though the quantities are not known (Raut and Sitaula 2012). Mineral fertilizers were available only in easily accessible areas, and hence, most of them were used by farmers in Kathmandu Valley. The response to nitrogen application on virgin soil was beneficial, and farmers became interested only in ammonium sulfate rather than superphosphate and muriate of potash. Thus, the demand for nitrogenous fertilizer increased, though the demand for other fertilizers stayed the same. Nitrogenous fertilizer was very popular among the farmers of Thimi and Bhaktapur. The unused phosphate and potassium fertilizer remained undistributed and later spoiled. Realizing farmers' interest and the increasing demand for fertilizer, the Government of Nepal created an agency called the Agriculture Marketing Corporation (AMC) in 1966 (which later became the Agriculture Input Corporation, AIC), to handle agricultural input such as seeds, fertilizers, and pesticides. AIC now handles fertilizers and plant protection chemicals while the Nepal Seed Company handles seeds.

The government provided a subsidy for fertilizer transport to remote, inaccessible districts so that fertilizer could be provided at the same price given to farmers in accessible districts. Various national and international non-government organizations also supported the transport of fertilizers to remote locations within districts. However, the government realized that the transport contractor misused the subsidy, and it was abolished in 1997/98, thus decreasing fertilizer demand. The government opened a fertilizer unit within the Ministry of Agriculture to regulate fertilizer procurement, though without the transport subsidy, the fertilizer remained in National Trading and Salt Trading stores for a long period. The fertilizer subsidy began again in March 2009.

The use of inorganic fertilizer increased in Nepal, particularly between 2003 and 2010, from 262,000 to 409,000 t. Much of this increase came from the Tarai (171,000–322,000 t), while no significant increase has been observed in the hills and mountains (Takeshima et al. 2016a). Urea and DAP are the two most used fertilizers. Only 28 kg of urea was used in the Tarai in 1995. The amount of urea used in 2003 was 38 kg ha^{-1}, though in 2010 its use increased to 63 kg ha^{-1}. Consumption of DAP increased from 17 kg ha^{-1} in 2003 to 47 kg ha^{-1} in 2010 (Takeshima et al. 2016b).

The government immediately regulated the subsidy in March 2009, increasing fertilizer demand in 2010. Based on a recent farm-level survey, fertilizer use exceeds 700,000 t at present. Fertilizer use is projected to increase to 1,500,000 t in 2022. Fertilizer import for 2018 was 330,000 t and was the actual supply of Urea, DAP, and MOP (Panta 2018). Total sale of urea in the year 2017/18 was 244,545.5 t. The quantity further decreased to 176,066.4 t in the year 2018/19. However, fertilizer use in the years 2017/18 and 2018/19 was 364,852.7 and 342,638.3 t, respectively. For

Table 9.8 World demand for fertilizer nutrients in 2014–2018 (in thousand tons)

Year	2014	2015	2016	2017	2018
Nitrogen (N)	113,147	115,100 (1.7%)	116,514 (1.23%)	117,953 (1.23%)	119,418 (1.24%)
Phosphate (P_2O_5)	42,706	43,803 (2.56%)	44,740 (2.14%)	45,718 (2.18%)	46,648 (2.03%)
Potash (K_2O)	31,042	31,829 (2.53%)	32,628 (2.51%)	33,519 (2.73%)	34,456 (2.8%)
Total ($N + P_2O_5 + K_2O$)	186,895	190,732	193,882	197,190	200,522

Source World fertilizer trends and outlook to 2018

the year 2019/20 until February 2020, fertilizer use was 296,950.4 t. Because the main fertilizer consuming season is approaching, fertilizer sales are expected to rise (Table 9.9). The data here differ between the Ministry and Panta (2018) mainly because the survey included fertilizer data from private handlers, whereas the ministry shows data from the official source of import.

Biofertilizers

Biofertilizers are substances with microorganisms that can colonize a rhizosphere and promote plant growth when applied to the soil or as a seed treatment. Biofertilizers can add nutrients by fixing atmospheric nitrogen, solubilizing phosphorus, or producing plant growth-promoting substances in soils. At present, biofertilizers are increasingly used in seed treatment or applied in soils to minimize harmful effects on biodiversity and the environment.

The host-specific rhizobium bacteria are isolated from the host plant roots under controlled conditions. They are multiplied in a growth medium and then transferred into carriers, which are mostly peat. Rhizobium is an inoculum to be inoculated into seeds using a sticky substance at the time of seeding. The seeds are dried in the shade, then sown into the field. As soon as the seeds come in contact with the soil, the bacteria from the inoculum colonize the rhizosphere and ultimately infest the plant roots. If it is N-fixing bacteria, then the N from the atmosphere (which can be air from the soil) transports the N into the plant roots in the form of nodules. The forming of nodules by the infested roots stimulates plant growth. The P solubilizing organisms solubilize phosphorus and make it available to plants. The Soil Science Division, Khumaltar, began producing and distributing rhizobium culture from 1970/71. The cultures were mainly for soybean crops, which were widely cultivated in and around Kathmandu Valley. Later, culture for other legumes such as moong bean, chickpeas, lentils, and other legumes was prepared and distributed to the different districts. Research on Azotobacteria, cyanobacteria (Anabaena), Phosphate Solubilizing Bacteria (PSB), and Azospirillum are also conducted but the application of these inoculum has not been widely practiced by the farmers.

In large cattle farms or even dairy farms, animal dung is collected every day from the yard and fed to earthworms to produce vermicompost, which producers claim as the best organic manure. Some farmers also practice bokashi and produce good quality compost. The Government of Nepal provides incentives to such producers, but their quality should be assured. At present, several private companies produce such fertilizers and sell them on the local market. It is good that organic manure is added to the soil and improves soil physical properties, but the way the government buys these fertilizers and distributes them to the farmers is in need of attention. There are more than 25 different FYM converting to organic fertilizers in Nepal, and all of them fail to meet the quality standard set by the Ministry of Agriculture (Amgai et al. 2017). Several commercial biofertilizer companies claim to include plant hormones, biofertilizers, and micronutrients in their products, but their ingredients and effectiveness are questionable.

9.6 Estimation of the Country's Fertilizer Requirement

While Nepal introduced chemical fertilizers around 1953/54, its organized supply began in 1965. A subsidy for fertilizers was introduced in 1973/74 to increase food production by encouraging farmers to use chemical fertilizer (Takeshima et al. 2016a). The subsidy policy included both a price subsidy on fertilizer throughout the country and a transport subsidy for selected districts in the Hills and mid-Hills (APROSC 1995). Nepal does not have the facilities to manufacture chemical fertilizers, and almost all of its inorganic fertilizers are imported. Seven types of fertilizers are mainly being used in Nepal, and include urea, diammonium phosphate (DAP), muriate of potash (MOP), ammonium sulfate (AS), single superphosphate (SSP), ammonium phosphate sulfate (APS), and NPK complex. Urea is the most consumed chemical fertilizer, followed by DAP and MOP. On average, farmers used 43 kg urea ha^{-1} and 29 kg DAP ha^{-1} in 2010/11 (Kyle et al. 2017). An estimated 70% of fertilizer use in Nepal is in the Tarai region, primarily for rice, as Tarai is the breadbasket of the country. About 52% of fertilizer is used for rice, followed by wheat (17%) and maize (15%) (Pandey 2015). Additionally, micronutrient deficiency issues including khaira disease (leaf bronzing) in rice due to Zn deficiency and sterility in wheat induced by an inadequate B supply (Panday et al. 2018) have also increased demand for micronutrients.

Table 9.9 Fertilizer distribution for 3 years in Nepal

Fiscal year	Urea (t)	DAP (t)	MOP (t)
2017/18	244,545.5	112,130.5	8176.7
2018/19	214,351.7	120,969.5	7317.1
2019/20 (Until February 2020)	176,066.4	115,744.2	5139.8

Source S. Siwakoti (Ministry of Agriculture, GON) personal communication 2020

Based on formal import data, fertilizer use in Nepal increased from less than 10,000 t in 1970 to over 90,000 t in 1994/95 and remained below that level until 2012/13. Fertilizer use thereafter increased again beginning in 2014/15. It is estimated that Nepal requires approximately 500,000–800,000 t of fertilizer per year (Pandey 2015). The supply of essential chemical fertilizer in Nepal is far below (around 30%) the present demand. As a result, farmers cannot obtain chemical fertilizers in a timely manner or end up paying a higher price. The government of Nepal needs to develop fertilizer manufacturing industries or reliable import and distribution systems to meet national demand, maintain the quality of supply, and reduce fertilizer costs for all farmers.

9.7 Constraints to Soil Fertility and the Way Forward

The mountains cover more than 80% of the total area of Nepal, which supports the lives of more than 50% of the population (Shrestha 1997). Terrace cultivation is practiced in most of the mountains, where cereal grain production is a crucial agricultural practice. The terraces where cereal grains such as rice, maize, wheat, barley, and buckwheat are grown are narrow, and holdings are scattered across several sites. Soils are often eroded due to heavy monsoon rains, and hence, soil fertility is the major constraint of crop production. Loss of basic cations increased soil acidity in most of the mountain soils except some river valleys. Mechanization is not possible in the narrow terraces, and human labor is used in all agricultural operations. Due to difficulties in transportation, lower amounts of fertilizer are applied to crops in the hills and mountains. The major constraints to soil fertility and crop production in Nepal are as follows:

Soil erosion: Soil erosion refers to the physical wearing down of the earth's surface and includes surface erosion and mass wasting. Erosion is greater in cultivated fields, grazing lands, heavy logging areas, construction (road construction) areas, and some recreational sites. It is caused by wind, water, glaciers, and gravity. Rainfall in Nepal ranges from 2,000 to 3,000 mm yr^{-1}, with most of the rain falling in the hills and mountains. This heavy rainfall erodes 1–12 tons of topsoil ha^{-1} yr^{-1} (Gardner and Gerrard 2003; Shrestha 1997). The rivers and rivulets loaded with sediments cut banks every year, making the watercourse wider (Gautam and Phaiju 2013). Soil erosion occurs when soil particles are detached from the group and transported. Soil erosion causes infertility and low productivity because fertile topsoil is often lost during erosion (Gardner and Gerrard 2003). Soil loss from the Himalayas and mountain regions is as high as 56 t ha^{-1} yr^{-1}, and even in degraded forest soils, it ranges from 1 to 9 t ha^{-1} yr^{-1}, with 8 t ha^{-1} yr^{-1} in grazing land (Shrestha 1997). Though wind erosion is high in the High Mountain and Himalayan regions, due to high-speed wind, detailed data on soil erosion is not available for windy areas such as the upper Mustang, upper Manang, Dolpa, and Solu districts. Wind in these areas is so high that some gravel is also blown away, and the wind speed increases with altitude (Quinn 2004). The sediment load from high-altitude to downstream is minimum (Watanabe 1994). Deforestation in the middle mountains is a major problem aggravating soil erosion, though with the implementation of the country's community forestry program, forest cover has increased and erosion has been minimized (Acharya 2002). Shrestha (1997), citing Gilmour (1991), also mentions the lack of reduction in forest cover in the Middle-Mountain regions, and emphasis is laid on cultivated terraces of the Middle Mountains for contributing major erosional sediment to the filling of canals and lakes.

In the High Mountains, glacial and preglacial erosion is common, though their quantification is highly challenging. An estimate in the High Himalayan region found that sediment deposition of debris depth is 0.1–2.4 m on average with a yearly sediment load of 5820 m^3 yr^{-1} (Heimsath and McGlynn 2008). There are 6000 rivers and rivulets in Nepal, with their watersheds running on different geomorphologic positions. Watersheds with granite rock generally have low to minimal erosion of 2.4 ± 0.9 mm ha^{-1} (Garzanti et al. 2007). Soil losses from cultivated fields in the Middle Mountains are 2.7–12 t ha^{-1} yr^{-1} (Gardner and Gerrard 2003). Similarly in the Siwaliks, because of the conglomerate geologic structure and unsettled landform, sheet erosion from forest and shrubs is 16 t ha^{-1} yr^{-1}, though Gully erosion is >14 t ha^{-1} yr^{-1} (Ghimire et al. 2013). Since the Tarai region consists of a flat plain and all the fields are leveled terraces, soil erosion is minimal.

Soil acidity: Soil salinity is a major problem in most of the upland soils in Nepal. In higher pH soils, calcium, magnesium, and sodium will be active, and micronutrient availability will be lower. Soil test results show that 53% of the soil tested in Nepal is acidic (Dawadi and Thapa 2015). Most of these acidic soils are found in the Middle Mountain region. Increased acidity is attributed to acidic parent materials such as sandstone, granite schist, and phyllites (Karki 1986; Schreier et al. 1994). To ameliorate soil acidity, ample amounts of well-decomposed FYM or compost, or a recommended dose of agricultural lime can be used. Ghimire and Bista (2016) also reported an increase in pH of highly acidic soils with the adoption of more diversified cropping systems. Integrating legume and other crops with low N demand minimizes fertilizer-induced acidity in cultivated soils.

Salinity: Soil salinity is a problem in hot arid and semiarid low rainfall areas of India, Pakistan, and Middle Eastern countries. Swampy areas developed through extensive irrigation where drainage is a problem also accumulate salts. Nepal does not have any coastline to salt wash from the sea or creeks with which to salinize the land. Rainfall in most of the areas is about 1000 mm yr^{-1}, which is enough to leach the salts developed on the surface. Arid and semiarid regions of Mustang and Manang can develop salinity problems, but the area is small, and melting snow often leaches or washes the salt. Therefore, Nepal does not have a significant soil salinity problem.

Soil Pollution: Heavy traffic on narrow Kathmandu roads emit various gasses, making the air unbreathable and causing respiratory diseases (Ale 2003; Giri et al. 2006; Hildebrandt and Pokhrel 2002). Brick kilns have worsened air quality, increasing human health problems (Pariyar et al. 2013). All of these emitted gasses fall onto the earth and pollute agricultural fields. Aichner et al. (2007) analyzed soil polycyclic aromatic hydrocarbons (PAHs) and polychlorinated biphenyls (PCBs) to study the pollution in farmers' fields. These pollutants are found in higher concentration near petrol stations and affect crop production. However, soil pollution is studied in Kathmandu Valley only. The sources of pollution are atmospheric deposition, plastic incineration, vehicle batteries, sewage sludge applied as manure, irrigation with sewage, brick factory fuel (low-quality coal and other solid fuel), burning of motor tiers, and fossil fuel burning, including fertilizers and manure/compost from city waste (Gautam et al. 2005; Karki 1995b). The main polluting metals are Cd, Co, Cr, Cu, Fe, Mn, Ni, Pb, and Zn. Gautam et al. (2005) conclude that soils around Kirtipur and Bagmati River sediments are high in Zn and Ni and other soils in the valley are below polluting levels. The groundwater of the Tarai is highly contaminated with arsenic. Even human hair has been found to contain a high amount of arsenic (Shrestha et al. 2003). Dahal et al. (2008) found that arsenic helped the growth of roots and stems in rice and vegetables but did not support crop production. More research on soil pollution, their impact on soil fertility and human health, and the development of policies to regulate soil pollutants are needed for healthy and sustainable food production in Nepal.

Forest fires: Forest fires are another cause of soil erosion in Nepal. There were 4,741 fire alerts in Nepal in the week from April 23 to April 30, 2019 (MODIS 2020). Forest fires affected 268,618 ha of forest from January to May 2016. Due to forest and bush fires, surface soil is burnt, and organic matter present in the surface of forest soils is lost during the pre-monsoon torrential rain (Atreya et al. 2008; Gardner and Gerrard 2003; Schreier et al. 1994). Mandal (2019) reported over 4000 fire alerts in the country every year, burning 1000 s of ha of land. Among these fire events, 40% are set in the hope of getting new grass for animals.

9.8 Conclusion and Policy Recommendations

The soil fertility status of Nepal is highly depleted. The root cause of low soil fertility is soil erosion along with soil acidity and organic matter depletion. Research is concentrated on crop responses to N, P, K management, and very little research has been conducted on nutrient removal during crop harvest, recycling of residual nutrients, and secondary and micronutrient availability and management. The recommended dose of mineral fertilizer, combined with an ample amount of organic manure, has benefited crop production. However, Nepal has a diverse climate and topography, and extensive region-specific research and policy recommendations based on research data is needed to improve soil fertility and sustain crop production. Agriculture policies should be formulated to address marginal farmers' needs, and poverty and food insecurity will only be alleviated when these needs are properly addressed.

There have been numerous studies pointing out to the problem of soil erosion, along with proposed solutions, but very few farmers showed interest in adapting the technology and methods. Furthermore, these studies have been conducted within the research community and have not been adequately communicated to planning bodies. Every few years, the government formulates new policies but has not addressed soil nutrient replenishment. The recent 20-year Agriculture Perspective Plan (APP, 1995–2015) was not successful; the plan aimed at integrating all concerned agencies such as road, irrigation, cooperatives, finance, and many others related to crop production under the coordination of the Ministry of Agriculture. However, other ministries did not cooperate with the Ministry of Agriculture at the district level. At present, Agriculture Policy 2004 and Agriculture Development Strategy (2015–2035) has been implemented, emphasizing organic agriculture. The government is promoting organic manure production by buying manure from the production unit, but the problem lies in distribution because the bulky bags of organic manure cannot be transported to scattered parcels of land in remote fields. In addition, organic manure and compost rates from native plants have not been optimized for various macro- and micronutrient needs of crops. Development of effective technology transfer mechanisms within federal, provincial, and local levels of government will improve technology adoption. The country needs a strong policy based around the problems, proven technology for addressing these problems, and firm implementation of its policy with frequent monitoring.

References

APROSC (1995) Nepal agriculture perspective plan. Kathmandu, Nepal: Agricultural Projects Services Centre, Washington DC: John Mellor Associates, Inc

Acharya KP (2002) Twenty-four years of community forestry in Nepal. Int for Rev 4:149–156

Acharya G, McDonald M, Tripathi B, Gardner R, Mawdesley K (2007) Nutrient losses from rain-fed bench terraced cultivation systems in high rainfall areas of the mid-hills of Nepal. J Land Degr 18:486–499

Adhikari C, Bronson K, Panuallah G, Regmi A, Saha P, Dobermann A, Olk D, Hobbs P, Pasuquin E (1999) On-farm soil N supply and N nutrition in the rice–wheat system of Nepal and Bangladesh. Field Crop Res 64:273–286

Adhikari B, Karki KB (2006) Effect of potassium on potato tuber production in acid soils of Malepatan, Pokhara. Nepal Agric Res J 7:42–48

Aichner B, Glaser B, Zech W (2007) Polycyclic aromatic hydrocarbons and polychlorinated biphenyls in urban soils from Kathmandu Nepal. Org Geochem 38:700–715

Ale B (2003) Fuel adulteration and tailpipe emissions. J Inst Eng 3:12–16

Amgai S, Paudel SR, Bista DR, Poudel SR (2017) Government intervention on organic fertilizer promotion: a key to enhancing soil health and environment. J Agric Environ 18:131–139

Andersen P (2007) A review of micronutrient problems in the cultivated soil of Nepal. Mt Res Dev 27:331–335

Atreya K, Sharma S, Bajracharya RM, Rajbhandari NP (2008) Developing a sustainable agro-system for central Nepal using reduced tillage and straw mulching. J Environ Manage 88:547–555

Atreya K, Sharma S, Bajracharya RM, Rajbhandari NP (2006) Applications of reduced tillage in hills of central Nepal. Soil Tilage Res 88:16–29

Bhattarai S, Maskey S, Shah R (1987) Use and promotion of biofertilizer in Nepal. In: First review/working group meeting on biofertilizer technology, Soil Science Division, Kathmandu, Nepal

Brown PH, Welch RM, Cary EE (1987) Nickel: A micronutrient essential for higher plants. Plant Physiol 85:801–803

Bäumler R, Zech W (1994) Soils of the high mountain region of Eastern Nepal: classification, distribution and soil forming processes. CATENA 22:85–103

Cassman K, Gines G, Dizon M, Samson M, Alcantara J (1996) Nitrogen-use efficiency in tropical lowland rice systems: contributions from indigenous and applied nitrogen. Field Crop Res 47:1–12

Celik I (2005) Land-use effects on organic matter and physical properties of soil in a southern Mediterranean highland of Turkey. Soil Tillage Res 83:270–277

Dahal N, Bajracharya RM (2011) Prospects of soil organic carbon sequestration; Implications for Nepal's mountain agriculture. J for Livelihood 9:1

Dahal N, Bajracharya RM (2012) Effects of sustainable soil management practices on distribution of soil organic carbon in upland agricultural soils of mid-hills of Nepal. Nepal J Sci Technol 13:133–141

Dahal BM, Fuerhacker M, Mentler A, Karki KB, Shrestha RR, Blum WEH (2008) Arsenic contamination of soils and agricultural plants through irrigation water in Nepal. Environ Pollut 155:157–163

Das S, Lee JG, Cho SR, Song HJ, Kim PJ (2019) Silicate fertilizer amendment alters fungal communities and accelerates soil organic matter decomposition. Front Microbiol 10:2950

Dattilo BF, Freeman RL, Zubovic YM, Brett CE, Straw AM, Frauhiger MJ, Hartstein AR, Shoemaker LM (2019) Time-richness and phosphatic microsteinkern accumulation in the Cincinnatian (Katian) Ordovician, USA: an example of polycyclic phosphogenic condensation. Palaeogeography 535:109362

Dawadi DD, Risal C, Maskey KH, Rijal BR, TB, T (2015) Mobl Soil Testing Laboratory (MSTL): Experience of soil management directorate to aware farmers about soil health. In: Karki KB, Tripathi BP, Manandhar R, Adhikari B, Vista HP (eds) Second national soil fertility research workshop, pp 6–15, Kathmandu

Dawadi D, Thapa M (2015). Soil fertility status of Nepal: Report from Laboratory Analysis of soil samples from 5 development regions. In: Karki KB, Tripathi BP, Manandhar R, Adhikari B, Vista SP (eds) Second national soil fertility research workshop, pp 42–52. Soil Science Division, NARC, Kathmandu, Nepal

de Queiroz Barcelos JP, de Souza Osório CRW, Leal AJF, Alves CZ, Santos EF, Reis HPG, dos Reis AR (2017) Effects of foliar nickel (Ni) application on mineral nutrition status, urease activity and physiological quality of soybean seeds. 11:184

Friel JK, Andrews WL, Matthew JD, Long DR, Cornel AM, Cox M, McKim E, Zerbe GO (1993) Zinc supplementation in very-low-birth-weight infants. J Pediatr Gastroenterol Nutr 17:97–104

Gardner RAM, Gerrard AJ (2003) Runoff and soil erosion on cultivated rainfed terraces in the Middle Hills of Nepal. Appl Geogr 23:23–45

Garzanti E, Vezzoli G, Andò S, Lavé J, Attal M, France-Lanord C, DeCelles P (2007) Quantifying sand provenance and erosion (Marsyandi River, Nepal Himalaya). Earth Planet Sci Lett 258:500–515

Gautam P, Blaha U, Appel E (2005) Integration of magnetism and heavy metal chemistry of soils to quantify the environmental pollution in Kathmandu, Nepal. Island arc 14:424–435

Gautam TP, Mandal TN (2013) Soil characteristics in moist tropical forest of Sunsari district Nepal. Nepal J Sci Technol 14:35–40

Gautam DK, Phaiju AG (2013) Community based approach to flood early warning in West Rapti River Basin of Nepal. IDRiM J 3:155–169

Ghasemi A, Ansari MR, Omidyar RJR (2016) Optimum concentration of boric acid and sodium molybdate for enhancing mungbean productivity in Nepal. 2:1

Gheibi M, Malakouti M, Kholdebarin B, Ghanati F, Teimouri S, Sayadi RJ (2009) Significance of nickel supply for growth and chlorophyll content of wheat supplied with urea or ammonium nitrate. J Plant Nutr 32:1440–1450

Ghimire R, Adhikari KR, Chen Z, Shah SC, Dahal KR (2012) Soil organic carbon sequestration as affected by tillage and crop residue management in the rice-wheat system. Paddy Water Environ 10:95–102

Ghimire R, Bista P (2016) Crop diversification improves pH of acidic soils. J Crop Improv 30:657–667

Ghimire KS, Higaki D, Bhattarai PT (2013) Estimation of soil erosion rates and eroded sediment in a degraded catchment of the Siwalik Hills Nepal. Land 2:370–391

Ghimire R, Norton U, Bista P, Obour AK, Norton JB (2017) Soil organic matter, greenhouse gases, and net global warming potential of irrigated cropping systems. Nutr Cycl Agroecosyst 107:49–62

Gilmour DA (1991) Trends in forest resources and management in the middle mountains of Nepal. In: Soil fertility and erosion issues in the middle mountains of Nepal, pp. 33–46

Giri D, Murthy K, Adhikary P, Khanal SJ (2006) Ambient air quality of Kathmandu valley as reflected by atmospheric particulate matter concentrations (PM 10). Int J Sci Tech 3:403–410

Graham RD, Welch RM, Saunders DA, Ortiz-Monasterio I, Bouis HE, Bonierbale M, De Haan S, Burgos G, Thiele G, Liria R (2007) Nutritious subsistence food systems. Adv Agron 9:1–74

Gupta UC, Gupta SC (2005) Future trends and requirements in micronutrient research. Commun Soil Sci plant Anal 36(1–3):33–45

Gupta UC, Wu K, Liang S (2008) Micronutrients in soils, crops, and livestock. Earth Sci Front 15:110–125

Gómez-Galera S, Rojas E, Sudhakar D, Zhu C, Pelacho AM, Capell T, Christou P (2010) Critical evaluation of strategies for mineral fortification of staple food crops. Transg Res 19:165–180

Havlin JL, Tisdale SL, Nelson WL, Beaton JD (2016) Soil fertility and fertilizers. Pearson Education India

Heimsath AM, McGlynn R (2008) Quantifying periglacial erosion in the Nepal high Himalaya. Geomorphology 97:5–23

Hildebrandt ML, Pokhrel S (2002) Climate and air quality: a case study of PM10 pollution in Kathmandu, Nepal. In: Papers and proceedings of applied geography conferences, vol 25, pp 27–35

Hänsch R, Mendel RR (2009) Physiological functions of mineral micronutrients (cu, Zn, Mn, Fe, Ni, Mo, B, cl). 12:259–266

Ichiyanagi K, Yamanaka MD, Muraji Y, Vaidya BK (2007) Precipitation in Nepal between 1987 and 1996. Int J Climatol 27:1753–1762

Imtiaz M, Rashid A, Khan P, Memon M, Aslam MJ (2010) The role of micronutrients in crop production and human health. Pak J Bot 42:2565–2578

Joshi D (1997) Soil fertility and fertilizer use in Nepal. Soil Science Division, Kathmandu

Joshy D (1997) Indigenous technical knowledge in Nepal. In: Technical Issues, pp 45–61. International Board for Suastainable Land Management (IBSRAM), Bangkok Issues on Sustainable Land Management

Karki KB (1986) Assessment of land degradation and formulation of productivity improvement strategies in Southern Lalitpur District of Nepal. Partial Fulfillment for the degree Central Luzon State University, The Philippines

Karki KB (2003) Status of potassium in intensively cultivated soils of Kathmandu valley. Nepal J Sci Technol 5:83–89

Karki KB (2008) Soil fertility evaluation and environmental nexus in mid-western regional Nepal. IFAD, Rome

Karki KB, Karki AB (2001) Response of biogas slurry application in maize and cabbage production in Lalitpur District (Fanal Report). Alternative Energy Promotion Centre (ARPC), Kathmandu

Karki KB (1995b) Influence of organic and inorganic fertiliser in sustaining crop productivity in the central hills of Nepal. A Ph D, Dissertation, Universitaet fuer Bodenkultur, Vienna, Austria (unpublished)

Karki KB, Joshi D (1993) Soil fertility and fertilizers use in Nepal. In: HLS T (ed), Soil fertility and fertilizer use in South Asia. Development and Consultation Organization, New Delhi, India

Karki R, Schickhoff U, Scholten T, Böhner J (2017) Rising precipitation extremes across Nepal. Climate (Open access) 5:4. https://doi.org/10.3390/cli5010004

Karki KB (2006) City waste compost and sustainability of rice-wheat cropping system. Nepal Agric Res J 7:49–53

Karki KB (2004) Estimation of plant nutrients content in biogas slurry and losses during handling. In: Joshi BK, Joshi SL, Paudel KP (eds), Agricultural research for enhancing livelihood of Nepalese people, vol 30, pp 330. NARC/SAS, Kathmandu

Karki KB (2015) Soil fertility and fertilizer use in Nepal: past, present and furure. In: Karki KB, Tripathi BP, Manandhar R, Adhikari BH, Vista SP (eds), Second national soil fertility research workshop, pp 16–25. Soil Science Division, NARC, Kathmandu Nepal

Karki KB, Tuladhar J, Uprety R, Maskey S (eds) (2005) Distribution of micronutrients available to plants in different ecological regions of Nepal, pp 1–17–29. ICIMOD, Kathmandu, Nepal

Karki KB (1995a) Effect of fertilizers and boron on yield and boron uptake in wheat. In: Rawson HM, Subedi KD (eds) Sterility in wheat in subtropical Asia: extent, causes and solutions, pp. 39–43. ACIAR, LumJe Agricultural Research Centre, Pokhara, Nepal

Khor GL, Misra S (2012) Micronutrient interventions on cognitive performance of children aged 5–15 years in developing countries [online]. Asia Pac J Clin Nutr 21:576–486

Kyle J, Resnick D, Karkee M (2017) Improving the equity and effectiveness of Nepal's fertilizer subsidy program, vol 1685. International Food Policy Research Institute

Lang BF, Laforest MJ, Burger G (2007) Mitochondrial introns: a critical view. Trends Genet 23(3):119–125

Leigh RA, Wyn Jones R (1984) A hypothesis relating critical potassium concentrations for growth to the distribution and functions of this ion in the plant cell. New Phytol 97:1–13

Luyckx M, Hausman J-F, Lutts S, Guerriero G (2017) Silicon and plants: current knowledge and technological perspectives. Front Plant Sci 8:411

MODIS (2020) May 2, 2019—Fires, snow and smoke in Nepal, vol 2020. Moderate Resolution Imaging Spectroradiometer MODIS

Mandal CK (2019) Wildfires single biggest threat to tens of thousands of hectare forest cover. In Kathmandu Post, https://kathmandupost.com/national/2019/04/29/wildfires-single-biggest-threat-to-tens-of-thousands-of-hectareforest-cover, Kathmandu

Marschner H (2011) Marschner's mineral nutrition of higher plants. Academic Press, Heidelburg, Germany

Maskey S, Bhattarai S, Peoples M, Herridge D (2001) On-farm measurements of nitrogen fixation by winter and summer legumes in the Hill and Tarai regions of Nepal. Field Crop Res 70:209–221

Maskey S, Bhattarai S, Karki K (2004) Long-term effect of different sources of organic manures on wheat soybean rotation. SAARC J Agric 2:243–256

McLaren RG, Cameron KC (1996) Soil sciences, ustainable production and environmental protection. Oxford University Press Auckland

Mengel K (2015) Chapter 4. Potassium. CRC Press, New York

Mengel K, Kirkby E (eds) (1987) Principles of plant nutrition. Bern, pp 1–695. International Potash Institute, Bern, Switzerland

Nepal AK, Gelal B, Mehta K, Lamsal M, Pokharel PK, Baral N (2014) Plasma zinc levels, anthropometric and socio-demographic characteristics of school children in eastern Nepal. BMC Res Notes 7:18

Pal D, Srivastava P, Durge S, Bhattacharyya T (2001) Role of weathering of fine-grained micas in potassium management of Indian soils. Appl Clay Sci 20:39–52

Panday D, Maharjan B, Chalise D, Shrestha RK, Twanabasu B (2018) Digital soil mapping in the Bara district of Nepal using kriging tool in ArcGIS. PloS One 13(10):e0206350

Pandey SP (2015) The role of fertilizer in transforming the agricultural economy of Nepal

Panta H (2018) Supply chain of subsidized chemical fertilizers in Nepal. Inst Agric Animal Sci 35:9–20

Pariyar SK, Das T, Ferdous TJ (2013) Environment and health impact for brick kilns in Kathmandu valley. Int J Sci Technol Res 2:184–7

Paudel S, Sah JP (2003) Physiochemical characteristics of soil in tropical sal (Shorea robusta Gaertn.) forests in eastern Nepal. Himal J Sci 1:107–110

Paudel S, Timilsina YP, Lewis J, Ingersoll T, Jnawali SR (2015) Population status and habitat occupancy of endangered river dolphins in the Karnali River system of Nepal during low water season. Mar Mamm Sci 31:707–719

Quinn M (2004) The effects of wind and altitude in the 400-m sprint. J Sports Sci 22:1073–1081

Quinton JN, Catt JA, Hess TM (2001) The selective removal of phosphorus from soil. J Environ Qual 30:538–545

Rahman H, Sabreen S, Alam S, Kawai S (2005) Effects of nickel on growth and composition of metal micronutrients in barley plants grown in nutrient solution. 28:393–404

Raut N, Sitaula BK (2012) Assessment of fertilizer policy, farmers' perceptions and implications for future agricultural development in Nepal. Sustain Agric Res 1(526-2016-37838)

Rawal N, Chalise D, Khatri N (2017) Omission plot experiment in wheat to determine the native nutrient supplying capacity of soil in Bhairahawa Nepal. Res Agric Vetenary Sci 1:108–115

Regmi A, Ladha J, Pasuquin E, Pathak H, Hobbs P, Shrestha L, Gharti D, Duveiller E (2002) The role of potassium in sustaining yields in a long-term rice-wheat experiment in the Indo-Gangetic Plains of Nepal. Biol Fert Soils 36:240–247

Rerkasem B, Rawson H, Subedi K (1996) A general survey of the incidence of wheat sterility. In: Sterility in wheat in subtropical Asia: extent, causes and solutions, pp 8–12

Sah S, Brumme R (2003) Altitudinal gradients of natural abundance of stable isotopes of nitrogen and carbon in the needles and soil of a pine forest in Nepal. J for Sci 49:19–26

Sah G, Shah S, Sah S, Thapa R, McDonald A, Sidhu H, Gupta R, Sherchan D, Tripathi B, Davare M (2014) Tillage, crop residue, and nitrogen level effects on soil properties and crop yields under rice-wheat system in the Tarai region of Nepal. Glob J Biol Agric Health Sci 3:139–147

Salwen MJ (2017) Vitamins and trace elements. In: Henry's clinical diagnosis management by laboratory methods, pp 416, e-Book

Sapkota A, Yang H, Wang J, Lu Z (2013) Role of renewable energy technologies for rural electrification in achieving the Millennium Development Goals (MDGs) in Nepal. ACS Publications

Schreier H, Shah P, Lavkulich L, Brown S (1994) Maintaining soil fertility under increasing land use pressure in the Middle Mountains of Nepal. Soil Use Manage 10:137–142

Shah P, Schreier H (1995) Maintaining soil fertility in agriculture and forestry. In: Schreier HH, Shah PB, Brown S (eds) Challenges in mountain resources management in Nepal: processes, trends and dynamics in middle mountain. ICIMOD, Kathmandu, Kathmandu

Sherchan DP, Gurung GB (1998) Effect of long term application of chemical fertilise and manure on crop yields and soil chemical properties under rice-wheat cropping pattern (the results of 8 cropping cycles), PAC Technical Paper No. 183. Pakhribas (Nepal): Pakhribas Agricultural Research Centre

Shrestha DP (1997) Assessment of soil erosion in the Nepalese Himalaya: a case study in Likhu Khola Valley, Middle Mountain Region. Land Husbandry 2:59–80

Shrestha RR, Shrestha MP, Upadhyay NP, Pradhan R, Khadka R, Maskey A, Maharjan M, Tuladhar S, Dahal BM, Shrestha K (2003) Groundwater arsenic contamination, its health impact and mitigation program in Nepal. J Environ Sci Health Part A 38:185–200

Shrestha S, Rymer C, Jayaswal M, Lama N, Neupane K, Jha V (2005) Supplementation of selenium: a strategy to increase fertility in goats. In: Andersen JKTP, Karki KB, Maskey SL (ed) Micronutrients in South South East Asia, pp 43. SSD/NARC/ICIMOD/Bergen Uni/NORAGRI, Kathmandu, Nepal

Sillanapaeae M (1982) Micronutrients and the nutrients status of soils: a global study. FAO, Rome

Sillanpää M, Jansson H (1992) Status of cadmium, lead, cobalt and selenium in soils and plants of thirty countries. Food & Agriculture Org

Skerry C, Moran K, Calavan K (1991) Four decade of development: the history of US assistance to Nepal 1951–1991. United States Agency for International Development (USAID) in Kathmandu, Kathmandu Nepal

Sroor A, El-Dine NW, El-Shershaby A, Abdel-Haleem A (2003) Major and trace elemental analysis in milk powder by inductively coupled plasma-optical emission spectrometry (ICP-OES) and instrumental neutron activation analysis (INAA). J Environ Sci 15:570–576

Steinnes E (2009) Soils and geomedicine. Environ Geochem Health 31:523–535

Stevens CJ, Quinton JN, Bailey A, Deasy C, Silgram M, Jackson D (2009) The effects of minimal tillage, contour cultivation and in-field vegetative barriers on soil erosion and phosphorus loss. Soil Tillage Res 106:145–151

Subedi SKA (2015) Domestic biogas production and use in Nepal: a simple, reliable, clean, and cost-effective solution to provide energy security to the rural households. Massey University, Palmerston North, New Zealand

Subedi K (1993) Adhatoda vasica (asuro): a superior indigenous green manure species for rice in the hills of Nepal. Int Rice Res Notes 18

Sumner ME, Noble AD (2003) Soil acidification: the world story. Marcel Dekker, New York

Takeshima H, Adhikari RP, Kaphle BD, Shivakoti S, Kumar A (2016a) Determinants of chemical fertilizer use in Nepal: insights based on price responsiveness and income effects. International Food Policy Research Institute

Takeshima H, Adhikari RP, Kaphle BD, Shivakoti S, Kumar A (2016b) Determinants of chemical fertilizer use in Nepal: Insights based on price responsiveness and income effects, vol 1507. International Food Policy Research Institute

Tandan RP, Rai R, Karki KB (2015) Soil nutrition distribution in Eastern Tarai of Nepal: a case Study of Morang District of Nepal. In: Karki KB, Tripathi BP, Manandhar K, Adhikari B, Vista SP (eds) Second national soil fertility research workshop, pp 368–373. Soil Scince Division, NARC, Kathmandu Nepal

Uchida R (2000) Plant nutrient management in Hawaii's soils, approaches for tropical and subtropical agriculture. In: Plant nutrient management in Hawaii's soils. In: U. o. H. a. M. College of Tropical Agriculture and Human Resources (ed), pp 25

Upadhyaya H, Begum L, Dey B, Nath PK, Panda SK (2017) Impact of calcium phosphate nanoparticles on rice plant. J Plant Sci Phytopathol 1:1–10

Wagley MP (ed) (1995) Status of watershed management in Nepal, pp 1–99–118. PWMTA, Wallingford

Watanabe T (1994) Soil erosion XE "Soil erosion" on yak-grazing steps in the Langtang Himal Nepal. Mount Res Dev 14:171–179

Welch RM, Graham RD (2004) Breeding for micronutrients in staple food crops from a human nutrition perspective. J Exp Bot 55:353–364

Westarp S, v., Sandra Brown, H. S., and Shah, P. (2004) Agricultural intensification and the impacts on soil fertility in the Middle Mountains of Nepal. Can J Soil Sci 84:323–332

Xue T, Hartikainen H, Piironen V (2001) Antioxidative and growth-promoting effect of selenium on senescing lettuce. Plant Soil 237:55–61

Yamada T, Sharma C (1993) Glacier lakes and outburst floods in the Nepal Himalaya . IAHS Publ Publ Int Assoc Hydrol Sci 218:319–330

Yong W, Link B, O'Malley R, Tewari J, Hunter CT, Lu CA, Carpita NC (2005) Genomics of plant cell wall biogenesis. Planta 221(6):747–751

Land Degradation

10

Krishna Bahadur Karki and Roshan Babu Ojha

Abstract

Land degradation is one of the major issues facing by Nepalese agriculture and can mostly be attributed to the loss of soil coupled with heavy rainfall during the monsoon season. Furthermore, soil nutrient mining, soil organic carbon depletion, floods, and chemical degradation are other forms of soil degradation. Based on the literature and our own observations, we elucidate the land degradation status in Nepal, around 12% of which is affected by the degradation process. Forest land (38%), Pasture/rangeland (37%), and agricultural land (10%) are under the threat of serious degradation for which urgent land management for restoration is required. Soil erosion by water solely accounted for 45.5% of the area under degradation, followed by wind erosion (4% of total area), chemical degradation (0.3% of total area), and physical degradation (0.2% of total area). Rill-erosion, inter-rill erosion, and gully erosion are serious threats facing the Mountain regions. Sheet erosion and floods are serious threats facing the Tarai region. Physical degradation such as soil compaction, deterioration of the soil structure, and increase in bulk density are overlooked researchable issues in Nepal. Chemical degradation such as pesticide pollution in soils, nutrient imbalance and mining, and the leaching of storing herbicides are other issues that need addressing. The yearly loss of life and land due to floods and landslides are irrecoverable and directly affect the country's economy. It is estimated that on average 6 t ha^{-1} yr^{-1} of soil is lost annually from Nepal, which strongly indicates the need for proper land management.

Keywords

Deforestation • Floods • Land degradation • Soil acidity • Soil erosion

K. B. Karki (✉) · R. B. Ojha
Nepal Agricultural Research Council, Kathmandu, Nepal

10.1 Land Degradation

Environmental degradation and its adverse impact on soil nutrition status is a major concern facing Nepalese agriculture, particularly in the hills and mountains. Its adverse effect is the outcome of several factors, including physical, biological, and chemical changes in soil properties developed by land-use practices, including the socio-economic condition of the farming community. Deterioration in the physical, biological, and chemical properties of soil determines the productive capacity of the land (Barrow 2010). The FAO (1976) states that even imbalanced use of fertilizers can lead to nutrient deficiencies and degrade the land. Salinity and other structural degradation are alarming in most tropical countries, though in Nepal soil acidity caused by massive soil erosion and the development of infertility are major causes of soil degradation (Gardner and Gerrard 2002; Karki 1986).

Degradation of land can be wholly or partly natural but can also be a complex phenomenon. Therefore, appropriate measures for assessing land degradation are critical (Johnson and Lewis 1995). In developed nations soil compaction caused by modern agricultural implements is the main reason for land degradation (Blum 1998; Kayombo and Lal 1994), whereas in developing nations soil erosion, imbalance in plant nutrient application, and nutrient mining are major factors of land degradation (Dumanski et al. 1991; Lal 2001).

The fertility level of Nepalese soil is declining mainly due to multiple cropping practices adopted by farmers that mine heavy nutrients which are not replenished (Karkee 2004; Karki 1986; Karki and Joshi 1993). In addition, soil erosion in the hills and mountains is a serious problem that washes away a huge amount of applied plant nutrients (Carson 1992). The use of fertilizer nutrients in Nepal is 70 kg ha^{-1}, whereas the nutrients harvested by a single maize crop are more than 70 kg ha-1 (Ghani and Brown 1997). Moreover, farmers apply fertilizer (NPK) in a ratio of 1:0.37:0.03 (Joshy 1997), indicating higher use of nitrogenous fertilizer. As a result, Nepalese soils are highly acidic in nature and

R. B. Ojha and D. Panday (eds.), *The Soils of Nepal*, World Soils Book Series,
https://doi.org/10.1007/978-3-030-80999-7_10

low in organic matter, nitrogen, phosphorous, potassium and cation exchange capacity (SSD 1998).

The physiographic and climatic contrast to Nepal's small area has added extensive complexity to this problem. The climate changes drastically across the country and can range from subtropical monsoons to Alpine Tundra as the elevation gradually increases from Tarai to the High Himalayas. Moving from lowland Tarai to the high Himalayas, an array of land uses, along with corresponding problems become apparent in watersheds. Mountains and hills constituting three-fourths of the area in Nepal comprise 4 macro, 24 meso, and more than 6,000 micro watersheds, with diverse biophysical conditions and biological deterioration under the influence of ongoing human activities (Wagley 1995). Climate and human pressure further accelerate land degradation processes (Gardner and Gerrard 2002; Paudel and Thapa 2004). Four major river systems drain the nation: the Sapta Koshi, the Narayani, the Karnali, and the Mahakali. All four rivers have their headwaters (water towers) in the north of the Great Himalayan range, and the first three originate in the Tibetan Plateau (Paudel and Thapa 2004).

Overall, land degradation has been taken as a major threat to economic and rural development. However, systematic information is lacking from the country as a whole (Karkee 2004). The physical form of land degradation in Nepal has been highlighted more adequately, but the chemical and other aspects of land degradation have not been well studied (Karki 1986). The inherent properties of soils such as light texture and shallowness have been considered major constraints due to their susceptibility to erosion. However, specific management problems are associated with soil in different physiographic regions. The development of sound land management practices to address land degradation problems is lacking in the country. While there have been some efforts at addressing land degradation, these efforts require careful understanding and attention to the country's complex farming system and their interactive roles and thus have been lacking. Here, we would like to briefly discuss the status, types, and causes of land degradation in Nepal.

10.2 Land Degradation Causes and Status in Nepal

Land degradation problems in Nepal have been documented in several publications (Acharya and Kafle 2009, 2002; Karkee 2004; Karki 1986). While many types of land degradation exist in Nepal and the most prevalent are deforestation, mass wasting, erosion, and flooding. In addition, the steepness of the terrain, high-intensity rainfall during monsoons, unplanned and scattered settlement patterns, and improper land use practices in the hill and mountain slopes are other causes of watershed degradation. More recently, rapid growth in population, slow economic growth and transformation, unplanned construction of infrastructure, and the degradation of the natural environment have posed further threats to watershed degradation (Karki 1986; Wagley 1995). Common examples of human-induced degradations include accelerated soil erosion and downstream sedimentation, landslides, floods, desertification, forest fires, and droughts. For many researchers working in Nepalese soils, water-induced soil erosion is considered land degradation (Chalise et al. 2019; Gardner and Gerrard 2002; Shrestha et al. 2004) but there are numerous other causes such as soil acidity, nutritional deficit, waterlogging, overgrazing, compactness or soil structural degradation, elemental toxicity, and many more. Limited studies are available, though their methodologies vary greatly, and they give inconsistent information. The following causative factors are recognized in our soil to accelerate land degradation, as explained below.

Population: The population of Nepal has been increasing ever since the population census began. The national population report published by the Ministry of Population and Environment of Nepal shows that the population in 1961 was 9,412,996 with a growth rate of 1.64% per annum (pa), but that growth increased at the rate of 2.06% pa and the population reached 11,555,983 in 1971. Ten years later in 1981, the growth rate was 2.62% pa. In 1991 the growth was 2.08% pa and the population had reached 18,491,097. In 2001 the population was 23,151,423 and growing at a rate of 2.25% pa, though by 2011 the growth rate had fallen to 1.35% and the total population has risen to 26,494,504. The population growth rate to 2.62% and 2.08% in 1981 and 1991 could be the result of labor being imported from Bangladesh and India as the brick and garment industries mushroomed. The migrant laborers did not return to their respective countries, and rapid urban growth increased 412% encroaching 31% of agricultural land (Ishtiaque et al. 2017). The decreased rate of population growth from 2.25 to 1.35 in 2011 and the continual decrease is due to a 1.92 M absentee population of Nepalese who have migrated in search of better opportunities for themselves and their families (MOPE 2017; UNFPA 2017).

The increasing population rate of 2.25% is one of the major hurdles in sustaining Nepal's natural resources for the livelihoods of people, as the country has more than 200 people per km^2 making it one of the world's highest densities per hectare of land. The distribution between population and arable land in each ecological zone is also unbalanced; this is seen in the heavy human population pressure in the Hill and Tarai regions. Despite its large geographical area, the percentage of arable land available in the mountain region is very low in terms of sustaining many people. Most people in Nepal own very small parcels of land. It is estimated that for a healthy livelihood, 0.5 ha of land per capita under intensive cultivation is needed (Pimentel et al. 2010). Demographic pressure on natural resources is a worldwide problem and the rate at which the present population—especially in the developing world—is growing, there will be tremendous pressure on land to feed this ever-growing population (Timah et al. 2008). However, in Nepal, the average land holding per capita is 0.18 ha (MOPE, 2000), and the pressure on the marginal lands in the mid–hills of Nepal is higher than is required to fulfill the basic demand. This has led to the intensive cultivation—and hence, degradation—seen today (Blaikie and Brookfield 2015).

Land Use: The total area of Nepal is 14,718,000 ha, 21% (3,091,000 ha) of which is capable of cultivation. While Joshi (1997) considers the land under cultivation to be fully occupied, there remains 7% (10,30,000 ha) of total land that could be brought under cultivation. The pressure on the land is said to be one of the highest in the region, indicating that the land is intensively cultivated. Poudel and Shaw (2015) demonstrated that the rapid population growth has put tremendous pressure on natural resources, along with the construction of unplanned and haphazard infrastructure and no investment in conservation, all of which have led to depleting resources and decreased agricultural productivity. Urbanization is another problem as urban areas encroach on productive agricultural land, thus pushing agriculture to marginalization (2009).

In the Tarai and lower river valleys three crops are usually grown each year, whereas in the Middle Mountain and higher altitudes two or three crops are grown. Where vegetables are grown farmers raise three to four crops in one year. The land is so intensively cultivated that it has insufficient time to regain its optimal level of fertility, and therefore its crop-bearing sustainability (Atreya et al. 2008). It is notable that due to the intensity of Nepal's soil cultivation an in-depth study of soil physico-chemical and biological properties has yet to be carried out. There are some studies indicating that the mining of Nepalese soil is heavy and there is a negative balance of nutrients (Ghani and Brown 1997; Joshi 1997; Karki 2003). If this mining continues, Nepal's presently cultivated land will eventually turn to the desert (Zdruli et al. 2017).

Rainwater erosion: Rainfall is an active agent causing erosion at a rate dependent upon its distribution, amount, and intensity. This is a great threat to Nepalese crop production. It has been estimated that in Nepal the monsoon rains wash away 240 M m^3 of soil annually, thereby raising the bed level of rivers in the plains by 15 to 30 cm (Joshi 1985). This leads to a heavy loss of plant nutrients, which is deleterious to crop production (Acharya et al. 2007); Gardner and Gerrard 2002; Tiwari et al. 2010). Based on the information on sediment yields gained from surveys of the main Himalayan rivers, it is estimated that the average soil erosion rate in Nepal is 20 to 30 t ha^{-1} yr^{-1}. Balla (1983) estimated the average soil loss to be 11.6 t ha^{-1} yr^{-1} in the Phewa Tal Watershed using the Universal Soil Loss Equation (USLE). Quantification of the rate of erosion varies, as erosion plot experiment results depend on the slope, type of soil, plot size, and location of the experimental site (Mawdesley et al. 1998). Schreier et al. (2000) estimated the annual soil erosion rates in Bari (upland) fields at an average of around 20 t ha^{-1} yr^{-1}. Putting aside the debate on the rate of deforestation, all studies have noted that soil erosion and land degradation is higher in the hills and cultivated uplands. Some reports are available on a watershed basis, indicating 60 t ha^{-1} yr^{-1}; 33 t ha^{-1} $yr^{-1}$from the Andhi Khola watershed whereas the Tinau watershed is 18 t ha^{-1} yr^{-1}. Very different figures are reported from the Nakkhu Khola watershed, namely 44 t ha^{-1} yr^{-1} (Maharjan and Tamrakar 2011).

Accelerated (human-induced) erosion is another type of physical wearing down of the surface of the earth that is being increasingly felt in Nepal. Acceleration of erosion is characterized by the loss of topsoil by infrastructural engineering processes with an associated sheet and rill erosion followed by a decrease in soil fertility in allied cultivated areas (Merz et al. 2006; Mulmi 2009; Paudel and Thapa 2004; Shrestha 2009). Agriculture extended to marginal land due to infrastructure occupying land aggravates soil erosion and environmental degradation, adding loads of sediment to rivers and streams (Gentle et al. 2014; Raut et al. 2010).

Seasonal differences in soil loss in agricultural lands have been found to be significantly high. Different authors report different studies in various sites and find variations in results. Annual rates of erosion from forests, as recorded by most authors, are similar, though results from degraded and grazing lands differ. This could be due to variation in soil type and rainfall intensity and canopy cover, including the type of vegetation (Awasthi 2004; Gardner and Gerrard 2003; Sherchan and Gurung 1995). Average soil loss calculated by the mentioned authors for each land use category in the watershed showed the least annual soil loss of 2.7 t ha^{-1} from dense forest and the highest soil loss of 63.1 t ha^{-1} from unprotected grazing land.

About 68% of total soil loss occurs during pre-monsoon tilling and the early monsoon period in agricultural plots (Chalise and Khanal 1997). Carver and Nakarmi (1995) reported that 50 to 90% of soil erosion is pre-monsoon and 30% of total rainfall received during the pre-monsoon season produced 60% soil loss in the mountain watershed. The reason for the higher soil erosion in the pre-monsoon season is that during this period the vegetative cover in cultivated land is at a minimum (Sharma 1996).

Landslides, mass wasting, and flooding: The Himalayan mountain range is the most fragile mountain system in the world. Because of its fragile nature, it suffers from monsoon-triggered landslide problems every year, and during these periods a great number of people throughout the country are affected by large- and small-scale landslides. Triggering landslides is a result of the construction of non-engineered roads, irrigation canals, and dams/reservoirs without considering natural hazards (Thapa 2015). The annual loss due to landslide damages alone in this region is estimated at hundreds of human fatalities and economic losses of more than 1 B (billion) US dollars in repair costs. Casualties due to landslides and floods in Nepal from 1971 to 2014 totaled 8758 deaths, with many more missing or unaccounted for and millions left homeless with the loss of their property. This total loss of property was accounted to be Nepalese Rs. 26,271,251,281 (Adhikari et al. 2016). Flooding from 19 to 21 July 1993 was caused by high-intensity rain of 65 mm h^{-1} and a total of 540 mm rain in 24 h in the Bagmati river basin, which cost the lives of 1336 people and injured 136 injured more, with a property loss of NRs 4.9 B (Gautam and Pokherel 2004). Unplanned and haphazard construction of rural roads has further aggravated these problems in rural areas causing heavy loss of lives and property. Studies indicate that the loss due to landslides and related problems in the Himalayan region alone constitutes about 30% of the world's total landslide-related damage (Li 1990; Wang et al. 2019; Zhang et al. 2016).

Lamsal (2014) listed triggers and causes of landslides in Nepal as cloud outbursts (200 to 1000 mm of rain in 24 h), the uncontrolled flow of water on slopes, toe cutting, earthquakes, and flash flooding due to GLOF. Anthropogenic causes include deforestation, rock blasting and quarrying, hill cutting, the continuous irrigation of paddy fields in the hills and mountains, water storage ponds, tunneling, and vibration due to vehicles and heavy machines. The erosion process blocks natural slopes, leading to the downward movement of high-volume water and forming gullies and other geological conditions. Landslides in the mountains can cause highways to be blocked for days (Dahal and Hasegawa 2008; Hasegawa et al. 2009). The prime cause of rockslides is deemed to be intense fracturing and the alteration of rocks along the road corridors. Major types of landslides include rockslides, shallow linear slides, debris flow, debris slides, and bank cutting. The damages to roads come from bank cutting, debris flow, and subsidence (Kayastha et al. 2013). Soil erosion and landslides from upstream watersheds during the monsoon period lead to floods downstream, as every year streams and rivers running north to south carry debris from the hills into the thousands of hectares that make up the fertile plain of the Tarai, causing hundreds of hectares of damage as fertile land is covered with debris and rocks (Joshi et al. 1998).

Various authors have expressed different views on whether landslides are natural or anthropogenic. Carson (1985) states that landslides on Nepalese slopes are natural, whereas Laban (1978) estimated that 74% of landslides occur under natural conditions but 26% seem to occur due to the activities of human beings. This includes about 5% of the country's landslide and erosion problems that are associated with roads and trails and thus more hazardous to human life. One human activity that causes landslides is quarrying (i.e., the extraction of stones for construction) (Basnyet 1989). As reported by Regmi (2001), out of 29 landslides in the Kanthe Khola watershed in the Baglung district, only eight are classified as natural, covering 23.94 ha, while 21 are classified as artificial, covering 173.8 ha. The author further states that from 1979 the amount of landslide area due to human pressure has doubled. Joshi et al. (1998), citing reports from the DPTC (1994) and MOH (1994), explained that in July 1993 flood and landslides destroyed 363 km of road and 277 bridges and damaged 60,000 ha of prime agricultural land. In that year landslides and floods affected 500,000 people and took the lives of 1475 people and 25,388 heads of livestock in the country.

Sedimentation load: Rivers in Nepal are generally characterized by steep gradients and high sediment loads, along with bank-cutting erosion and deposition. River gradients are steepest in the High Himalayan region, especially where the rivers cross mountain ranges. Sediment contribution represents the entirety of the erosion process in the watershed, including all types of slope failures. The difference in the annual sediment contribution rate depends on the rainfall and intensity. Sediment transport by all of the Himalayan river systems in Nepal is very high, especially during peak monsoon runoff periods (Nepal et al. 2014). The sediments are derived from bank erosion, bed load suspension, terrace erosion, and glacial outwash in addition to landslides. Estimates show that the Karnali River transports a mean annual total sediment load of about 170 M t, the major source of sediment being glacier sediments carried all the way from the Tibetan Plateau and mass wasting from gullying and landslides. In this manner, a large portion of the sediment load is derived from the Siwalik region (Shrestha 1985). Different views are expressed in the case of the Marsyangdi River, which is reported as mass wasting (Dingle et al. 2016; Shrestha 1985, 2000). The information on sediment yields based on suspended sediment measurements and erosion rates compiled by the WEC Secretariat (1987) and reported by Kayastha (1998) illustrated that the Tamur watershed has the highest erosion rate at 240 t ha-1 yr-1 followed by the Marshyangdi with 210 t ha^{-1} yr^{-1}, the Koshi with 169 t

ha^{-1} yr^{-1}, the Karnali with 153 t ha^{-1} yr^{-1}, and the Kali Gandaki with 148 t ha^{-1} yr^{-1}.

With the increase of global warming caused by human activities in recent decades, it is estimated that the beds of rivers in the Tarai are rising by 15 to 30 cm annually. This phenomenon is responsible for flooding as well as for the shifts in the river courses. As an example, the Koshi river carries an annual load of 119 times 106 m^3 of silt, which is equivalent to 2 mm of topsoil depth over its entire watershed. It has shifted its course about 112 km westward into the Bihar state of India within the last two centuries, leaving 15,000 km^2 of once fertile land buried beneath the sediment (Danish et al. 2013).

Sedimentation surveys have been carried out in lakes and reservoirs such as Phewa Lake and Kulekhani reservoir, plus Lake Begnas and Lake Rupa indicate that the average annual sediment contribution rate from a hectare of the watershed is about 17 m^3 in Phewa Lake (DSC 1994) and 35 m^3 in Begnas Lake, whereas, in Kulekhani the rate was 42 m^3 from 1979 to 1992 but 415,73 and 82 m^3 for 1993, 1994, and 1995, respectively (Sthapit 1989; Sthapit and Balla 1998). The annual sediment contribution rate from 1979 to 1993 can be considered as the average rate, whereas the rate for 1993 was due to the disastrous monsoon that occurred in that year. The rates for the years 1994 and 1995 can be considered as also being exaggerated by the disastrous monsoon of 1993.

There have been very few studies regarding the land made infertile every year in Nepal by these 6000 rivers and rivulets. Adhikari (2010), citing the work of Dixit (1995), adumbrated that 2 m depth of sedimentation was deposited by the river Lakhandehi in Rautahat District over 45 years but does not state that how much area was covered with this sedimentation. The Koshi river also destroyed 700 ha of land in Nepal, covering it with a meter of sand and silt and making it useless for cropping. While this was a one-off phenomenon, there are other rivers such as the Lal Bakaiya, Ratu, and others for which this is a yearly phenomenon depending on the magnitude of rainfall during the year.

Rainfall intensity: The enormous variation in the Nepalese Himalayan climate is mainly due to the range of altitudes within the very short north–south distance of 150 to 230 km. The presence of the east–west-trending Himalayan massifs to the north and the monsoonal alteration of wet and dry seasons greatly contribute to local variations in precipitation. Six climatic zones are recognized in Nepal based on altitude: the tropical and subtropical zone below 1,000 m in altitude, the warm temperate zone of 1,000 to 2,000 m, the cold zone of 2,000 to 3,000 m, the sub-alpine climatic zone of 3,000 to 4,000 m, the alpine zone of 4000 to 4,500 m, and the Arctic zone above 4500 m in altitude (Figs. 10.1 and 10.2). In terms of natural vegetation regimes and distribution patterns, altitude plays a significant role. Below 1,000 m, the dominant form of vegetation consists of tropical and subtropical rainforest that changes to tundra-type vegetation when the altitude exceeds 4000 m.

Altitude also affects annual rainfall and precipitation patterns. Up to about 3,000 m, the total annual rainfall increases as the altitude increases; thereafter, it diminishes with increasing altitude and latitude (Dhar and Rakhecha 1981). In addition to latitudinal differentiation in rainfall, two other patterns can be discerned. First, given the northwestward movement of the moisture-laden summer monsoons (June to September), the amount of annual rainfall generally decreases from east to west (Kansakar et al. 2004). Second, the horizontal extension of hill and mountain ranges creates a moist condition on the south- and east-facing slopes, whereas it produces a major rain shadow on the north-facing slopes. The aridity increases with altitude, especially on the northern slopes, and reaches its climax in the inner Himalayan region and on the Tibetan Plateau. Eastern Nepal receives approximately 2,500 mm of rain annually, the Kathmandu area about 1,420 mm, and western Nepal about 1,000 to 2000 mm.

During the summer monsoons, there is a strong flow of moist air from the southwest that follows the pre-monsoon season. Even though the arrival of the summer monsoons in Nepal can vary by as much as a month, it generally arrives in

Fig. 10.1 **a** Land degradation due to tree cutting, **b** Goat grazing in the public forest as the forest is pushed away, and **c** Sloping land cultivation without any scientific knowledge

Fig. 10.2 Farmers grow maize crops slopes such as this one, where pre-monsoon torrential rain washes away all the topsoil and manure applied as fertilizer

early June, is preceded by violent lightning and thunderstorms, and lasts through September, when it begins to recede. During the peak season, the intensity of rain reaches its climax. Sometimes the rainfall intensity ranges from 10.3 to 19.0 mm per day, with the maximum rainfall in a 24 h-duration reaching 456 mm. Rainfall measured for the ten year-period of 1994–2003 is considered high-intensity rainfall (Gautam and Pokherel 2004). Dahal and Hasegawa (2008) calculated the intensity of rainfall with respect to landslides as $I = 73.90D - 0.79$, where I is the intensity and D is the duration. The plains and lower Himalayas receive more than 70% of their annual precipitation during the summer monsoon season. Although the success of farming is most dependent on the timely arrival of the summer monsoon, it periodically causes problems such as landslides and the subsequent loss of human lives and farmlands, along with the heavy flooding in the plains. Conversely, when prolonged breaks in the summer monsoons occur, severe drought and famine often result.

10.2.1 Forest Degradation

Nepalese farmers are highly dependent upon their local forests and get over 80% of their energy for cooking and heating from firewood (Baral et al. 2019). In addition, 8% of the rural population uses cattle/buffalo dung and agricultural residue as heating fuel. In the last several years, the rise in petroleum prices has pushed middle-class petroleum-based fuel users back to traditional energy sources, which puts increased pressure on forests. Nepal uses 48% of its energy in households and 38% in industries (Poudyal et al. 2019). Depending on the altitude, one person needs from 398 to 486 kg yr-1 of firewood, as higher altitudes demand higher amounts of fuelwood per year (Puri et al. 2017). Forest clearance for fuel wood, along with timber for construction and furniture, denuded forests at a rate of 1.3%, which is much lower than that 1.7% cited by Chaudhary et al. (2016). This reduction is due to the use of 365,000 biogas plants and the increasing use of LPG for cooking even in rural areas, which has reduced traditional family dependence on fuelwood (Bhatta et al. 2018; Poudyal et al. 2019). The present rate of forest clearance for agriculture and fuel wood, plus timber for construction and furniture, etc., comes to around 1.06% yr^{-1} (Bhatta et al. 2018; Poudyal et al. 2019).

The grazing of animals in public forests, especially goats, is a serious menace to forest maintenance. Goats in particular feed on the growing buds of forest plants, and as the growing buds are eaten, plant and tree growth are restricted (Fig. 10.1b). The effect of such feeding can be clearly observed in the Terai and Siwaliks forest. In addition, tree looping (Fig. 10.1a), shifting cultivation (Fig. 10.1c) are other problems, as farmers fail to consider the long-term effects of haphazardly lopping forest plants for fodder. On average, a single-family loops 3.3 t yr^{-1}, 30% of which comes from community forests (Gurung et al. 2009).

With Nepal forest cover gradually decreasing, waste shrub cover is increasing. The rate of decrease in forest cover was more rapid during the period 2000–2005, and the forestland in the Tarai region seems to be more slowly decreasing in size than the forestland in the hills, bringing overall deforestation as 1.06% in 2010. Based upon currently available data it is apparent that compared to 1970 about 11.68% of Nepalese forestland is now degraded through poor management by landowners and at the local and central government levels. Encroachment into forestlands for infrastructural rehabilitation following extreme weather or industrial actions and for agricultural and infrastructural encroachments accounts for 10% of forest degradation (Acharya et al. 2011). However, the type of degradation is different depending on the type of forest. The poorly managed forest is highly degraded with 5.3 M ha, followed by poorly managed sloppy terraces with 3 M ha, while the area damaged by flood and landslides is 11.6 M ha. This rate is further increasing due to the recent political instability in Nepal. A recent national forestry inventory report shows that Nepal's forest area has decreased from 38 to 29% of the country's land area during the last 16 years (Gautam et al.2004). This rate of forest loss dramatically shows that the gap between current forest consumption levels and the essential and sustainable forest stock capacity levels will further widen in the future. In addition, Fox et al. (2019) suggested the decline in total forest growth in Nepal may be caused more by the decrease in the overall forest stock by other means rather than due to physical decline in tree growth rate. Therefore, the Government of Nepal needs to formulate and rapidly implement appropriate policy instruments to rapidly narrow this gap if the Nepali forest stock is to return to a sustainable state to meet the needs of current and future generations.

10.2.2 Forest Fires

Forest fires are a key source of loss of national stock and are common during the dry season between October and May in Nepal's jungles and protected areas. Fires can be set intentionally by honey collectors, Saal seed collectors in the Tarai region, hunters chasing wild animals, firewood collectors, and land encroachers. Fire can also break out due to cigarette and bidi bits carelessly thrown by travelers, tourists, picnickers, or cowboys; or by sparks caused by rolling stones; sparks thrown by passing vehicles; or the burning of stubble by farmers. This can also be part of the "slash and burn" practice that farmers traditionally employ to gain better short-term vegetation and agricultural yields at the cost of long-term losses in soil quality. There are no areas safe from forest fires. Forests from the Tarai to High Himalayan regions are equally affected. Regionwide distribution of forest fires shows that 37% of fires occur in Siwaliks; 26% in the Hills, 20% in the Tarai, and 13% in the Middle Mountain Regions. The government of Nepal reported by Wagley et al. (2014) further notes that 58% of fires were caused deliberately, 22% by negligence, and 20% were accidental. Because the area damaged by fire varies every year, it is difficult to pinpoint the exact amount of forest affected by fires because it is unpredictable where and when fires will break out. Most of the damages are seen in the loss of biodiversity and soil forest timber loss, which cannot be calculated. In 2009 a total of 105,350 ha of land was damaged by forest fires that resulted in 43 human casualties, 12 people injured, 375 livestock killed, and 74 houses destroyed at an estimated economic loss of $2.4 M USD (UNDP 2009). Shrestha (2019), citing data from 1971 to 2016, communicated that there were 6,766 fire events that took the lives of 1,491 people, injured 1,699 people, and 313,816 people are affected (Sombai et al. 2018).

Wildfires denude the terrain, thereby providing sources of major erosion flooding along with increased river sediment-loading events, leading to a marked reduction in water quality. After a fire, runoff often increases due to several factors, including the occurrence of water repellence, changes in soil hydraulic properties, removal of vegetation, loss of biodiversity, and loss of litter cover. After a certain period, soil, vegetation, and water flow patterns recover to pre-fire conditions. Soil properties are highly deteriorated due to fire and 1.3 Mg ha^{-1} soil loss has been observed after a fire broke out (Fonseca et al. 2017). Similar effects on soil properties were affirmed by Alcañiz et al. (2018). The severity of the untraditionally long dry period fits the pattern of increasing extreme weather Nepal has witnessed in recent years. Meteorologists are worried that the rapidly changing climate will bring an annually increasing number of disasters in the years ahead (Bhujel et al. 2017).

Much is known about the occurrence of fire-induced soil water repellence, the change in soil nutrient status, and changes in soil structure and infiltration, but the information is missing that is suitable for use in process-based hydrological and erosion models. Most studies on the effect of fire on soils and hydrology focus on the plot- or hill-slope scale, but the catchment scale is little studied. Also, hydrological modeling is not a common tool in forest fire research because of the large spatial heterogeneity and the occurrence of water repellence (Stoof 2008). Soil hydrological properties affected by fire mainly occur due to the clogging of pore spaces, reducing infiltration, and increasing runoff (Stoof et al. 2016).

10.3 Soil Fertility and Land Degradation

The fertility level of Nepalese soil is declining mainly because of multiple cropping, especially nutrient-exhausting crop cycles such as the rice–wheat-maize sequence that heavily mines soil. Traditionally farmers do not replenish mined nutrients (Ghani and Brown 1997; Karki 1995; Karki and Dacayo 1990; Karki and Joshi 1993), and soil erosion in the hills and mountains is a serious problem as large amounts of applied plant nutrients are washed away (Carson 1992), along with the major sedimentary loads taken annually. Soil erosion not only washes away the fertile soil but also the base cations, leaving acid-forming cations on the soil surface and causing higher soil acidity. The present use of fertilizer nutrients in Nepal is about 70 kg ha^{-1}, and although it is targeted to reach 131` kg ha^{-1} (Panta 2018), a good single maize crop harvest requires the planting of nutrients at more than 400 kg ha^{-1} (Ghani and Brown 1997). The nutrient balance in some intensively cultivated land (three crops per year) is given in Table 10.1. Moreover, farmers apply fertilizer (N: P: K) in a ratio of 1:0.37:0.03, respectively (Joshy 1997), indicating a higher use of nitrogenous fertilizer. As a result, Nepalese soils are extremely to highly acidic in nature, and low in OM N, P, and K, including CEC (Dawadi and Thapa 2015; Karki 2002).

Anthropogenic causes of the exploitation of forest resources, such as overgrazing and the excessive collection of fuelwood, fodder, and timber, are accelerating soil erosion and landslides particularly in mid-mountain watersheds and leading to overall watershed degradation (Wagley 1995). The country's ever-increasing population and allied infrastructural growth coupled with rapidly changing climatic conditions are exacerbating these conditions. The global assessment of soil degradation (GLASOD) attempted to map the severity of degradation on a world scale and showed that 34% of land in Nepal is affected by waters erosion (FAO 2000), leaving infertile soil in the fields.

Acid soil covers approximately 35% of the total geographical area in Nepal (Dawadi and Thapa 2015). The acidity is an inherent property of soils in the hills and mountain regions mainly due to higher rainfall and acid-forming parent materials. However, concerns over soil acidification have featured prominently in recent literature (Shah and Schreier 1995) because it has an adverse impact on the availability of phosphorus and impairs organic matter decomposition, in addition to increased leaching of base cations. The use of acid-forming chemical fertilizers in some pockets of the country has also resulted in increased soil acidity. Without addition through external sources, the loss of calcium and magnesium due to leaching, as well as surface run-off of soil particles coupled with the removal of calcium and magnesium by crop harvests, will cause soil acidity to increase (Tripathi 1999). The loss of major plant nutrients and organic matter is also an important factor leading to chemical degradation through the loss of productive topsoils and leaching through the soil profile. This type of chemical degradation is the result of low-input agricultural production systems as well; for example, in the hills and mountains, the use of chemical fertilizers on main crops is far below 70 kg ha^{-1} on average.

Table 10.1 Nutrient losses under different land-use categories in Nepal

Land type/land use	Nitrogen, kg ha^{-1} yr^{-1}	Phosphorus, kg ha^{-1} yr^{-1}	Potassium, kg ha^{-1} yr^{-1}
Sherchan and Gurung (1995)			
Maize/millet bari land	55.00	2.53	7.88
Sole maize bari land	53.67	3.09	7.05
Maskey and Joshy (1991)			
Maize + soybean	0.43–0.83	0.002–0.003	0.07–0.17
Rainfed bench terrace fallow	3.80	5.0.00	10.00
Rainfed marginal land	15.00	20.00	40.00
Grazing land degraded	75.00	100.00	200.00

Topsoil degradation and loss are possibly the most serious processes affecting the sustainability of farming systems in the hills and mountains of Nepal. Throughout the hills, particularly at lower elevations where rainfall intensity is highest, erosion is a major contributor to soil fertility decline. The estimated soil and nutrient loss for the major production systems of Nepal below 1,000 m elevation are given in Carson (1992), which notes that rainfed bench terraces lose 0.4 mm of soil in depth creating soil loss of 5.0 t ha^{-1} yr^{-1} while rainfed marginal land loses 1.0 mm yr^{-1} and degraded grazing land loses 8.0 mm yr^{-1}. Loss of soil nutrients from degraded grazing land totaled 100 t soil, with 150 kg of organic matter, 75 kg of N, 100 kg of P, and 200 kg of K each year from a hectare of land.

The results of some studies indicate that the erosion from cultivated land is negligible while major erosion occurs on marginal land, abandoned land, and grazing land. Different types of intercultural operations promote soil erosion. Sherchan and Gurung (1995) showed in the eastern hills of Nepal that nitrogen lost from a maize/millet cropping sequence can total 55 kg ha^{-1} yr^{-1}, though Maske and Joshy (1991) report that growing soybean instead of millet creates minimal nitrogen loss. The authors measured higher nitrogen losses from marginal lands and degraded rangelands, showing losses of 75 kg ha^{-1} yr^{-1}. In both experiments, the loss of P and K were lower in cultivated fields, whereas the highest losses (that is, 100 kg P and 200 kg K) occurred in degraded grazing land (Table 10.1). Erosion studies conducted in the hills indicate that rainfall and its intensity, along with soil, are major factors causing annual erosion and the resultant heavy soil and nutrient loss from sloping terraces (Bari land). Cited by Acharya et al. (2007), reported that the loss of major plant nutrients through leaching has been found to be significantly higher than losses through runoff from the hills.

When nitrogenous fertilizer was first applied to Nepalese 'virgin soil' there was a tremendous yield response. The application of phosphorus and potassium fertilizer did not make any difference. From that point on Nepalese farmers trusted nitrogenous fertilizer (ammonium sulfate, later replaced by urea) while ignoring the continuous need for the application of organic manure. After several years, this practice resulted in the virgin fertile soil becoming unproductive due to its increased acidity. The work of Karki and Dacayo (1990), Karki and Joshi (1993), and Karki (2003) showed that more than 63% of soil samples measured in the Nepal Agricultural Research Council and Department of Agriculture were found to be acidic to strongly acidic (pH 5 to 6) (Table 10.2). The present condition of soils in Nepalshows that their percentage of nitrogen, available phosphorus, and organic matter is low. Only the available potassium in most soil samples was found to be high, though not significantly. Micronutrients, mainly boron, molybdenum, and zinc, are low. In some sites, these elements have posed serious problems, and without the application of these elements, sensitive crops such as barley, citrus, legumes, and vegetable yields are very low (Karki 2006).

10.3.1 Wind Erosion and Desertification

Wind-caused erosion is a serious problem, particularly in the high Himalayan Mustang, Manag, and Dolpa districts where a cold and arid climate exists. There has been a lack of systematic studies to quantify the rate and size of the problem; however, based on the limited studies that have been carried out, wind erosion has been calculated to have permanently degraded 10,000 ha of land in these areas to date. Due to its serious moisture and temperature limitations, the resilience of this land is very weak. In the Tibetan Plateau and High Himalayan areas of Nepal evidence of desertification is visibly in progress. In addition, low-rainfall areas such as the Dolpa, Mustang, Surkhet, and Dang districts are also affected by wind erosion, and further desertification is evident and progressing (Dhinwa 2018).

Table 10.2 Soil nutrient status in Nepal. The percentage coverage indicates the percentage of the total cropping area

Soil pH, % coverage	Alkaline—10.5; Neutral—26.6; Acidic—63.0		
Soil nutrient content	High	Medium	Low
Soil organic matter, % coverage	7.5	33.9	59.4
Soil phosphorous, % coverage	48.2	23.9	27.9
Soil potassium, % coverage	39.0	29.4	31.5
Soil nitrogen, % coverage	Low		
Soil micronutrients (B, Mo, and Zn)	The majority are low but Cu, Mn, and Fe in the aerobic condition is not a problem		

Source Calculated from SST/DOA 2000, Karki et al. (2005)

10.3.2 Waterlogging and Floods

Waterlogged areas are caused by either poor drainage or terrain susceptible to flooding during the high peak period of the monsoon season. Also, marshy land is observed in the Tarai region, south of the Bhabar zone where two lithological units with different porosity and permeability meet along with the change in elevation, resulting in the creation of springs, ponds, and lakes. The area affected by waterlogging during the monsoon season is estimated at about 7297 km^2 and has been mainly used for rice cultivation. This figure was arrived at via a rough estimation based on the land system reports of the Land Resource Mapping Project at Tarai and the Siwalik physiographic region, with 4.8% and 0.2% of the total land, respectively, being waterlogged (LRMP 1986b).

Flooding, which is an overflow of natural or artificial river banks, cause and threaten damages to property and life. Intense rainstorms have been the main reason for flooding in the low-lying valleys and Tarai plains. There are several rivers originating from Churia (Siwalik) that flood downstream in Tarai during monsoon season. During the dry season, most of these rivers are dry and there is a limited amount of water. When there is heavy rain during the monsoon season they flood and inundate the fertile land. Such heavy flooding is due to their narrow river width as these rivers cross the Indian border. Taking advantage of this narrowness, India has dammed these rivers, thus inundating the border towns and large areas of cultivated land. The total area of flooding and inundated land is more than 8987 km^2 (Adhakari 2013). The most damaging rivers due to flood and inundation in the border towns of Nepal lie in the East Tarai District: the Biring and Ratuwa rivers in Jhapa, the Bakraha and Lohendra in Morang, the Sunsary in Sunsari, the Khando in Saptari, the Balan and Gagan in Sirha, the Rato and Jangha in Mahottari, the Jhim and Lakhandehi including the Bagmati in Sarlahi, the Lal Bakaiya in Rautahat, the Pasaha in Bara, the Rohini in Rupandehi, the Khutiya in Kailali, and the Dhondha in Kanchanpur. Similarly, the Laxmanpur Dam on the Rapti River creates havoc in Nepalgunj in the Bake district, and the Jaleswor of Mahottari and the Gaur of Rautahat experience flooding every year due to damming at the border.

Through extreme climatic events, the path of the Koshi River shifted over 110 km from east to west during the period from 1731 to 1963 in Bihar. Besides destroying about 7800 km^2. of land in Bihar and 1300 km^2 in Nepal through sand deposits (Gole and Chitale 1966), the Koshi River has wiped out towns and villages during its shifting course that displaced 6.5 million people (Reiger 1976). Enormous amounts of runoff and sediment carried by the Koshi are responsible for its shifting nature. Moreover, the Multipurpose Projects on the Koshi River and the Ganda River were intended to provide irrigation to 61,640 ha, along with the Gandak west canal was intended to irrigate 40,800 ha of land in Nepal; however, these projects never received these amounts of irrigation, with only 35,200 ha in the Koshi west in Sirha and 10,000 ha in Nawalparasi. However, the gate is opened by the Indian side and these areas are inundated during the monsoon season, flooding rice crops and remaining wet during the winter, where they hinder winter crop plantation (Dhungel and Adhikari 2009; Pradhan 2009). These inundations are caused not only by the damming of the rivers at the border but also by heavy sand and stone quarrying upstream (Shrestha 2013). There is no permanently waterlogged land except for the area near the lakes, which is negligible.

Koshi River Floods in Sunsari and Saptari Office for the Coordination of Humanitarian Affairs (OCHA) Situation Koshi Flood Report, August 2008

The Koshi River was fully flooded and broke its eastern embankment on 18 August 2008 and has subsequently changed its course. The Sunsari district that lies east of the Koshi river was severely affected. The flood displaced at least 70,000 people. Most of the displaced are poor farmers or land laborers, many of them Maithili speakers. In addition to the persons displaced from VDCs in Sunsari District, the number of flood-displaced persons from India has been reported to be considerable. The Saptari district that lies west of the Koshi River hosts most of the displaced people from Sunsari from the east and some from neighboring India where the flood extended beyond the Indian border and caused significant damage and human suffering.

The East-West highway remained impassable for several months, as it had been broken at three points at least and was covered by the river waters. The Koshi and Mechi Zones thus remained disconnected from Nepal's road network. It took over nine months to repair and make the road passable. An alternative route through Bathanaha-Birpur-Bahantabari in Bihar, India, was used that had also reportedly become impassable in places, requiring deviations to be put in place, thus considerably lengthening the route. The national telephone service (NTC) was restored by the 21st of August but remained intermittent for some months.

The media reported four to six people might have been killed. For those living, 27 different camps were organized to provide shelter for those flood victims that remained active for four to five months. Now, several government and non-governmental organizations are active in rehabilitating the affected area and returning the displaced people to their original areas.

The East-West Highway is made passable with temporary bridges over flood-cut road, and transport is made possible. A full repair of the East-West Highway and the breached eastern embankment of the Koshi River has not yet been completed. Now that the monsoon season has already started, more floods. Floods are expected this year and the problem will be further aggravated. It is announced that the water is returning to its pre-flooding course and levels. As the upgrading/repair of the Koshi River barrage and embankment system, built in the 1950s, had been an issue of intense debate for several years, the discussion about a lasting solution to the recurring flooding by the Koshi River in the area is likely to run in parallel to relief and rehabilitation efforts.

10.3.3 Chemical Degradation

Chemical degradation is a topic for which a systematic study is lacking. Acid soil covers approximately 60% of the total geographical area in Nepal (Sherchan and Gurung 1995), but Dawadi and Thapa (2015) reported that over 70% of cultivated land is acidic. The acidity is an inherent property of soils in the hills and mountain region mainly due to higher rainfall and acid-forming rocks. However, concerns over soil acidification have featured prominently in the literature (Shah and Schreier 1995) as it has an adverse impact on phosphorus availability for crops and impairs organic matter decomposition, including the leaching of base cations. The use of acid-forming chemical fertilizers in some pocket areas of the country has also been found to be increasing, causing acidity that has resulted in the loss of calcium and magnesium (Shah and Schreier 1995; Tripathi 1999) in associated soils. In addition, basic cations are also removed by crops. Soil acidification has been found to be a major contributing factor to the decline in soil fertility in Nepal.

10.3.4 Shifting Cultivation and Watershed Degradation

Shifting cultivation had been a sustainable agro-ecosystem in the past, but it cannot serve as a model for the future. Regeneration of forests is crucial for the long-term productivity and sustainability of slash and burn agro-ecosystems, but many farmers are no longer able to leave their fields fallow for the necessary period of time (Pratap and Watson 1994). Karki and Joshi (1993) reported that farmers have shortened fallow periods due to increased population pressure for food. This action has also reduced the productivity of agricultural lands. Due to the short period (two to three yrs) rotation in regions that adopt the traditional slash and burn farming technique some good trees are destroyed and palatable grass species have been made patently extinct, while invasive shrubs such as *Eupatorium* sp. have grown to dominate the area. Each year additional natural land is put under slash and burn, which is causing a shortage of fuel, fodder, and timber in the villages in the mid-hills of central Nepal. Another source of land degradation arises with shifts in cultivation, the extent and coverage of which is not clearly known. The regular shifting of cultivation is practiced in the mid-hills of Nepal in different regional intensities, serving as one cause of land degradation in many parts of the region. Soil erosion during the monsoon season due to high runoff is a serious problem, and this process has further been accelerated due to the practice of shifting and sloping terrace cultivation in the hills and mountains of Central Nepal.

For example, Balla et al. (2000) reported that in the Kali Khola watershed in the Chitwan district, farmers have cleared their slopes up to 81° for cultivation and the area under agriculture and shifting cultivation has increased by about 13% from 1978 to 1999. Likewise, in another watershed in Tanahun, the increase is about 18%. In both watersheds, the shift in land use is from forest and shrubland to agriculture. It was concluded that landslides were increasingly common and the degree of land degradation increased in both watersheds was serious.

The term "natural or environmental resource degradation" covers deforestation, watershed destruction, loss of biodiversity, loss of soil fertility, soil erosion, groundwater depletion, over-grazing, and fuel and fodder shortages (Jodha 1998). Specifically, land degradation is the temporary or permanent lowering of the productive capacity of the land. The UNEP has found that the adverse change in soil quality resulting from the decline in the productive capacity of the land is due to processes induced mainly by human intervention (UNEP 1992).

The ICIMOD (1986) identifies six types of forces active in watersheds, which influence the quality of land resources: natural forces, demographic pressures, livestock pressures, development pressures, techniques and management practices, and socio-economic pressures. These forces directly or indirectly influence natural resource management or mismanagement practices, including forest resources in the mountains. However, Sharma (1996) identified not only natural forces, but deforestation, cultivation of marginal and non-arable land, and inappropriate construction of infrastructure (i.e., dams, canals, roads, etc.) as major human-induced factors of hazards in Nepal. Scarcity of arable land for agricultural expansion, maintenance of many unproductive livestock that rely heavily on forest and rangelands, improper and inadequate farming practices, widespread deforestation to satisfy basic forest product needs and agricultural expansion, and frequent occurrence of fires in much of the natural forest have greatly altered the natural vegetation of Nepal, exposing the soil for erosion. All of these factors have accelerated the process of soil erosion and are characterized by the loss of topsoil by sheet and rill erosion.

10.3.5 Livestock Farming and Land Degradation

At present, there are more than 22,135,000 heads of livestock in Nepal. Due to the increased feed supply required for the animals, their population increase can become a major threat to land degradation in the absence of proper range and pasture management. Plowing of marginal lands for agriculture together with over-exploitation of vegetation for fodder and fuelwood has resulted in land degradation, severe erosion, and the drying of water sources. In some areas, the results are already being shown in families' low farming productivity and poor animal health. Farmers must spend more time collecting water, fuelwood, and fodder, as well as leaf litter for manure, all of which are essential components of rural subsistence agriculture. This resulting unbalanced environmental status has resulted in low socio-economic conditions for subsistence farmers and instability (Sthapit 1989).

However, the livestock sector is a major performer in the agricultural economy, and sustainable management is, therefore, an important aspect. The land available for raising livestock and feed availability according to these requirements is in a state of unbalance. Only 56% of cattle and buffalo are raised in stalls; the remaining livestock is left in public forests or on public lands for grazing (Karki 2006). These animals feed on their preferred species of grasses and fodder leaving non-preferred species to increasingly cover the watersheds. As per the 2017 data, there are 7.3 M cattle, 5.1 M buffalo, 0.8 M sheep, and 10.9 M goats in Nepal. Among them, only 1.0 M cattle and about 1.0 M buffalo are milking (Nirmal 2017), while the rest are unproductive animals. This shows that the number of unproductive cattle and buffalo is double that of the productive cattle, and it is the unproductive animals that are left to graze on public forest and degrade the land.

The current stocking density is exceeding the carrying capacity of the agricultural land, with the exception of the alpine meadows region. The situation is more alarming in the mid-hills and open grassland. The work of Karki and Dacayo (1990) shows that the availability of grazing land in the mountains is high (63%) for only 13% of the livestock, whereas the population of grazing animals is much higher (52%). For this higher number, the availability of grazing land is only 14%, while the Tarai has 34% of the country's livestock population but only 22% of the available grazing land (Table 10.3). The higher numbers of livestock depending on the smaller area indicates higher pressure on grazing, leaving the land overgrazed and placing it in the high degradation category.

Land availability for crop production, livestock husbandry, and other uses is vital. Human population pressure is leading to the fragmentation of agricultural lands, eventually increasing the number of marginal and small farmers throughout the hills and mountains (Pratap 1995). Moreover, in the present context, in many areas, the contribution of richer groups to resource degradation is greater than that of the poor, with the poorest districts having the lowest natural resource degradation (ICIMOD 1986, 1997). The carrying capacity of mid-hills is 0.31 livestock units, but the number of livestock units in the mid-hills is 4.08, meaning this livestock do not have enough feed and fodder. Similarly, the carrying capacity of the alpine meadow is 1.42 livestock unit ha^{-1}, whereas available grazing land is only 0.64 livestock unit ha^{-1} (Table 10.3). There is a clear relationship between livestock, soil fertility, and natural resources in the mountain farming system, and the improper management of this relationship results in land degradation.

There has been a general decline in the number of livestock particularly in the mid–hill area, where there is intense pressure and the decline per household is most prominent. This is due to the decreased size of holdings, reduced livestock feed resources, increasing population pressure, and deepening shortage of human labor. The reduced livestock number results in reduced farmyardmanure (FYM) since an

Table 10.3 Percentage of livestock population and grazing land availability in different ecological regions

Ecological region	Livestock population, %	Grazing land, %
Mountains	13.4	63.8
Hills	52.2	13.7
Tarai	34.4	22.5

Source Agriculture Statistic of Nepal, 2019

adult animal produces 1140 kg FYM yr^{-1}. Thus, in the high-pressure hill/mountain areas sustainable soil management is a critical issue due to deteriorating soil fertility and the consequent decline in crop productivity (Tulachan and Neupane 1999).

For hill farming systems to be sustainable primarily requires a net transfer of fertility from the forest through fodder and leaf-litter to stall-fed animals. The dung produced is mixed with leaf-litter to obtain the compost, which is transferred to the fields and generally comprises the only means available for maintaining an already low level of soil fertility. The declining availability of forest-produced fodder and leaf-litter means that nutrient levels and soil structure cannot be maintained, thus resulting in a decline in land productivity. Given the present rate of soil and nutrient loss from the hills, the productivity of hilly agricultural land will continue to decline even if forests are restored (Upadhya 1994). The declining land to man ratio is the most fundamental problem of development facing the hill region, and declining farm productivity resulting from this declining ratio is the root cause of the hill farmers' problems.

10.4 Conclusion

Soil erosion, nutrient mining, soil acidification, soil compaction, aggregate destabilization, floods, soil organic carbon depletion, and nutrient imbalance are land degradation issues faced by Nepalese soils. Of these issues, soil erosion (water and wind erosion) solely accounted for 49.5% of total land degradation. Steep topography coupled with unscientific and unplanned land use management practices accelerated soil degradation. The extent of degradation is severe in 12% of the land that requires urgent attention for restoration. Most of the forest area is under degradation, followed by range/pastureland and cropland. The vulnerability of forestland and range/pastureland to degradation is mainly due to a lack of scientific forest- and range/pastureland management accompanied by high grazing intensity. Similarly, agricultural land degradation is mainly caused by an imbalanced input to output ratio, minimal or no residue recycling, erosion-led nutrient mining, and a depleting level of soil organic carbon. In Tarai, more than 95% of agricultural land is under the critical limit of SOC (2%). Furthermore, soil compaction is an emerging issue in the soils of Tarai due to extensive tillage and the operation of heavy tillage equipment. As a result, we have highlighted the most attentive issues of land degradation with the goal of seeking policy enforcement and well-planned land management strategies at both the national and local scales.

References

Acharya AK, Kafle N (2009) Land degradation issues in Nepal and its management through agroforestry. J Agric Environ Earth Sci 10:133–143

Acharya G, McDonald M, Tripathi B, Gardner R, Mawdesley K (2007) Nutrient losses from rain-fed bench terraced cultivation systems in high rainfall areas of the mid-hills of Nepal. Land Degrad Dev 18:486–499

Acharya K, Dangi R, Acharya M (2011) Understanding forest degradation in Nepal. Unasylva 62:238

Adhakari BR (2013) Flooding and inundation in Nepal Tarai XE "Tarai": issues and concerns. Hydro Nepal J Water Energy 12:59–65

Adhikari J (2010) Food insecurity, conflict and livelihood threats in Nepal. In: Upreti BR, Müller-Böker U (eds) Livelihood insecurity social conflict in Nepal, pp 73–130. South Asia Regional Coordination Office, Swiss National Centre of Competence in Research (NCCR) North-South Kathmandu

Adhikari S, Basyal G, Chaudhary S, Adhikari H, Thapa I (2016) Disaster inventory and information management system in Nepal

Alcañiz M, Outeiro L, Francos M, Úbeda X (2018) Effects of prescribed fires on soil properties: a review. Sci Total Environ 613–614:944–957

Atreya K, Sharma S, Bajracharya RM, Rajbhandari NP (2008) Developing a sustainable agro-system for central Nepal using reduced tillage and straw mulching. J Environ Manage 88:547–555

Awasthi KD (2004) Land use change effects on soil degradation, carbon and nutrient stocks and greenhouse gas emission in mountain watersheds. Agric Univ Nor Oslo

Balla M, Awasthi K, Shrestha P, Sherchan D, Poudel D (2000) Degraded lands in mid-hills of central Nepal: a GIS appraisal in quantifying and planning for sustainable rehabilitation. In: LI-BIRD, Pokhara, Nepal. LI-BIRD, Pokhara Nepal

Balla MK (1983) Developing a scheme for application of the USLE in Nepal. UGA, Georgia, USA

Baral S, Basnyat B, Gauli K, Paudel A, Upadhyaya R, Timilsina YP, Vacik H (2019) Factors affecting fuelwood consumption and CO_2 emissions: an example from a community-managed forest of Nepal. Energies 12:4492

Barrow C (2010) How is environmental conflict addressed by SIA? Environ Impact Assess Rev 30:293–301

Basnyet SK (1989) Micro-level environmental management: observation on public and private responses in kakani panchayat. ICIMOD, Kathmandu

Bhatta BR, Banskota K, Giri D (2018) Household fuel choice in urban Nepal: a multinomial logistic regression analysis. J Bus Soc Sci Res 3:65–90

Bhujel KB, Maskey-Byanju R, Gautam AP (2017) Wildfire dynamics in Nepal from 2000–2016. Nepal J Environ Sci 5:1–8

Blaikie P, Brookfield H (2015) Land degradation XE "Land degradation" and society. Routledge

Blum W (1998) Soil degradation caused by industrialization and urbanization. Towar Sustain Land Use 755–766

Carson B (1985) Erosion and sedimentation process in the Nepalese Himalaya, Kathmandu

Carson B (1992) The land, the farmer and the future. In: ICIMOD discussion occasional paper no. 21. ICIMOD, Kathmandu

Carver M, Nakarmi G (1995) The effect of surface conditions on soil erosion and stream suspended sediments. In: Challenges in mountain resources management in Nepal, Kathmandu (HS Schreier, PB, edition). ICIMOD, Kathmandu

Chalise D, Kumar L, Kristiansen P (2019) Land degradation XE "Land degradation" by soil erosion in Nepal: a review. Soil Syst 3:12

Chalise SR, Khanal NR (1997) Erosion processes and their implications in sustainable management of watersheds in Nepal Himalayas, regional hydrology: concepts and models for sustainable water resource management. In: Proceedings of the Postojna, Slovenia conference, vol 246, IAHS. Chapman & Hall London and UNEP, pp 261

Chaudhary RP, Uprety Y, Rimal SK (2016) Deforestation in Nepal: causes, consequences and responses. Biological Environmental Hazards, Risks, Disasters, pp 335–372

Dahal RK, Hasegawa S (2008) Representative rainfall thresholds for landslides in the Nepal Himalaya XE "Nepal Himalaya." Geomorphology 100:429–443

Danish M, Gupta P, Alam J, Muzammil M (2013) River Kosi, sorrow of India: an overview. In: Proceedings of 2nd international conference on emerging trends in Engineering and technology, vol 12, pp 13

Dawadi D, Thapa M (2015) Soil fertility status of Nepal: report from laboratory analysis of soil samples from 5 development regions. In: Karki KB, Tripathi BP, Manandhar R, Adhikari BH, Vista SP (eds) Second national soil fertility research workshop. Soil Science Division, NARC, Kathmandu, Nepal, pp 42–52

Dhar O, Rakhecha P (1981) The effect of elevation on monsoon rainfall distribution in the central Himalayas. Monsoon Dyn 253–260

Dhinwa P (2018) Desertification and land degradation in Indian subcontinent: issues, present status and future challenges. In: Climate variability impacts on land use and livelihoods in Drylands. Springer, Berlin, pp 181–201

Dhungel DN, Adhikari BR (2009) Inundation at the southern border. In: The Nepal–India water relationship: challenges. Springer, Berlin, pp 269–293

Dingle EH, Sinclair HD, Attal M, Milodowski DT, Singh V (2016) Subsidence control on river morphology and grain size in the Ganga Plain. Am J Sci 316:778–812

Dixit A (1995) Resource endowment and associated uncertainty of water resources. Water development-Nepalese perspective. Konark Publishers Pvt. Ltd., Main Vikas Marg, Delhi, India

DSC (1994) Sedimentation survey of Phewa lake, Jan 1994. Department of soil conservation, Kathmandu

Dumanski J, Eswaran H, Latham M (1991) Proposal for an international framework for evaluating sustainable land management. In: Evaluation for sustainable land management in the developing world: proceedings of the international workshop on evaluation for sustainable land management in the developing world, Chiang Rai, Thailand. International Board for Soil Research and Management, Bangkok, Thailand, pp 15–21

FAO (1976) A framework for land evaluation. FAO, Rome

FAO (2000) Types of land degradation. In: Land degradation (FAO nations, edition). Rome, Italy

Fonseca F, de Figueiredo T, Nogueira C, Queirós A (2017) Effect of prescribed fire on soil properties and soil erosion in a Mediterranean mountain area. Geoderma 307:172–180

Fox J, Saksena S, Hurni K, Van Den Hoek J, Smith AC, Chhetri R, Sharma P (2019) Mapping XE "Mapping" and understanding changes in tree cover in Nepal: 1992 to 2016. J for Livelihood 18:1

Gardner RAM, Gerrard AJ (2002) Relationships between runoff and land degradation on non-cultivated land in the middle hills XE "Hills" of Nepal. Int J Sust Dev World 9:59–73

Gardner RAM, Gerrard AJ (2003) Runoff and soil erosion on cultivated rainfed terraces in the middle hills XE "Hills" of Nepal. Appl Geogr 23:23–45

Gautam DK, Pokherel A (2004) Extreme floods in Bagmati River basin. In: Hydroinformatics (in 2 volumes, with CD-ROM). World Scientific, pp 1800–1807

Gentle P, Thwaites R, Race D, Alexander K (2014) Differential impacts of climate change XE "Climate change" on communities in the middle hills region of Nepal. Nat Hazards 74:815–836

Ghani A, Brown MW (1997) Improvement of soil fertility in Nepal through balanced application of fertilizer. Land Management Group, AgResearch, Raukura Agricultural Research Centre, Hamilton, New Zealand

Gole CV, Chitale SV (1966) Inland delta building activity of Kosi River. J Hydraul Div 92:111–126

Gurung B, Nelson KC, Smith JL (2009) Impact of grazing restrictions on livestock composition and husbandry practices in Madi valley, Chitwan National Park. Nepal Environ Conserv 36:338–347

Hasegawa S, Dahal RK, Yamanaka M, Bhandary NP, Yatabe R, Inagaki H (2009) Causes of large-scale landslides in the lesser Himalaya of central Nepal. Environ Geol 57:1423–1434

ICIMOD (1986) Managing mountain watersheds. In: International workshop in watershed management in the Hindu Kush-Himalaya region. ICIMOD, Kathmandu, Chengdu, China

ICIMOD (1997) Districts of Nepal: indicators of development. ICIMOD, Kathmandu

Ishtiaque A, Shrestha M, Chhetri N (2017) Rapid urban growth in the Kathmandu Valley, Nepal: Monitoring land use XE "Land use" land cover XE "Land cover" dynamics of a himalayan city with landsat imageries. Environment 4:72

Jodha N (1998) Poverty and environmental resource degradation: an alternative explanation and possible solutions. Econ Polit Wkly 2384–2390

Johnson D, Lewis L (1995) Land degradation: creation and destruction. Oxford, UK: Blackwell Scientific Publishers, pp 335

Joshi D (1997) Soil fertility and fertilizer use in Nepal. Soil Science Division, Kathmandu

Joshi J, Bhattarai TN, Sthapit KM, Omura H (1998) Soil erosion XE "Soil erosion" and sediment disaster in Nepal–a review XE "Agriculture." J Fac Agric 42:491–502

Joshi M (1985) Country paper on the state of art in atershed management in Nepal. In: International workshop on watershed management in the Hindu-Kush Himalayan region, held at ICIMOD, Chengdu, China

Joshy D (1997) Indigenous technical knowledge in Nepal. In: Technical issues. International Board for Suastainable Land Management (IBSRAM), Bangkok Issues on Sustainable Land Management, pp 45–61

Kansakar SR, Hannah DM, Gerrard J, Rees G (2004) Spatial pattern in the precipitation XE "Precipitation" regime of Nepal. Int J Climatol J R Meteorol Soc 24:1645–1659

Karkee K (2004) Land degradation XE "Land degradation" in Nepal: A menace to economy and ecosystems. Lund University, Lund, Sweden

Karki K (1995) Effect of fertilizers and boron on yield and boron uptake in Wheat In: Rawson HM, Subedi KD (eds) Sterility in wheat insSubtropical asia: extent, causes and solutions. ACIAR, LumJe Agricultural Research Centre, Pokhara, Nepal, pp 39–43

Karki K (2003) Status of potassium XE "Potassium" in intensively cultivated soils of Kathmandu valley. Nepal J Sci Technol 5:83–89

Karki KB (1986) Assessment of land degradation and formulation of productivity improvement strategies in Southern Lalitpur district of Nepal. Partial Fulfillment for the degree Central Luzon State University, The Philippines

Karki KB (2002) Response of biogas effluent to maize and cabbage in Lalitpur district. In: Proceedings of the international seminars on mountains, Tech. S 6.5. RONAST, Kathmandu, pp 55–56

Karki KB (2006) Soil degradation in Nepal and some measures of rehabilitation. Agric Environ 6:32–36

Karki KB, Dacayo JB (1990) Assessment of land degradation in Southern Lalitpur of Nepal. In: 14th congress of international society of soil science. Internation Society of Soil Science, Kyoto, Japan

Karki KB, Joshi D (1993) Soil fertility and fertilizers use in Nepal. In: Soil fertility and fertilizer use in South Asia. T. HLS edition. Development and Consultation Organization, New Delhi, India
Karki KB, Mentler A, Blum WEH (2005) Crop productivity and food security in kathmandu Valley, Nepalese J Dev and Rural Stud, 2:14–22
Kayastha GP (1998) Reservoir sedimentation, its prediction and mitigation measures, DPTC, Lalitpur. In: Lecture notes, training on disaster prevention
Kayastha P, Dhital MR, De Smedt F (2013) Application of the analytical hierarchy process (AHP) for landslide susceptibility mapping: a case study from the Tinau watershed, West Nepal. Comput Geosci 52:398–408
Kayombo B, Lal R (1994) Chapter 13—responses of tropical crops to soil compaction. In: Soane BD, van Ouwerkerk C (eds) Developments in agricultural engineering, vol 11. Elsevier, pp 287–316
Laban P (1978) Field measurements on erosion and sedimentation in Nepal. IWM/WP/05 (cited by LRMP, 1986), Department of Soil Conservation and Watershed Management. FAO/UNDP
Lal R (2001) Soil degradation by erosion. Land Degrad Dev 12:519–539
Lamsal A (2014) Major landslides in Nepal. Kathmandu University, Kathmandu
Li T (1990) Landslide management in the mountain area of China. In: Occasion paper, vol 15. ICIMOD, Kathmandu, pp 50
LRMP (1986b) Land systems report: the soil landscapes of Nepal. Land Resources Mapping Project, HMG/N, Kathmandu, Nepal
Maharjan B, Tamrakar NK (2011) Deterioration of the Nakhu River: impairment on dynamic and recreational functions of the river. J Nepal Geol Soc 43:225–240
Maskey R, Joshy D (1991) Soil and nutrition loses under different soil management practices in the middle mountains of Central Nepal. In: The workshop proceedings of soil fertility and erosion issues in the middle mountains of Nepal held in Kathmandu. Sponsored by IDRC, Canada
Mawdesley KJ, Tripathi BP, Gardner RAM (1998) Quantitative indicators of soil erosion on upland bariland in the mid-hills of Nepal. In: LARC seminar paper. Lomle, Kaski, Nepal
Merz J, Dangol P, Dhakal M, Dongol B, Nakarmi G, Weingartner R (2006) Road construction impacts on stream suspended sediment loads in a nested catchment system in Nepal. Land Degrad Dev 17:343–351
MOPE (2000) Nepal's state of the environment Nepal. Ministry of Population and Environment (MOPE), HMG
MOPE (2017) National population report. Kathmandu
Mulmi AD (2009) Green road approach in rural road construction for the sustainable development of Nepal. J Sustain Dev 2:149–165
Nepal S, Flügel W-A, Shrestha AB (2014) Upstream-downstream linkages of hydrological processes in the Himalayan region. Ecol Process 3:19
Nirmal B (2017) Strategy to enhance the production and productivity of buffaloes for the growth of Livestock Gross Domestic Product (LGDP) of Nepal. In: Livestock sector improvement program. Department of Livestock Services, Kathmandu
Panta H (2018) Supply chain of subsidized chemical fertilizers in Nepal. J Inst Agric Anim Sci 35:9–20
Paudel GS, Thapa GB (2004) Impact of social, institutional and ecological factors on land management practices in mountain watersheds of Nepal. Appl Geogr 24:35–55
Pimentel D, Whitecraft M, Scott ZR, Zhao L, Satkiewicz P, Scott TJ, Phillips J, Szimak D, Singh G, Gonzalez DO (2010) Will limited land, water, and energy control human population numbers in the future? Hum Ecol 38:599–611
Poudel S, Shaw R (2015) Demographic changes, economic changes and livelihood changes in the HKH region. Mountain hazards and disaster risk reduction. Springer, Kathmandu, pp 105–123
Poudyal R, Loskot P, Nepal R, Parajuli R, Khadka SK (2019) Mitigating the current energy crisis in Nepal with renewable energy sources. Renew Sustain Energy Rev 116:109388
Pradhan BK (2009) Personal reflections: Nepal-India water relations. In: The Nepal–India water relationship: challenges. Springer, Berlin, pp 243–267
Pratap T (1995) High value cash crops in mountain farming: mountain development processes and opportunities. ICIMOD, Kathmandu
Pratap T, Watson H (1994) Sloping agriculture land technology (SALT). In: ICIMOD occassional paper 23. ICIMOD, Kathmandu
Puri L, Nuberg I, Ostendorf B (2017) The adequacy of community forest operational plans for estimating fuelwood supply and consumption in Nepal. Aust for 80:317–327
Raut N, Sitaula BK, Bajracharya RM (2010) Agricultural intensification: linking with livelihood improvement and environmental degradation in mid-hills of Nepal. J Agric Environ 11:83–94
Regmi B (2001) Inventory of landslides: past and present trend in Kanthe Khola sub-watershed in Baglung district, F Pokhara. NRMP, SAP/IO, Pokhara
Reiger HC (1976) Floods and droughts: the Himalayas and the Ganges plain as an ecological system. Tribhuvan University Press
Schreier H, Shah P, Nakarmi G, Shrestha B (2000) Prevention, degradation and rehabilitation: a new dilemma in Himalayan watersheds. In: Environmental reconstruction in headwater areas. Springer, Berlin, pp 147–159
Shah P, Schreier H (1995) Maintaining soil fertility in agriculture and forestry. In: Schreier HH, Shah PB, Brown S (eds) Challenges in mountain resources management in Nepal: processes, trends and dynamics in middle mountain. ICIMOD, Kathmandu
Sharma BP (1996) Sustainable management of soils in rainfed cropping systems in the mid-hills of Nepal: Soil and Water Management, APROSC/SDC, Kathmandu, Nepal
Sherchan DP, Gurung GB (1995) Soil and nutrient losses under different crop husbandry practices in the hills of East Nepal. In: Soil fertility and erosion issues in the middle mountains Nepal. ISS Kathmandu, Kathmandu
Shrestha AB, Wake CP, Dibb JE, Mayewski PA, Whitlow SI, Carmichael GR, Ferm M (2000b) Seasonal variations in aerosol concentrations and compositions in the Nepal Himalaya XE "Nepal Himalaya." Atmos Environ 34:3349–3363
Shrestha BR (2019) An assessment of disaster loss and damage in Nepal. Geogr Base 6:49–57
Shrestha CL (1985) River sedimentation and remedial measures in Nepal. Water Int 10:14–17
Shrestha D, Zinck J, Van Ranst EJC (2004) Modelling land degradation in the Nepalese Himalaya 57:135-156
Shrestha H (2013) Sand mining as a flood hazard mitigation measure in Nepal. Int J Landslide Environ 1:87–88
Shrestha HR (2009) Harmonizing rural road development with mountain environment: green roads in Nepal. SCAEF, Nepal
Shrestha S (2000) Sediment handling facilities for the middle Marsyangdi hydroelectric project. In: Proceedings of the first international conference silting problems in hydro power plants. CRC Press, New Delhi, India
Sombai IG, Karakatsoulis J, Gardner W, Gautam AP, Sharma SP, Adhikari B (2018) Forest governance in Nepal: rationale for centralised forest and wildfire management. J Manag Dev Stud 28:16–35
SSD (1998) Annul report 1998. Soil Science Division (SSD), Khumaltar, Lalitpur Nepal
Sthapit KM (1989) Soil conservation status in Nepal, Kathmandu. Department of Soil Conservation and Watershed Management, Watershed Management and Conservation Education. Ministry of Forests, Kathmandu (Nepal)

Sthapit KM, Balla MK (1998) Review of runoff and soil loss studies: A teching material. Institute of Forestry (IOF), Pokhara, Nepal

Stoof C (2008) Fire effects on soil water movement. Land Degradation and Development, Group Wageningen, The Netherlands

Stoof CR, Gevaert AI, Baver C, Hassanpour B, Morales VL, Zhang W, Martin D, Giri SK, Steenhuis TS (2016) Can pore-clogging by ash explain post-fire runoff? Int J Wildland Fire 25:294–305

Thapa PB (2015) Occurrence of landslides in Nepal and their mitigation options. J Nepal Geol Soc 49:17–28

Timah EA, Ajaga N, Tita DF, Ntonga LM, Bongsiysi IB (2008) Demographic pressure and natural resources conservation. Ecol Econ 64:475–483

Tiwari KR, Sitaula BK, Bajracharya RM, Børresen T (2010) Effects of soil and crop management practices on yields, income and nutrients losses from upland farming systems in the middle mountains XE "Mountains" region of Nepal. Nutr Cycl Agroecosyst 86:241–253

Tripathi BP (1999) Review on acid soil and its management in Nepal. In: Lumle seminar paper. Lumle Agriculture Center, Lumle Pokhara Nepal

Tulachan PM, Neupane A (1999) Livestock in mixed farming systems of the Hindu Kush-Himalayas: trends in sustainability. ICIMOD, Kathmandu, Nepal

UNDP (2009) Nepal distaster report 2009:the hazards and vulnarebility, vol 2020. Ministry of Home Affiar GON Nepal, Kathmandu, Nepal

UNEP (1992) World atlas of desertification. London

UNFPA (2017) Population situation analysis of Nepal. United Nation Fund for Population Activity

Upadhya M (1994) Restoring mountainous region: a strategy at the local level. BWMP, Kathmandu

Wagley MP (1995) Status of watershed management in Nepal. PWMTA, Wallingford, pp 1–99–118

Wagley MP, Poudel A, Shakya PB, Gautam N (2014) Assessing climate change impacts and vulnerabilities and identifying adaptatation of options/measures in selected landscapes within the Tarai Arc landscapes. NARMA Consultancy Private Limited, Kathmandu, WWF Nepal

Wang Y, Wu N, Kunze C, Long R, Perlik M (2019) Drivers of change to mountain sustainability in the Hindu Kush Himalaya. In: The Hindu Kush Himalaya assessment. Springer, Berlin, pp 17–56

Zdruli P, Lal R, Cherlet M, Kapur S (2017) New world atlas of desertification and issues of carbon sequestration, organic carbon stocks, nutrient depletion and implications for food security. In: Carbon management, technologies, and trends in mediterranean ecosystems. Springer, Berlin, pp 13–25

Zhang J-Q, Liu R-K, Deng W, Khanal NR, Gurung DR, Murthy MSR, Wahid S (2016) Characteristics of landslide in Koshi River Basin, Central Himalaya. J Mt Sci 13:1711–1722

Soils and Humans

11

Dinesh Panday, Shree Prasad Vista, Dikshit Poudel, Arjun Chhetri, and Bharat Sharma Acharya

Abstract

Soils have been vital to humankind since time immemorial. They provide food, fiber, energy, and nutritional security. In Nepal, soils are intricately linked to cultures and traditions. However, in the recent decades, soil degradation and food shortages are also evident due to marginalization of agriculture, land use change, and out-migration of rural people. In this chapter, we explored different soil functions, and trends and repercussions of land cover change. We also outlined different soil and water quality issues resulting from heavy metals, organic pollutants, and pathogens, and provided brief recommendations to guide and improve soil health and quality.

Keywords

Culture • Erosion • Human nutrient • Land cover • Soil health • Urbanization

11.1 Soils in Culture and Traditions

Soils have been vital to humankind in several ways since time immemorial. Human existence and well-being rests on healthy soils. Vedic Sanskrit Scriptures (1500 BC) also noted "Upon this handful of soil our survival depends. Husband it and it will grow food, our fuel, and our shelter and surround us with beauty. Abuse it and the soil will collapse and die, taking humanity with it" (Oliver and Gregory 2015). This undeniably highlights the interdependence of soils and human society since ancient civilization, and function of soils as basis of all life on Earth. Indeed, soils support life through multiple assets, functions, and services to humankind. In exploring the inter-linkages between soils and human, this article brings examples from Asia, particularly Nepal.

In Nepal, soils are largely referred to as "mother" and represent one of the five life supporting elements of nature (Panchatatwa) in Hindu Mythology. The other four elements are "*Akash*" (space or sky), "*Vayu*" (air), "*Jal*" (water), and "*Agni*" (fire). Soils are intricately linked to culture and traditions. For instance, red soils of Nepal are generally regarded best and most suitable for plastering or painting earthen walls and floors in mud homes. Earthen plaster adds to the attraction and protects houses from insects, rain splash, and erosion. This is one of the oldest soil conservation technologies practiced in Nepal. Soil's use is also different among various ethnic and cultural groups. However, most ethnic groups in Nepal use soils in child naming ceremony, birthday, and wedding celebration, and other spiritual conventions. "*Ashadh* 15" (Ashadh or Asar is the third month in Nepali calendar that corresponds to the end of June) is celebrated as the Ropai Festival, rice-planting day, which enjoys soil and human relationships. Further, soils are being used for medicines in healing wounds and treating diseases, and in spas and beauty therapies. Thus, soil-based cultural references embark upon multiple roles of soils and humans in all ancient and modern societies.

11.2 Role of Soils

11.2.1 Food Production and Food Security

Soils and humans are related by way of numerous soil physical, biological, and chemical properties and processes.

D. Panday (✉) · A. Chhetri
Department of Biosystems Engineering and Soil Science, The University of Tennessee-Knoxville, Knoxville, TN, USA

S. P. Vista
Nepal Agricultural Research Council, Kathmandu, Nepal

D. Poudel
Department of Agricultural and Applied Economics, University of Georgia, Athens, GA, USA

B. S. Acharya
Department of Mines, Oklahoma City, OK, USA

R. B. Ojha and D. Panday (eds.), *The Soils of Nepal*, World Soils Book Series,
https://doi.org/10.1007/978-3-030-80999-7_11

The basic process of growing crops in soils and producing consumable food starts with photosynthesis. Food production inputs comprise of soil, fertilizer, seed, labor, equipment, and energy. From an economic standpoint, soils have the potential to control all these variables. For example, a superior seed or the finest labor is worthless, if soils have lower quality.

Soils are the major determinant of food security and nutrition. Food production and consumption in Nepal is, however, among the lowest in the world. As a result, the country suffers from food and nutritional insecurity. Further, the uneven distribution of produced and imported food products has increased hunger and malnutrition. Land degradation and food shortages are observed in many geographical areas like hills and mountains due to marginalization of agriculture and out-migration of rural people. The effects of recent COVID-19 pandemic are yet to be observed in the food security of the many marginalized groups.

Soil conservation and management is crucial to improve agricultural production for food and nutritional security and enhance soil health and quality. Generally, soil management improves soil quality by ameliorating soil's physical, chemical, and biological properties, which increases yield and farm profitability. Food and Agriculture Organization (2015) reported that sustainable soil management could increase agricultural yield by up to 58%. Such management involves crop residue management, crop diversification, reduced tillage, irrigation management, nutrient management, and precision farming, and effects may vary with soil type, climate, vegetation, and technology available, among others. For example, sandy soils are light, warm, and dry, and generally lower in nutrient content. Thus, nutrient and drainage management in sandy soils, such as in far-eastern part of Nepal can be utilized for melons and coconuts production. Similarly, irrigation management in clay soils in Terai is better suited for rice production.

11.2.2 Human Nutrition Supply

Soils are an important source of nutrients in food and medicines. Soil affects human health directly and indirectly —directly through ingestion, inhalation, and absorption of soil constituents, and indirectly through the quantity and quality of food (Steffan et al. 2018). Passage of toxic substances or pathogenic organisms through the human food chain can be detrimental to human health (Brevik and Burgess 2013). Intuitively, human health and soil health depend upon each other. Humans' diet and nutrition are derived hugely from soils and in turn, humans nurture soils (Brevik et al. 2017; El-Ramady et al. 2019). Overall, human health depends on healthy food. Therefore, improving soil quality is central to the challenge of reducing hunger and increasing food provision and nutritional security.

11.3 Land Cover and Its Change

Land Resource Mapping Project in Nepal classified Nepal into five physiographic regions: Terai plain (14% of total land cover), Siwalik low hill (13%), Middle Mountain (30%), High Mountain (20%), and High Himal (23%) (Paneru 2013). However, Nepal has three major ecological belts; Mountain, Hill, and Terai, all of which extend from east to west. Mountain, Hill, and Terai covers 5,151,300 (35%), 6,181,560 (42%), and 3,385,140 (23%) ha, of total land area in Nepal, respectively. Of the total land area, cultivated and uncultivated arable land covers 3,091,000 (21%) and 1,030,000 ha (7%), respectively. Forest, shrubland, grassland and pasture, and water covers 4,268,000 ha (29.1%), 1,560,000 ha (10.6%), 1,766 ha (12.7%), and 383 ha (2.6%), respectively (MoALD 2020). Each of the ecological belts has different climate (details in Chap. 3) and soil types with significant variations in land cover and cultivation practices (MoALD 2020). Moreover, microclimatic regions are prevalent within each belt, which contribute to the rich biodiversity (MoALD 2020).

Fourteen soil groups within four soil orders (Entisols, Inceptisols, Mollisols, and Alfisols) are dominant in Nepal (Paneru 2013); however, soil orders such as Spodosols, Histosols, Ultisols, and Aridosols are also occasionally found. Mollisols are the dominant soil group in the southern sal (*Shorea robusta*) forest (also known as broad leaf forest) and south western grassland. Alfisols, commonly present in the sloping lands of mountain regions, are formed by the subsurface accumulation of silicate clays. Within the Great groups of Alfisols, Rhodustalf are found in the western part of Nepal, and Haplustalfs and Hapludalfs in the Godavari area of the Kathmandu Valley. Ultisols are present in the Tars of central and eastern parts of Nepal as well as in the Jhikhu Khola (east of Kathmandu valley). Aridisols are dominant in the arid regions of Nepal, such as Mustang district, where annual precipitation is very less, i.e., around 300 mm.

Land Use and Land Cover (LULC) of Nepal is changing overtime. Various anthropogenic activities together with climatic factors have influenced LULC changes (Paudel et al. 2016), eventually leading to alterations in soil properties. Rural migration to cities and urban areas is increasing, which has resulted in farmlands of hills and mountains being abandoned (Chaudhary et al. 2020). Similarly, many forest lands have been converted to croplands, and croplands to urban lands in the last 30 years (1978/79–2010). Such land use changes along with poor soil management increase risks of soil erosion and degrade soil quality. In the following section, we discuss different land covers and their change over time.

11.3.1 Croplands

In Nepal, the highest cultivated land recorded was 59.27% in 2001/2002, but recently this has declined up to 21% (Chaudhary et al. 2020). Generally, rice, wheat, corn, millet, and barley are the major cereal crops grown and consumed as staple food in Nepal. Tarai is considered as the grain basket of Nepal where rice, wheat, and maize are primarily grown. Importantly, around 48% of total cultivable land in Tarai is suitable for growing rice (Bhandari et al. 2015). Besides cereal crops, fruits like mango (*Mangifera indica*), banana (*Musa paradisica*), pineapple (*Ananans comosus*), and litchi (*Litchi chinensis*) are grown in this region. The diverse cropping system, rice–wheat, maize–mustard, and rice–fallow rotation is followed in recent alluvial plain landforms in Terai where the soils range from fine loamy to loamy texture (Dijkshoorn and Huting 2009).

Likewise, Middle Mountain is suitable for cereals such as corn, wheat, and millet, vegetables, and fruits. In this region, irrigated terrace cultivation (rice–wheat), non-irrigated terraced cultivation (maize–mustard), rice or wheat, maize or mustard are commonly practiced. Irrigated and non-irrigated terrace cultivation is done in the moderately to steeply sloping mountainous terrain with loamy skeletal loamy textured soil. However, rice or wheat cultivation is common in alluvial plains and fans with fragmental sandy textured soil. Maize or mustard is common in tars with loamy textured soil. In high Mountain, barley and millet are the major cereal crops; potato is a major vegetable, and apple and walnut are major fruits. In this region, irrigated rice is cultivated in alluvial plains rich in loamy textured soil. Similarly, irrigated maize–wheat in high Mountain is found in past-glaciated mountainous terrain rich in loamy skeletal soil.

Similar to Terai, in the Hills, maize–mustard, rice–fallow, and rice–wheat cropping system is followed. Rice–wheat crop rotation is practiced in active alluvial lands whereas rice–fallow rotation in the depositional basins. Maize–mustard is cultivated either in depositional basins or in fans, aprons, and ancient river terrains.

In Nepal, variations in climate, soils, and altitude have created numerous productive ecological niches and provided opportunities to feed the country throughout the year. However, urbanization, poor soil management, limited technologies, and incoherent links between research and extension in soil science have contributed to decline in agricultural productivity. Cultivable lands in many urban and semi-urban areas exhibit rapid and unplanned settlements, constructions, and loss of agricultural lands. Best examples of urbanization are Pokhara and Kathmandu in middle Mountain, and Butwal, Bharatpur, Biratnagar in Terai. Ishtiaque et al. (2017) reported that urban cover in Kathmandu has increased by 412% in past three decades, and cultivable land has decreased by 31.3%. If this rate of land transformation continues, no cultivable land will likely be left to feed future generations. Similarly, poor soil management has degraded soil quality in Nepal. Heavy tillage operation coupled with imbalanced chemical fertilizer application is widely reported across the country, which impairs soil quality. Gami et al. (2009) reported significantly higher soil organic carbon (SOC) in a soil treated with farm-yard manure (FYM) compared with soils treated with NPK and control plots under rice–wheat rotation in Terai. This research suggests that the soil of Terai districts, where livestock is an integral part of cropping system, have maintained SOC, and therefore soil quality. As such, minimal tillage, residue management, and integrated nutrient management are important to maintain and enhance soil health (Tiwari et al. 2009).

Soil erosion is one of the most serious agricultural and water quality problems in Nepal. Steep slopes in the middle Mountains and high Mountains of Nepal together with concentrated rainfall (80% of annual precipitation) during monsoon season transport fertile soils to downstream. Human activities such as conversion of marginal land to agricultural land, intensive cultivation practices with heavy tillage operation in sloppy land, and infrastructure development such as roads construction are among the many factors increasing erosion risks, and other soil quality issues in Nepal. This calls for sustainable soil management practices to enhance soil health and quality. For example, in middle and high Mountain of Nepal, terrace farming, strip farming, and cover crop could be implemented to reduce erosion risks and manage/sustain important soil nutrients. However, more studies on crop rotation, fertilizer management (farmyard manure, chemical fertilizer, bio-fertilizer, organic fertilizer), and irrigation management are necessary to better understand and improve crop production, soil health, and sustainability.

11.3.2 Forest Lands

Nepalese people rely on forest for timber and non-timber forest products such as fodder, litters, and bedding materials. Besides, forest helps to maintain natural environment, and conserve biodiversity, soil, and watersheds. Mostly, those forests are managed as community forest, collaborative forest, leasehold forest, religious forest, protected forest, and government managed forest (Baral 2011). With the amendment of the Forest Act 1977, Panchayat Forest (PF) and Panchayat Protected Forest (PPF) came into existence, which largely fueled the community forestry programs and policies (Kanel et al. 2006).

According to the Forest Act (1993), the local community forest users' group (CFUGs) with their active participation utilize, manage, and protect the forest products (Kanel et al. 2006). Community's participation has helped to strengthen local institutions in managing forest for biodiversity improvement, hazard management, and forest products harvesting (Acharya 2004). These forests further play a major role within the local hydrological cycles and in mitigating the global climate change through sequestering atmospheric carbon into the soils.

The forest cover in Nepal has changed overtime; forest cover was 45.3% in 1990, 41.6% in 2000, and 42.1% in 2010 (Uddin et al. 2018). Khair, sisso forest, dry sal forest, mixed sal and bel forest, and tropical mixed hard wood forest is the characteristic of Tarai forest (Dijkshoorn and Huting 2009). Khair, sisso forest are present in active alluvial deposited loamy textured soil. Dry sal forest, sal plusbel forest, and tropical hard wood forest are found in alluvial fan apron complex of landform, where dry sal and tropical mixed hard wood forest are present and sal plusbel forest in dominant soil. Sal forest in Siwaliks is present in fans, aprons, and ancient river terraces landforms with loamy textured dominant soil. Degraded pine forest and degraded sal forest are the characteristics of middle Mountain forest. Both type of forests lie in steeply to very steeply sloping mountainous terrain with loamy skeletal textured dominant soil. Likewise, in high Mountains, blue pine forest and abies forest in past-glaciated mountainous terrain are the dominant forest types in loamy skeletal textured soil (Carson et al. 1986). Thus, different forests and functional types have soil anchoring functions and help in reducing runoff, erosion, and sediment export.

A study conducted in Likhu khola of middle Mountains in Nepal under different land cover: grass, sal (*Shorea robusta*) forest, and mixed broadleaf forest, and soil types: loams, silty loams, and silty clay loams demonstrated least runoff from grass cover followed by leaf litter-covered sal forest, and highest runoff from degraded forest (Gardner and Gerrard 2002). Results also highlight the importance of sal forest or broadleaf forest cover in middle Mountains in reducing runoff and erosion risks by improving soil properties like aggregate stability and SOC content. Similar study in Dhading district also attributed runoff to differential land cover and land use (Tiwari et al. 2009). Thus, forest type together with land topography and climatic factors such as precipitation determine soil's physical, chemical, and biological properties, and ecohydrological processes. Broad leaf forests with high litter covers have higher SOC than the degraded forest and needle leaf forests. Similarly, steep slopes in middle and high Mountain accentuates runoff, thereby erosion and sediment export, which in due course results in soil acidity, sedimentation, and pollution of water bodies.

11.3.3 Grazing Lands

In 1986, grasslands and pastures covered 11.81% of total land cover in Nepal (LRMP 1986). Small fluctuation of grasslands and pastures cover was observed in 2019, where total grasslands and pastures covered 1,766,000 ha (11.99%) area (MoALD 2020). Shrublands and pasture in Nepal are mostly found in high Himalayas (Uddin et al. 2018); however, grasslands are distributed in Tarai, Siwalik, and High Mountain regions. Soils in grassland of Tarai and Siwalik are characterized by active alluvial deposition with sandy texture. Tarai grasslands are also prone to severe river flooding and wind erosions (Dijkshoorn and Huting 2009).

In high Mountains, grazing lands are characterized by past-glaciated mountainous terrain with soil dominant in loamy skeletal texture. Grazing lands are important part of farming system in Nepal, which are characterized by the integration of crop and livestock. The byproducts of crop and livestock systems benefit each other. Livestock are used for draft, milk, meat, and manure. Manure from them act as the fertilizer source to grow crops, while livestock browse crop residue and helps in nutrient cycling. Typically, people from high Mountain depend on livestock and thus on grazing land for their livelihood. However, these lands are vulnerable to degradation and soil loss. Heavy stocking density, increased grazing intensity, and longer grazing durations could potentially reduce soil cover, alter soil properties, and thus trigger soil degradation.

11.3.4 Wetlands

Wetlands are perennial water bodies that originate from underground sources of water or rainfall. They refer to swampy areas with either flowing or stagnant fresh or salt water. They also include marshy lands, riverine floodplains, lakes, ponds, water storage areas, and agricultural lands. Wetlands are either natural or manmade, and permanent or temporary (MoFE 2018).

Nepal has a total of 29 freshwater wetlands, where 19 are natural and 10 are manmade inland wetlands. Wetlands in Nepal extends from low land Tarai to glacial lakes in high Himalayas (Siwakoti and Karki 2009), occupying almost 5% of total land area in Nepal (MoFE 2018). Wetland provides a wide range of ecosystem services such as food, fuel, fodder, domestic water supply, non-timber forest products, nutrient retention, flood control, groundwater recharge, C sequestration, biodiversity, and recreational values (Sharma et al. 2015). Consequently, they support livelihood of many subsistence farmers and large ethnic communities (Siwakoti and Karki 2009).

Wetlands in Nepal are also classified based on their proximity to protected areas: (a) wetlands within protected areas and/or buffer zones, (ii) wetlands outside protected areas, and (iii) wetlands in private lands (NLC 2020). Soils in such wetlands vary spatially but are largely hydric permitting storage of water and buffering landscapes against extreme climatic events and variability such as flood and drought. Ironically, many wetlands in Nepal are degraded, converted, or lost to human disturbance and interventions; thereby changing soil properties and ecological processes, and jeopardizing valuable ecosystem functions and services. Moreover, sediments deposition in wetlands is a great threat to wetland loss in Tarai.

Human disturbance in wetlands can be direct or indirect. Vegetation removal, stream channelization, construction of dams, and drainage of wetlands for agriculture, urban, and rural development are few examples of direct disturbances reported in Nepal. Disturbances have also occurred indirectly through discharge of municipal and industrial wastes. One or many such activities reportedly change soil physical, chemical, and biological properties, and biogeochemical cycles; thereby increasing overland flow, flooding, and greenhouse gaseous emission. More recently, cultivation of wetlands and conversion to rice farms and fishponds are reported in lowlands of Nepal (Siwakoti and Tiwari 2007), which may result in loss of biodiversity and alteration of soil C cycling depending on cultivation age and type, hydrology, and antecedent soil OM (Xu et al. 2019). An improved understanding of soil dynamics and biogeochemical cycles in wetland will serve valuable to conserving existing wetlands and to restoring degraded ones.

11.4 Soil Pollution

11.4.1 Heavy Metals

Heavy metals constitute an ill-defined group of inorganic chemical hazards, and those most found at contaminated sites include lead (Pb), chromium (Cr), arsenic (As), zinc (Zn), cadmium (Cd), copper (Cu), mercury (Hg), and nickel (Ni) (Adriano 2001). There is an increasing concern towards human health hazards due to heavy metals in soils entering human food cycles. Fertilizers containing heavy metals as impurities are the primary sources of heavy metals, and straight fertilizers are reported to add more metals compared to mixed fertilizers. Use of agricultural pesticides is also common, which aggravate heavy metals contamination and associated soil microbial processes; thereby affecting human health and well-being.

Multiple studies have investigated heavy metals accumulation in soil and plants and their entry into food chains. For example, Shrestha (2003) observed an increasing accumulation of Cd in leafy vegetables compared to other crops. Most of the areas within the riverbanks are found to be heavily contaminated with Cd compared to agricultural lands. Yet, other metals in prime agricultural lands are common due to increasing use of industrial wastes (Khanal et al. 2014). The effluents of battery industries, leather factories, and dye factories are directly dumped into the river system of urban areas in Nepal, which results in Cd accumulation in the riverbank soils. Similarly, the addition of sewage water and sludge has increased a risk for accumulation of Zn, Cu, Pb, Cr, and Cd in both plants and soils (Kayastha 2014; Khanal et al. 2014). Compared to Cd and Pb, As contamination of soil and groundwater is of serious concern, especially in Tarai (20 districts) and Hilly (4 districts) regions in Nepal (FAO 2004). The widespread use of As-contaminated groundwater in irrigation for a prolonged period could elevate its concentration in surface soil and eventually enter into crops, for example, in rice straw and grain. A study carried out by Shrestha et al. (2017) reported that the level of As concentration in groundwater of southern district of Nepal, Nawalparasi, was higher than the national (0.05 mg L^{-1}) and WHO (0.01 mg L^{-1}) standards. Overall, heavy metals accumulation in soils through metals-contaminated irrigation-water is increasing in different parts of the country; affecting soil and water quality, which deserves considerable attention.

11.4.2 Organic Chemicals

The application of numerous biosolids (e.g., livestock manures, composts, and municipal sewage sludge) to land inadvertently leads to the accumulation of heavy metals in the soil. In Nepal, compost or FYM is generally applied to the field for crop production as a nutrient supplement. More recently, poultry manure is gaining popularity in Nepal due to flourishing poultry industry. Although most manures are seen as valuable fertilizers, the As laden poultry health products and Cu and Zn supplemented poultry and pig diets have potential to cause metal contamination of soils, if applied as manure.

The availability and utilization of many organic fertilizer products is growing across the country. Vista and Shrestha (2019) reported 78 different soil agrochemicals such as micronutrients, PGR/Hormones/Enzymes/Soil Conditioners are available in Kathmandu valley. Since the Government of Nepal is prioritizing organic farming, the number of organic fertilizers producing companies have increased recently. A few laboratory reports suggest that such organic fertilizers have permissible limits of heavy metal. Indeed, they have lower health risks, and can be valuable for agricultural application, but more studies are needed.

11.4.3 Soil Pathogens

Every year, Nepal faces severe crop loss from diversified groups of pathogens including fungi, bacteria, nematodes, and viruses borne in soil, air, seed, water, etc. The common soil-borne pathogens found in Nepal are *Rhizoctonia, Fusarium, Pythium, Phytophthora, Sclerotium, Sclerotinia,* etc. There are some institutions [for e.g., Nepal Agricultural Research Council (NARC), International Maize and Wheat Improvement Center (CIMMYT)] which run studies on plant pathogens and management techniques focused on staple crops. For example, the introduction of soil solarization technique was investigated by NARC to determine the controls of soil-borne pathogens. The response of local cultivars like Lumle Kalo, a local potato cultivar in Eastern Nepal, was found to be resistant against late blight disease (Shrestha et al. 2019). Other management techniques for soil-borne pathogens include using appropriate crop rotations, seed treatments, and biocontrol agents. Importantly, plant protection institutions in Nepal, such as NARC, should expand research activities and prepare them for current and future epidemic and pandemic situations that could negatively affect plant civilization and soil services.

11.4.4 Water Quality

Water quality deterioration in surface and groundwater bodies is an increasing concern in Nepal. It has been linked to N, P, heavy metals, sediment, and pathogens loading into water bodies from agricultural, commercial, residential, and industrial areas. Agricultural use of fertilizers and pesticides is a major non-point source of water pollution in Nepalese ecosystems (Dahal et al. 2008; Sharma et al. 2015). Burning and/or disposal of solid waste are reported in areas adjacent to water bodies across the country, which have increased risks of contaminants transport or runoff. Manfredi et al. (2010) reported burning and disposal of solid wastes in open dumps near water bodies in Sagarmatha National park and buffer areas causing water quality degradation, particularly, during heavy rainfall events. Land use change, urbanization, and industrialization are other soil degradation issues, which adversely affect water resources. One of such effects was observed and reported in the Bagmati River in the Kathmandu valley (Karn and Harada 2001).

Soil erosion is one of the major threats to land degradation in Nepal. It reduces soil and water quality and exacerbates crop productivity (Chalise et al. 2019; Uddin et al. 2018). Soil erosion is being observed across middle hills, high mountains, and Tarai region, such as in Siwalik Hills, Koshi basin, Aringale Khola watershed, Sagarmatha National Park, and Kulekhani, among others (Chalise et al. 2019), and have resulted in preferential loss of nutrients and soil thickness at spatial and temporal scales. The detachment, transportation, and deposition of soil particles and sediment are strongly linked to monsoons, concentrated rainfall, undulated topography, and human exploitation of natural resources (Morin et al. 2018; Chalise et al. 2019). Clearing of forest, intensive tillage, over grazing, and crop residue removal are largely prevalent in remote and hilly areas of Nepal, which increase erosional effects and degradation of water quality (Chalise et al. 2018, 2019). Thus, management of agricultural and forestry systems through precision farming, conservation tillage, residue management, cover cropping, and agroforestry are recommended to improve soil health, water quality, and long-term soil productivity.

11.5 Conclusion and Recommendations

Soils are the basis of all life on Earth. They deliver multiple functions and services including food and nutritional security. However, marginalization of agriculture and land use change have triggered soil degradation, food shortages, and water pollution in Nepal. For example, urban expansion into prime agricultural lands is common, which has brought soil, air, and water quality issues, and diminished crop yields. Herein, we recommended different suite of measures to sustain soil quality and functions. Integrated nutrient management, agroforestry, crop residue management, conservation tillage, grazing management, mulching, intercrops, hedgerows, grass strips, and terracing provide soil protection against runoff and erosion. They reduce the availability, detachment, and transport of pollutants. Vegetative buffers, regulating and managing of agricultural and silvi-cultural activities, storm water management, and wastewater treatments are important to conserve wetlands and underlying soils. Utilization of fertilizers, biosolids, and industrial wastes must consider soil contaminants such as heavy metals and associated human health risks. Management of soil-borne pathogens using appropriate crop rotations, resistant cultivars, seed treatments, biocontrol agents, and soil solarization technique is equally important. Overall, more studies are needed to unravel complex interactions between humans and soils, and improve strategic planning and management of soil health and quality at local to national levels.

References

Acharya KP (2004) Does community forests management supports biodiversity conservation? Evidences from two community forests from the mid hills of Nepal. J forest livelihood 4(1):44–54

Adriano DC (2001) Bioavailability of trace metals. In: Trace elements in terrestrial environments, Springer, New York, NY, pp. 61–89

Baral Jamarkattel S (2011) Mapping carbon stock using high resolution satellite images in sub-tropical forest of Nepal. Master's thesis, University of Twente

Bhandari NB, Bhattarai D, Aryal M (2015) Cost, production and price spread of cereal crops in Nepal: a time series analysis 2071/2072 (2014/2015). Kathmandu, Department of Agriculture

Brevik EC, Steffan JJ, Burgess LC, Cerdà A (2017) Links between soil security and the influence of soil on human health. In: Global soil security, pp 261–274. Springer, Cham

Brevik EC, Burgess LC (eds) (2013) Soils and human health. CRC Press, Boca Raton, FL

Carson B, Maharjan PL, Shah PB (1986) Land systems report: the soil landscapes of Nepal. Kenting Earth Sci Limited

Chalise D, Kumar L, Kristiansen P (2019) Land degradation by soil erosion in Nepal: A review. Soil Syst 3(1):12

Chalise D, Kumar L, Shriwastav CP, Lamichhane S (2018) Spatial assessment of soil erosion in a hilly watershed of Western Nepal. Environ Earth Sci 77(19):1–11

Chaudhary S, Wang Y, Dixit AM, Khanal NR, Xu P, Fu B, Yan K, Liu Q, Lu Y, Li M (2020) A synopsis of farmland abandonment and its driving factors in Nepal. Land 9(3):84

Dahal BM, Fuerhacker M, Mentler A, Karki KB, Shrestha RR, Blum WEH (2008) Arsenic contamination of soils and agricultural plants through irrigation water in Nepal. Environ Pollut 155 (1):157–163

Dijkshoorn JA, Huting JRM (2009) Soil and terrain database for Nepal (1.1 million) (No. 2009/01). ISRIC-World Soil Information

El-Ramady H, Alshaal T, Elsakhawy T, Omara AED, Abdalla N, Brevik EC (2019) Soils and humans. In: The soils of Egypt, pp 201–213. Springer, Cham

FAO (2004) Arsenic threat and irrigation management in Nepal, Rome

Gami SK, Lauren JG, Duxbury JM (2009) Soil organic carbon and nitrogen stocks in Nepal long-term soil fertility experiments. Soil Tillage Res 106(1):95–103

Gardner RAM, Gerrard AJ (2002) Relationships between runoff and land degradation on non-cultivated land in the Middle Hills of Nepal. Int J Sustain Dev World Ecol 9(1):59–73

Ishtiaque A, Shrestha M, Chhetri N (2017) Rapid urban growth in the Kathmandu Valley, Nepal: monitoring land use land cover dynamics of a himalayan city with landsat imageries. Environments 4 (4):72

Kanel KR, Poudyal RP, Baral JC (2006) Current status of community forestry in Nepal. Accessed 10 Sep 2020

Karn SK, Harada H (2001) Surface water pollution in three urban territories of Nepal, India, and Bangladesh. Environ manage 28 (4):483–496

Kayastha SP (2014) Heavy metal pollution of agricultural soils and vegetables of Bhaktapur district Nepal. Sci World 12(12):48–55

Khanal BR, Shah SC, Sah SK, Shriwastav CP, Acharya BS (2014) Heavy metals accumulation in cauliflower (Brassica oleracea L. var. Botrytis) grown in brewery sludge amended sandy loam soil. Int J Agric Sci Technol 2:87–92

Manfredi EC, Flury B, Viviano G, Thakuri S, Khanal SN, Jha PK, Maskey RK, Kayastha RB, Kafle KR, Bhochhibhoya S, Ghimire NP (2010) Solid waste and water quality management models for Sagarmatha National Park and Buffer Zone Nepal. Mount Res Dev 30(2):127–142

MoALD (2020) Statistical Information on Nepalese Agriculture 2018/2019. Ministry of agriculture and livestock development. Singh Durbar, Kathmandu, Nepal. https://s3-ap-southeast-1.amazonaws.com/prod-gov-agriculture/server-assets/publication-1595229368881-0dc12.pdf. Accessed 25 Sep 2020

MoFE (2018) National Ramsar Strategy and Action Plan, Nepal (2018–2024). Ministry of Forests and Environment. Singh Durbar, Kathmandu, Nepal. http://d2ouvy59p0dg6k.cloudfront.net/downloads/national_ramsar_strategy_and_action_plan__nepal___2018_2024_.pdf. Accessed 25 Sep 2020

Morin GP, Lavé J, France- C, Rigaudier T, Gajurel AP, Sinha R (2018) Annual sediment transport dynamics in the Narayani basin, Central Nepal: assessing the impacts of erosion processes in the annual sediment budget. J Geophys Res Earth Surf 123(10):2341–2376

NLC (2020) Classification of wetlands and Management Policy. Nepal Law Commission. http://www.lawcommission.gov.np/en/archives/8192. Accessed 31 July 2020

Oliver MA, Gregory PJ (2015) Soil, food security and human health: a review. Eur J Soil Sci 66(2):257–276

Paneru H (2013) Major soils of Nepal. https://hrpaneru.wordpress.com/2013/08/06/major-soils-of-nepal/#: ~ :text=The. Countrywide survey carried by, and Aridisols are occasionally found. Accessed 25 Sep 2020

Paudel B, Zhang YL, Li SC, Liu LS, Wu X, Khanal NR (2016) Review of studies on land use and land cover change in Nepal. J Mt Sci 13 (4):643–660

Sharma B, Rasul G, Chettri N (2015) The economic value of wetland ecosystem services: evidence from the Koshi Tappu Wildlife Reserve Nepal. Ecosyst Serv 12:84–93

Shrestha HD (2003) Heavy metals pollution in the environment of Kathmandu. J Phys IV France 107

Shrestha RK, Kafle BP, Gurung A, Regmi D (2017) Arsenic in groundwater, soil and rice grain of Nepal. Int J Agric Environ Biores 2(6)

Shrestha S, Manandhar HK, Shrestha SM, Karkee A (2019) Response of local potato cultivars to late blight disease (Phytophthora infestans (mont.) De bary) under field and laboratory conditions at Pakhribas, Dhankuta, Nepal. Adv Cytol Pathol 4(1):10–13

Siwakoti M, Karki JB (2009) Conservation status of Ramsar sites of Nepal Tarai: an overview. Botan Orient J Plant Sci 6:76–84

Siwakoti M, Tiwari S (2007) Emerging needs of wetlands protection for the conservation of wild rice biodiversity in Nepal: a case study from Lumbini area. Sci World 5(5):95–99

Steffan JJ, Brevik EC, Burgess LC, Cerdà A (2018) The effect of soil on human health: an overview. Eur J Soil Sci 69(1):159–171

Tiwari KR, Sitaula BK, Bajracharya RM, BØrresen T (2009) Runoff and soil loss responses to rainfall, land use, terracing and management practices in the Middle Mountains of Nepal. Acta Agric Scand Sect B-Soil Plant Sci 59(3):197–207

Uddin K, Abdul M, Maharjan S (2018) Assessment of land cover change and its impact on changes in soil erosion risk in Nepal. Sustainability 10(12):4715

Vista SP, Shrestha A (2019) Availability of micronutrients, PGRs, soil conditioners, enzymes, hormones, etc. in Kathmandu valley of Nepal. J Pharmacog Phytochem 8(1):542–548

Xu S, Liu X, Li X, Tian C (2019) Soil organic carbon changes following wetland cultivation: a global meta-analysis. Geoderma 347:49–58

Soil Issues and Future Perspective

12

Roshan Babu Ojha, Krishna Bahadur Karki, and Dinesh Panday

Abstract

In the previous chapters, it has been pointed out that the soils of Nepal are young due to active geology and weathering. Being fragile and loosely aggregated, soils of the plain, hill, and mountain regions are on the verge of degradation and thus require proper care and management. With continued cultivation, low prioritization by government soil conservation programs, and weak policy regulations coupled with natural forces, the soils of Nepal face many challenges. In this chapter, we discuss the issues of Nepalese soils that are either inherent, overlooked, and emergent or likely to emerge soon. These issues were envisioned based on a review of all the chapters of this book, other relevant literature, our own on-the-ground observations, and personal communications with concerned stakeholders. There are numerous soil issues that have arisen due to a lack of proper land management strategy. Soil erosion, soil acidity, flooding, and nutrient depletion are year-round problems. Soil organic carbon depletion is an emerging issue in the Tarai region due to an imbalanced organic input-output ratio. Similarly, soil compaction in the Tarai is another emerging issue due to a multitude of factors such as heavy tilling equipment, a low level of soil organic carbon, frequent flooding-drying cycles, continuous use of chemical fertilizer, and less residue recycling. Cropland abandonment is an emerging issue in the Hills, and land degradation is pronounced in the first five years in this abandoned cropland. Vegetation succession in abandoned land later reduces degradation to some extent, but studies are still required to examine how soil properties are changing in abandoned cropland. In addition, very few studies have been carried out in soil biodiversity, soil pollution, and landscape-scale biogeochemical cycling of nutrients in Nepal. There are numerous potentialities camouflaged in Nepalese soil that are both challenges and opportunities for present and future generations which is discussed in this chapter.

Keywords

Challenges • Nepal • Opportunities • Soil issues • Soil perspective

12.1 Introduction

The United Nations Food and Agriculture Organization identified ten major soil threats in its voluntary guidelines for sustainable soil management (Baritz et al. 2018): soil erosion, low soil organic matter content, imbalanced soil nutrients and cycles, soil salinization and alkalinisation, soil contamination, soil acidification, soil biodiversity loss, soil sealing, soil compaction, and depleting soil moisture. Apart from soil salinization, these threats are prevalent in the soils of Nepal on different scales (Table 12.1). Further, new issues regarding soil are arising in Nepal; for instance, abandoned agricultural soil, shallow rooting depth due to erosion and rock outcrops, microplastic pollution, soil biodiversity change, and pesticide soil pollution, among others.

The government of Nepal has endorsed some of these soil issues in national policy documents. The National Agriculture Policy (NAP) 2004 in Nepal included in the Agriculture Perspective Plan (APP 1995–2015) is a cornerstone shaping agriculture strategies and policies in the country. The policy mostly focused on increasing agricultural production through agricultural input (seed and fertilizer) maximization and identified soil issues such as erosion, nutrient mining, and chemical pollution by agrochemicals as key problems of land degradation requiring attention. At present, the Agriculture Development Strategy (ADS 2015–2035), a

R. B. Ojha (✉) · K. B. Karki
Nepal Agricultural Research Council, Kathmandu, Nepal

D. Panday
Department of Biosystems Engineering and Soil Science, The University of Tennessee-Knoxville, Knoxville, TN, USA

R. B. Ojha and D. Panday (eds.), *The Soils of Nepal*, World Soils Book Series,
https://doi.org/10.1007/978-3-030-80999-7_12

Table 12.1 Major soil threats of Nepal

Agro-ecological region	Soil threats	Severity	References
Tarai	Low soil organic carbon—more than 95% croplands are below the critical level of soil organic carbon concentration (i.e., 2%)	High	(Yadav et. al. 2000; Ojha et al. 2019)
	Soil acidification in the Eastern region (more than 55% cropping area is affected by soil acidity) and soil alkalinity in the Western region	High to moderate	(Bajracharya and Sherchan 2009; Brown et al. 2006; Gautam and Chettri 2020)
	Soil sealing and compaction (due to heavy tillage equipment, low organic inputs, less residue recycling, and continuous use of chemical fertilizer)	Moderate	(Shrestha and Kafle 2020; Takeshima and Justice 2020)
	Flooding and sedimentation	High	(Dingle et al. 2020; Kafle et al. 2020)
	Nutrient imbalance	Moderate	(Regmi et. al. 2002; Brown et al. 2006; Karna and Bauer 2020)
	Lowered water table	Low to moderate	(Gautam and Prajapati 2014; Nepal et al. 2021)
	Arsenic toxicity in ground water (used as drinking and irrigation water)	Moderate	(Maharjan et al. 2007; Thakur et al. 2010)
Siwaliks/ Mid-Hills	High rate of natural erosion due to sloping terrain and fragile geology	High	(West et al. 2015)
	Soil nutrient mining due to erosion, less input, and no residue recycling	Moderate	(Chalise et al. 2019, 2020)
	Depletion of spring water sources (used as drinking and irrigation water)	Moderate to high	(Adhikari et al. 2020; Thapa et al. 2020)
	Low quality of organic manure (only contains 0.3 to 0.7% N)	Moderate	(Gami et. al. 2001; Karki 2003; Maskey and Mihara 2020; Ojha et al. 2014)
	Abandoned croplands (soil properties change in a degraded abandoned land and natural vegetation succession)	High	(Chaudhary et al. 2020; Paudel et al. 2018)
	Microplastic pollution	Low to moderate	(Yukioka et al. 2020)
	Agro-chemical soil pollution	Moderate to high	(Atreya 2008; Bhandari 2014; Pokhrel et al. 2018)
	Low water retention capacity (most soils are sandy loam)	Moderate	(Bajracharya and Sherchan 2009)
High Hills/ Himalayas	Shallow soil depth (rock outcrop)	High	(Baumler and Zech 1994)
	Microplastic pollution	Low to Moderate	(Napper et al. 2020)
	Rooting depth limitation	High	(Sitaula et al. 2004)
	Slow decomposition of soil organic matter due to low temperature and low soil moisture	High	(Drollinger et al. 2017; Sitaula et al. 2004)
	Low soil nutrient status	Moderate to high	(Drollinger et al. 2017)
	Glacial and wind erosion	Moderate	(Gabet et al. 2008; Wagnon et al. 2013)
	Soil pollution (toxic trace elements, persistent organic pollution, poly-aromatic hydrocarbons) in the upper Himalayas	Low to moderate	(Guzzella et al. 2011; Tripathee et al. 2018, 2016)

long-term strategy by the Nepal government to increase agricultural productivity and sustainability, identifies soil sustainability as a key indicator of sustainable agriculture growth, though many aspects of the soil threats listed in Table 12.1 are overlooked in the ADS. Thus, policy reform and intensive soil conservation programs should be a national priority.

12.2 Soil Issues/Threats Frame

Different soil threats result from either being overlooked or exist as inherent or emerging threats in the soil system (Fig. 12.1). These issues need to be addressed in government developmental and research programs and might require reforming the existing policy supported by research findings.

12.2.1 Inherent Issues

Some of the soil issues have arisen due to inherent soil characteristics. These issues can be attributed mostly due to natural causes, but human interventions accelerate the problem. These soil issues have been identified at a national scale and are widely reported in government and research reports.

Soil erosion: Due to steep topography, soil erosion is one of the inherent soil issues of Nepal. The natural rate of soil erosion is higher than the accelerated rate of soil erosion (West et al. 2015), and forestland is more degraded than cropland in Nepal (Chalise et al. 2019). This is one of the most discussed soil issues, and policymakers are well informed about it. Recent studies have shown that rates of erosion are increasing which requires proper care and management (Chalise et al. 2019). This scenario demands effective soil conservation work in the affected areas.

Soil acidity and alkalinity: These issues particularly arise due to the acidic/alkaline nature of parent materials. However, the imbalanced use of chemical fertilizer also accelerates the soil acidity, seen particularly in the Tarai region. Refer to Chaps. 7 and 8 for the cause and remedies of soil acidity and alkalinity issues in Nepal.

Flooding and sedimentation: Every year flooding and sedimentation degrades thousands of hectares of fertile land in the Tarai region. This is a year-round problem that is widely

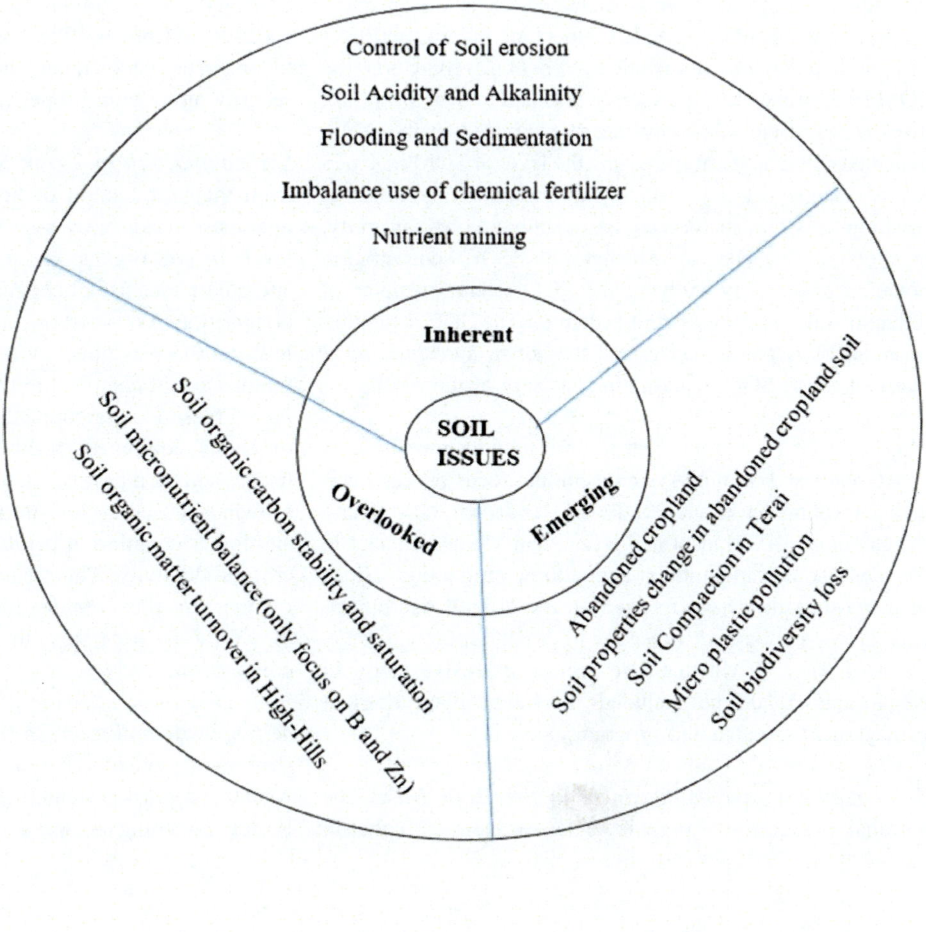

Fig. 12.1 Framing of soil threats or soil issues within the Nepalese perspective

reported in several studies and reports. The re-utilization of flooded or sediment-covered land is our prime concern. Most often than not, it is impossible to prevent floods, however, we need to develop strategies for the restoration of these lands.

Plant nutrient imbalance and nutrient mining: There is heavy nutrient mining in Nepalese soils due to continuous cropping without applying the recommended fertilizer amount in the field crops. Fertility management practices by farmers either rely only on chemical input or insufficient manure with low nutrient quality. This further deepens the nutrient imbalance in the field and triggers nutrient mining. Refer to Chap. 8 for the cause of soil fertility imbalance and its management through integrated nutrient management strategies.

12.2.2 Overlooked Issues

Soil organic carbon stability and saturation: The importance of SOC has recently been identified in the Agriculture Development Strategy (ADS 2015 to 2035). Previously, the Agriculture Perspective Plan (APP 1995 to 2015) had been endorsed by the government, which primarily focused on the chemical fertilizer distribution system and chemical fertilizer subsidy scheme and failed to integrate organic fertilizers into the subsidy program. Due to increased focus on chemical fertilizer and insufficient replenishment of organic fertilizer lost due to tillage or soil organic matter (SOM) oxidation, the SOM level of the cropland sharply declined to near 1% (ADS 2015). This decline has now been recognized in the ADS, which has set a target of increasing the level of SOM to 4% in 10 years' time as a key sustainability indicator. The target of reaching 4% SOM in 10 years is not supported by any study, and for this, we need to investigate the SOC sequestration potential of soils by studying the SOC saturation point of different soil types across agro-ecozones. The SOC saturation potential of Nepalese soil is still unknown, and thus, ADS overlooked the SOC sequestration capacity of these soils.

Micronutrients: Micronutrient use has now been realized in vegetable-based farming systems, but the focus is only on B and Zn (Shrestha et al. 2020a, b). Andersen (2007) and Panday et al. (2018) identified B, Zn, and Mo as deficient in Nepalese soils and limiting agricultural production. Since then, most studies have focused on these three micronutrients. However, there are other soil micronutrients such as Fe, Cu, Co, Ni, and Mn that also require extensive study in Nepal and should be included in future plant nutrient management research and programs.

Soil organic matter management in the High Hills: The regional distribution of organic matter seems to be higher in the High Hills, but low temperature and less soil moisture retard the decomposition rate. Low decomposition results in less availability of plant nutrients and is linked to low soil fertility and productivity. Research is needed to find ways to increase the decomposition rate to make nutrients available from the organic matter in the soil, including the rate of mineralisation from organic and inorganic sources.

12.2.3 Emerging Issues

Abandoned cropland: Of the 27% cultivable areas of the country, around one-third of the land is abandoned (Gautam 2004). Cropland abandonment is now an emerging and alarming problem in Nepal due to internal migration and labor out-migration (Ojha et al. 2017; Sunam and McCarthy 2016), causing significant land degradation problems, including soil fertility decline, soil organic matter reduction, gully erosion, mass movement, loss of topsoil, and vegetation cover reduction, all of which arise in the first few years of abandonment (Chaudhary et al. 2020; Jaquet et al. 2015; Khanal and Watanabe 2006). Similarly, the abandoned land goes to secondary natural succession after few years (Pathak 2015). As such, an extensive study is necessary to understand the changes in soil properties that occur with the passage of time in abandoned land, along with the ecosystem co-benefits of secondary natural succession. Also, retrospective and prospective econometric studies are necessary to understand the impact of abandoned cropland on food security along with proper re-utilization pathways.

Soil compaction in Tarai: Soil compaction in Tarai results from the introduction of heavy machinery for agricultural operation (Takeshima and Justice 2020). In addition, low levels of soil organic carbon, no or low residue recycling, and continuous use of chemical fertilizers further trigger soil compaction. Low carbon in soil reduces the bulk density, low residue recycling lowers soil organic matter, and continuous use of chemical fertilizer increases soil cementation. All of these factors coupled with the weight and drag force of heavy equipment result in massive soil structure and hence soil compaction. Compact soil restricts plant root growth and causes low infiltration, low microbial activity, and low acquisition of plant nutrients, which directly retards crop productivity. This issue is scantly reported in technical reports, and thus a widescale survey on soil compaction in the Tarai is necessary to understand the extent of the compaction.

Microplastic pollution: In the last decade, there has been a shift to a vegetable-based cropping system from a rice–wheat cropping system in Nepal (Shrestha et al. 2020b). Plastic mulching is common in vegetable farming and is commercially growing (Atreya et al. 2019). Plastic mulching is an effective measure to control weeds, temperature, and soil-borne pathogens. However, the improper management of plastic can lead to microplastic pollution. Some of the studies report traces of microplastic in the higher elevations of Nepal (Napper et al. 2020), and hence, future studies should examine the spatial extent of microplastic pollution in different eco-zones in Nepal.

Soil pollution (*Heavy metals, pesticide pollution, and organic pollutants*): Insecticides were first introduced to Nepal in 1954 when the Malaria Eradication Program was launched (Aryal et al. 2016; Sharma et al. 2012). Thereafter, many types of insecticides, fungicides, and herbicides have been imported. The highest amount of plant protective chemicals is applied to vegetable crops, including cereal crops. Recognized intensively vegetable-cultivated pockets in Nepal are Thimi of Bhaktapur, Nallah of Banepa, Kabhre, Panchkhal Valley of Kabhre, Trishuli of Nuwakot, and Naubishe of Dhading. These pockets supply vegetables to the Kathmandu valley. Later, Bara, Makawanpur, and many districts of the Tarai adopted vegetable farming to supply Kathmandu, Pokhara, and other densely populated cities. These vegetable-growing farmers have been using both organic and inorganic pesticides on their crops, including organochlorines, organophosphates, carbamates, formamidines, thiocyanates, organoioctines, dinitrophenol, synthetic pyrethroids, and many types of antibiotics, fungicides, and herbicides. When sprayed onto plants, 99.99% of these chemicals are washed onto the soil (Pimentel 1995). When the sprayed chemicals fall onto the ground they affect the beneficial microorganisms that are available on the soil surface. They can also enter the soil when the treated plant residues are incorporated into the soils. Some of these chemicals, in the form of dust or granules, are applied directly to the soil (Edwards 1966). The spray also enters the plant through leaf cuticles and the stomata (Mitra and Raghu 1998).

Pesticides, fungicides, and herbicides are increasingly applied to protect crops in Nepal, and it has become a habit for farmers to spray chemicals to protect their plants whether there are pests or diseases infestations. In general, farmers are advised to spray 2–3 times per crop season but farmers have been reported to spray between 2–10 times (Sapkota et al. 2020). Therefore, the surrounding environment of the intensive vegetable-cultivating pockets of Nepal is highly contaminated (Atreya 2008; Bhandari 2014; Pokhrel et al. 2018; Sharma et al. 2012). Obsolete pesticides and persistent organic pollutants are contaminating chemicals that have been brought to the country and kept in storehouses in various places across the country (e.g., Amlekhgunj) rather than being used, and because of unscientific storage methods, some of them have leaked into the soil and the open environment (Shah and Devkota 2009). Residues of these chemicals have been found above the tolerable limits in Nepalese soils (Kafle et al. 2015). More dangerous persistent organic pollutants such as organochlorine pesticides, Polycyclic aromatic hydrocarbons have been found in the soils of Kathmandu valley (Pokhrel et al. 2018). Sharma et al. (2012) mentioned that significant levels of pesticides have been found to contaminate vegetable-cultivating soil. Now is the time to start assessing soil and human health linked to pesticide residue levels in the soil. There is also an urgent need to set national standards/critical limits of soil pollution parameters to monitor our soil health status.

12.3 Future Perspectives

Soil biodiversity map: Occupying merely 0.1% of total global land area, Nepal is 31st in the world and 10th in Asia in terms of natural biodiversity (MoAD 2017). This indicates the richness of biodiversity in Nepal but does not include soil biodiversity. However, this national assessment of biodiversity suggests the potentiality of soil biodiversity richness too. To date, scattered studies have been conducted on soil bacteria, fungi, and earthworms. Most of them are related to soil pathogens that are harmful to crops. Studies related to soil microorganisms are limited to rhizobium cultures and their effect on legume production. However, the availability of different beneficial organisms and their potential needs to be studied and mapped. A national inventory of soil biodiversity is necessary to understand its status.

Establishment of soil study reference sites: The soils of Nepal are young, dynamic, and developing. Nepal could be a potential study site for researchers around the world to study the dynamic soil formation processes of pedogenesis. For this to occur, identifying the reference landscapes and establishing a reference study site is necessary. Inceptisols and Entisols, which are often referred to as young soils, are abundantly available in Nepal, and they undergo biological-physical–chemical changes to develop into other soil types. Researchers can harness the study of these active soil formations. Non-volcanic Andisols are also present in Eastern Nepal (Baumler and Zech 1994), and this presents Nepal as a potential place for the study of the dynamics of soil-forming factors with different soil types.

Soil atlas of Nepal: We need to make soil understandable to the public. A non-technical soil map is necessary to advocate the importance of soil resources to policymakers, farmers, and agro-entrepreneurs. Nepal joined the Global Soil Partnership program of the Food and Agriculture Organization of the United Nations and is now working to prepare the soil atlas of Asia. This atlas is necessary to scale up on a national scale and will be an important national document for planning and public education. It will be an important visual document that provides information about soils in plain language and an important resource for raising awareness to protect the soils.

12.4 Conclusion

Nepal generally has a rich body of policies for addressing general agricultural problems as well as different levels of soil-threatening conditions; however, there exists a wide gap between policy formulation and its implementation. In addition, soil biodiversity and soil micronutrient management, among other issues, were overlooked from a research point of view. Hence, the future of soils in Nepal should depend not only on policy reform and intensive soil conservation programs but on more data and facts that can be produced for evidence-based and effective implementation. All three levels of government—federal, provincial, and local—as well as the communities themselves, should give priority to the future of our soils and our food system.

References

Adhikari S, Gurung A, Chauhan R, Rijal D, Dongol BS, Aryal D, Talchabhadel R (2020) Status of springs in mountain watershed of western Nepal. Water Policy

Andersen P (2007) A review of micronutrient problems in the cultivated soil of Nepal. Mt Res Dev 27:331–335

Aryal KK, Neupane S, Lohani GR, Jors E, Neupane D, Khanal PR, Jha BK, Dhimal M, Shrestha BM, Bista B (2016) Health effects of pesticide among vegetable farmers and the adaptation level of integrated pest management program in Nepal, 2014. Nepal Health Research Council

Atreya K (2008) Probabilistic assessment of acute health symptoms related to pesticide use under intensified Nepalese agriculture. Int J Environ Health Res 18:187–208

Atreya PN, Kafle A, Suvedi BD, Shrestha SB (2019) Precision and protected horticulture in Nepal. Last Accessed 1 July 2019

Bajracharya RM, Sherchan DP (2009) Fertility status and dynamics of soils in the Nepal Himalaya: a review and analysis. Soil Fertil 111–135

Baritz R, Wiese L, Verbeke I, Vargas R (2018) Voluntary guidelines for sustainable soil management: global action for healthy soils. In: International yearbook of soil law and policy 2017. Springer, Berlin, pp 17–36

Baumler R, Zech W (1994) Soils of the high mountain region of Eastern Nepal: classification, distribution and soil forming processes. CATENA 22:85–103

Bhandari G (2014) An overview of agrochemicals XE "Agrochemicals" and their effects on environment in Nepal. Appl Ecol Environ Sci 2:66–73

Brown S, Schreier H, Shah PB, Lavkulich LM (2006) Modelling of soil nutrient budgets: an assessment of agricultural sustainability in Nepal. Soil Use Manag 15:101–108

Chalise D, Kumar L, Kristiansen P (2019) Land degradation by soil erosion in Nepal: a review. Soil Syst 3:12

Chalise D, Kumar L, Sharma R, Kristiansen P (2020) Assessing the impacts of tillage and mulch on soil erosion and corn yield. Agronomy 10:63

Chaudhary S, Wang YK, Dixit AM, Khanal NR, Xu P, Fu B, Yan K, Liu Q, Lu YF, Li M (2020) Spatiotemporal degradation of abandoned farmland and associated eco-environmental risks in the high mountains of the Nepalese Himalayas. Land 9

Dingle EH, Creed MJ, Sinclair HD, Gautam D, Gourmelen N, Borthwick AGL, Attal M (2020) Dynamic flood topographies in the Tarai XE "Tarai" region of Nepal. Earth Surf Proc Land 45:3092–3102

Drollinger S, Müller M, Kobl T, Schwab N, Böhner J, Schickhoff U, Scholten T (2017) Decreasing nutrient concentrations in soils and trees with increasing elevation across a treeline ecotone in Rolwaling Himal, Nepal. J Mt Sci 14:843–858

Edwards C (1966) Insecticide residues in soils. In: Residue reviews. Springer, Berlin, pp 83–132

Gabet E, Burbank D, Prattsitaula B, Putkonen J, Bookhagen B (2008) Modern erosion rates in the high Himalayas of Nepal. Earth Planet Sci Lett 267:482–494

Gami S, Ladha J, Pathak H, Shah M, Pasuquin E, Pandey S, Hobbs P, Joshy D, Mishra R (2001) Long-term changes in yield and soil fertility in a twenty-year rice-wheat experiment in Nepal. Biol Fertil Soils 34:73–78

Gautam B, Chettri MK (2020) Physicochemical properties of tropical forest top soil in relation to depth in western Nepal. Banko Janakari 30:39–48

Gautam D, Prajapati RN (2014). Drawdown and dynamics of groundwater table in Kathmandu valley, Nepal. Open Hydrol J 8:17–26

Gautam G (2004) Abandonment of cultivable land: farmers' dependency on imported cereals. (in Nepali)

Guzzella L, Poma G, De Paolis A, Roscioli C, Viviano G (2011) Organic persistent toxic substances in soils, waters and sediments along an altitudinal gradient at Mt. Sagarmatha, Himalayas, Nepal. Environ Pollut 159:2552–2564

Jaquet S, Schwilch G, Hartung-Hofmann F, Adhikari A, Sudmeier-Rieux K, Shrestha G, Liniger HP, Kohler T (2015) Does outmigration lead to land degradation? Labour shortage and land management in a western Nepal watershed. Appl Geogr 62:157–170

Kafle BK, Pokhrel B, Shrestha S, Raut R, Dahal BM (2015) Determination of pesticide residues in water and soil samples from Ansikhola watershed, Kavre, Nepal XE "Geology." Int J Geol Earth Environ Sci 5:119–127

Kafle KR, Khanala SN, Dahalb RK (2020) Impacts of long-term flood-induced sedimentation on agricultural land: case study of the 2008 Koshi flood in Eastern Nepal. Kathmandu Univ J Sci Eng Technol 14

Karki K (2003) Status of potassium XE "Potassium" in intensively cultivated soils of Kathmandu valley. Nepal J Sci Technol 5:83–89

Karna RD, Bauer S (2020) Analyzing soil nutrient balances on small-scale farms in the mid-hills of Nepal: do socio-economic factors matter for sustainable land use XE "Land use" ? Land Degrad Dev 31:3014–3023

Khanal NR, Watanabe T (2006) Abandonment of agricultural land and its consequences. Mt Res Dev 26:32–40

Maharjan M, Watanabe C, Ahmad SA, Umezaki M, Ohtsuka R (2007) Mutual interaction between nutritional status and chronic arsenic toxicity due to groundwater contamination in an area of Tarai XE "Tarai", lowland Nepal. J Epidemiol Community Health 61:389–394

Maskey S, Mihara M (2020) Current conditions and constraints in management of organic manure in mid-hills of eastern Nepal. Int J Environ Rural Dev 10:68–75

Mitra J, Raghu K (1998) Pesticides-non target plants interactions: an overview. Arch Agron Soil Sci 43:445–500

MoAD (2017) The state of Nepal's biodiversity for food and agriculture. Ministry of Agricultural Development, Kathmandu, Nepal

Napper IE, Davies BFR, Clifford H, Elvin S, Koldewey HJ, Mayewski PA, Miner KR, Potocki M, Elmore AC, Gajurel AP, Thompson RC (2020) Reaching new heights in plastic pollution—preliminary findings of microplastics on Mount Everest. One Earth 3:621–630

Nepal S, Neupane N, Belbase D, Pandey VP, Mukherji A (2021) Achieving water security in Nepal through unravelling the water-energy-agriculture nexus. Int J Water Resour Dev 37:67–93

Ojha HR, Shrestha KK, Subedi YR, Shah R, Nuberg I, Heyojoo B, Cedamon E, Rigg J, Tamang S, Paudel KP, Malla Y, McManus P (2017) Agricultural land XE "Agricultural land" underutilisation in the hills of Nepal: investigating socio-environmental pathways of change. J Rural Stud 53:156–172

Ojha RB, Lamichhane S, Adhikari K (2019) Bioclimatic factors associated with soil organic carbon accrual in subtropical Indo-gangetic croplands of Nepal. In: Soil organic matter in a stressed world, 7th internationational symposium on soil organic matter. Hilton Adelaide, South Australia. http://www.som2019.org/wp-content/uploads/2019/10/Poster-Presentation-Abstract-Book.pdf

Ojha RB, Shah SC, Pande KR, Dhakal DD (2014) Residual Effect of Farm Yard Manure on Soil Properties XE "Soil properties" in Spring Season, Chitwan, Nepal. Int J Sci Res Agric Sci 1:165–171

Panday D, Maharjan B, Chalise D, Shrestha RK, Twanabasu B (2018) Digital soil mapping in the Bara district of Nepal using kriging tool in ArcGIS. PloS One 13:e0206350

Pathak HN (2015) Secondary succession in abandoned crop fallows: a review. Int J Environ 4:166–176

Paudel B, Zhang Y, Li S, Liu L (2018) Spatiotemporal changes in agricultural land cover XE "Land cover" in Nepal over the last 100 years. J Geog Sci 28:1519–1537

Pimentel D (1995) Amounts of pesticides reaching target pests: environmental impacts and ethics. J Agric Environ Ethics 8:17–29

Pokhrel B, Gong P, Wang X, Chen M, Wang C, Gao S (2018) Distribution, sources, and air–soil exchange of OCPs, PCBs and PAHs in urban soils of Nepal. Chemosphere 200:532–541

Regmi A, Ladha J, Pathak H, Pasuquin E, Bueno C, Dawe D, Hobbs P, Joshy D, Maskey S, Pandey S (2002) Yield and soil fertility trends in a 20-year rice–rice–wheat experiment in Nepal. Soil Sci Soc Am J 66:857–867

Sapkota KR, Sapkota S, Katuwal K (2020) Pesticides handling practices among potato growers in Kavrepalanchok, Nepal. J Agric Nat Resour 3:77–87

Shah BP, Devkota B (2009) Obsolete pesticides: their environmental and human health hazards. J Agric Environ 10:60–66

Sharma D, Thapa R, Manandhar H, Shrestha S, Pradhan S (2012) Use of pesticides in Nepal and impacts on human health and environment. J Agric Environ 13:67–74

Shrestha S, Becker M, Lamers JPA, Wimmer MA (2020a) Boron and zinc fertilizer applications are essential in emerging vegetable-based crop rotations in Nepal. J Plant Nutr Soil Sci 183:439–454

Shrestha S, Becker M, Lamers JPA, Wimmer MA (2020b) Diagnosis of zinc and boron availability in emerging vegetable-based crop rotations in Nepal. J Plant Nutr Soil Sci 183:429–438

Shrestha S, Kafle G (2020) Variation of selected physicochemical and hydrological properties of soils in different tropical XE "Tropical" land use systems of Nepal. Appl Environ Soil Sci 2020:8877643

Sitaula BK, Bajracharya RM, Singh BR, Solberg B (2004) Factors affecting organic carbon dynamics in soils of Nepal/Himalayan region—a review and analysis. Nutr Cycl Agroecosyst 70:215–229

Sunam RK, McCarthy JF (2016) Reconsidering the links between poverty, international labour migration, and agrarian change: critical insights from Nepal. J Peasant Stud 43:39–63

Takeshima H, Justice SE (2020) Evolution of agricultural mechanization in Nepal. IFPRI book chapters, pp 285–325

Thakur JK, Thakur RK, Ramanathan A, Kumar M, Singh SK (2010) Arsenic contamination of groundwater in Nepal—an overview. Water 3:1–20

Thapa B, Pant RR, Thakuri S, Pond G (2020) Assessment of spring water quality in Jhimruk River Watershed, Lesser Himalaya, Nepal. Environ Earth Sci 79

Tripathee L, Guo J, Kang S, Paudyal R, Huang J, Sharma CM, Zhang Q, Rupakheti D, Chen P, Ghimire PS (2018) Concentration and risk assessments of mercury along the elevation gradient in soils of Langtang Himalayas, Nepal. Hum Ecol Risk Assess Int J

Tripathee L, Kang S, Rupakheti D, Zhang Q, Bajracharya RM, Sharma CM, Huang J, Gyawali A, Paudyal R, Sillanpää M (2016) Spatial distribution, sources and risk assessment of potentially toxic trace elements and rare earth elements in soils of the Langtang Himalaya, Nepal. Environ Earth Sci 75

Wagnon P, Vincent C, Arnaud Y, Berthier E, Vuillermoz E, Gruber S, Ménégoz M, Gilbert A, Dumont M, Shea JM, Stumm D, Pokhrel BK (2013) Seasonal and annual mass balances of Mera and Pokalde glaciers (Nepal Himalaya XE "Nepal Himalaya") since 2007. Cryosphere 7:1769–1786

West AJ, Arnold M, Aumaître G, Bourlès D, Keddadouche K, Bickle M, Ojha T (2015) High natural erosion rates are the backdrop for present-day soil erosion in the agricultural Middle Hills XE "Hills" of Nepal. Earth Surf Dyn 3:363

Yadav RL, Dwivedi BS, Prasad K, Tomar OK, Shurpali NJ, Pandey PS (2000) Yield trends, and changes in soil organic-C and available NPK in a long-term rice–wheat system under integrated use of manures and fertilisers. Field Crop Res 68:219–246

Yukioka S, Tanaka S, Nabetani Y, Suzuki Y, Ushijima T, Fujii S, Takada H, Van Tran Q, Singh S (2020) Occurrence and characteristics of microplastics in surface road dust in Kusatsu (Japan), Da Nang (Vietnam), and Kathmandu (Nepal). Environ Pollut 256:113447

Index

A

Active alluvial, 58, 64, 68
Agricultural land, 42, 45, 47, 49, 99
Agriculture, 1, 2, 5–9, 11, 12, 69, 97, 99, 101, 106, 120, 121, 123, 129, 135, 138, 151, 154, 155
Agrochemicals, 105, 147, 151
Agro-ecozones, 22
Alfisol, 59
Alluvial, 1
Alpine, 1, 18, 23, 59, 128
Anthropic, 1, 70, 75, 76
Anthropogenic, 58, 130, 134
Anthrosols, 1, 90
Azolla, 106, 107
Azotobacter, 102, 105–107

B

Biofertilizers, 106, 121
Black soils, 63, 67
Blue green algae, 106
Bulk density, 92, 95

C

Cambisols, 68–71
Cambrian, 31, 34, 56, 65
Calciorthids, 71, 77
Carbon, 11
Chemical degradation, 127, 137
Clay, 2, 5, 31, 34, 54, 55, 59, 63, 66, 67, 71–75, 77, 92–95, 98, 113, 114, 117, 146
Climate, 1, 15, 17, 19, 22, 23, 48, 56, 57, 64, 66, 67, 112, 128
Climate change, 2, 15, 17, 20, 23, 24, 48, 49, 146
Climatic, 1, 15–22, 24, 29, 57–59, 61, 66
Climatic trend, 20
Cold waves, 15, 17, 20, 22, 23
Colluvial, 1
Cropland abandonment, 151, 154
Cryorthod, 77
Cryumbrepts, 59, 60, 71, 77
Cryustept, 77
Cultivation, 1, 6, 18, 19, 63, 90, 95, 100, 105, 116, 122, 129, 136, 137, 147, 151

D

Desertification, 135
Droughts, 2, 17, 21, 22, 128
Dun Valleys, 59
Dystrochrepts, 59, 60, 68–70, 72, 75, 75

E

Endoaquept, 68
Entisols, 1, 53, 59, 60, 63, 90, 155
Eocene, 2, 29, 31, 32, 34, 56, 65, 66
Epipedon, 1, 58
Eutrochrepts, 59, 60, 68, 70
Eutrudepts, 68, 72, 76

F

Farming, 1, 9, 16–18, 22, 25, 58
Feldspar, 31, 117
Fertilizer, 3, 9, 11, 101–104, 120–122
Floods, 2, 15, 17, 20, 21, 24, 48, 68, 94, 127, 128, 130, 136, 139, 154
Fluvaquent, 76
Fluvisols, 68–70
Food security, 143
Forest degradation, 132
Forest fires, 123, 133
Fossil, 31, 56, 61, 123
Frosts, 22

G

Geology, 1, 2, 9, 31, 53–55, 59, 64, 151, 152
Geomorphology, 35, 55, 64
Gleysols, 68, 69
Grassland, 18, 24, 45, 48, 105, 138, 146
Gravimetric water content, 92
Greenhouse gas, 49, 105, 116
Green manure, 116

H

Haplaquents, 60
Haplaquepts, 60, 68, 69, 71
Hapludalfs, 59, 60, 75

R. B. Ojha and D. Panday (eds.), *The Soils of Nepal*, World Soils Book Series,
https://doi.org/10.1007/978-3-030-80999-7

Haplumbrept, 76
Haplustalf, 74
Haplustept, 73, 75, 76
Haplustoll, 72, 73
Heavy metals, 147, 155
Higher Himalaya, 2, 29, 31–33, 37
High Himalaya, 15, 22, 36, 37, 45, 56, 57, 66, 71, 77
High Mountain, 2, 15, 18, 20–22, 29, 34, 35, 43, 45, 58, 63, 65, 66, 70, 76, 94, 112, 122, 146
Hills, 15, 17, 22, 29, 34–37, 64, 99, 100, 102, 105, 108, 114, 118, 121, 133, 138, 148, 151, 152, 154
Histosols, 1, 53, 59, 60, 90
Human nutrition, 144

I

Inceptisols, 1, 53, 59, 60, 63, 90, 155
Indo-Gangetic Plain, 2, 29, 31, 33, 120
Intermontane basins, 29, 34, 37

K

Karakoram phase, 29
Kumaon Himalaya, 30

L

Land cover, 2, 3, 41, 42, 45, 47–50, 143, 146
Land degradation, 127, 128
Landform, 23, 58
Land systems, 58, 68
Land tenure system, 2, 41, 42, 48, 49
Land use, 2, 9, 24, 29, 35, 37, 41–43, 45, 47–49, 67, 70, 99, 105, 108, 112, 127–129, 134, 137, 139, 143, 146, 148
Lesser Himalaya, 2, 29, 31–33, 37, 55, 56, 58
Livestock farming, 138

M

Macronutrients, 100
Mahabharat, 34, 43, 55–58, 65
Main Boundary Thrust, 31, 32, 37, 55
Main Central Thrust, 32, 37, 55
Main Frontal Thrust, 31, 37, 55
Malla Johar phase, 29
Mapping, 66–68, 136
Micronutrients, 2, 11, 91, 97, 99, 101, 104, 107, 108, 111, 114, 115, 117, 119, 121, 147, 154
Microorganism, 105
Microplastic pollution, 151, 155
Middle Mountain, 15, 18, 20, 22, 29, 34–36, 43, 58, 64, 65, 75, 112, 122, 129, 133
Miocene, 29–31, 111
Monsoon, 19, 29, 35, 57, 64, 66, 112, 128, 131, 132, 148
Mountains, 1, 15, 16, 29, 37, 55–57, 59, 60, 63, 65, 70, 111, 113, 122, 128, 138, 146

N

Nepal Himalaya, 15, 19, 20, 29–37, 41, 42, 48, 49
N-fixing bacteria, 105–107, 121
Nitrogen, 100, 116, 120, 134
Nival, 19
North-East Frontier, 30
Nutrient management, 111

O

Ochrepts, 60
Organic chemicals, 147
Organic matter, 92, 95, 99, 101, 105, 114
Oxisols, 1, 59, 60, 90

P

Parent material, 98
Particle density, 92
Phosphorus, 100, 117, 134
Physical properties, 91
Physiographic regions, 1, 2, 15, 18, 20, 29, 34, 57, 58, 60, 64, 90, 95, 96, 112, 128
Physiographic subdivision, 34
Physiography, 29
Piedmont, 63, 64, 68
Piedmont soils, 63
Pleistocene, 30, 31, 34, 54, 64
Porosity, 92
Post-glaciated, 70, 71
Post-monsoon, 19
Potassium, 95, 98, 100, 104, 105, 108, 117, 118, 120, 128, 135
Precambrian, 31, 32, 56, 65, 66, 111
Precipitation, 2, 15–17, 19–22, 24, 25, 57, 59, 66, 102, 112, 114, 131, 132, 146
Pre-monsoon, 19
Psammaquents, 60
Psamments, 60, 68, 70
Punjab Himalaya, 30

Q

Quartz, 31

R

Recent alluvial, 68
Red soils, 63
Rhdustalf, 73
Rhizobium, 105, 106, 117, 121
Rhodustalf, 60, 75
Rocks, 2, 29, 31–35, 37, 53–56, 58, 59, 65, 93, 111, 117, 130, 137
Role of soils, 143

S

Sediment, 2, 31, 34, 59, 61
Sedimentary rocks, 32–35, 37, 54, 56
Sedimentation load, 130
Shifting cultivation, 137
Shrubland, 45, 48
Sikkim-Bhutan Himalaya, 30
Sirmurian phase, 30
Siwalik phase, 30
Siwaliks, 1, 2, 9, 29, 31, 34–37, 43, 55, 56, 63–65, 68, 70, 73, 122, 133, 146, 152
Soil acidity, 97, 98
Soil aeration, 95, 105
Soil biodiversity, 155
Soil biological properties, 101, 103
Soil chemical properties, 95
Soil classification, 2, 54, 63, 66–68
Soil color, 93
Soil compaction, 154

Soil consistency, 93
Soil crust, 94
Soil erosion, 24, 91, 94, 108, 122, 127, 130, 134, 137, 139, 148, 151, 153
Soil fertility, 6, 113, 115, 120, 122, 134
Soil formation, 55, 58
Soil genesis, 54
Soil in traditions, 143
Soil issues, 153
Soil microorganisms, 101, 114
Soil organic carbon, 24, 103, 127, 139, 151, 152, 154
Soil pathogens, 148
Soil pH, 97–100, 117
Soil plasticity, 93
Soil pollution, 123, 147
Soil properties, 91
Soil salinity, 122, 123
Soil sealing, 94, 152
Soils in culture, 143
Soil structure, 92, 94
Soil survey, 58, 59, 63, 67
Soil temperature, 67, 94
Soil temperature regime, 67
Soil texture, 92, 113, 114
Soil threats, 151, 153
Soil types (USDA), 68–71
Soil types (WRB), 68–71
Sub-alpine, 1, 18, 67
Subtropical, 1, 17, 64, 67

T
Tarai, 1, 2, 7, 9, 12, 15–17, 19–24, 29, 34, 36, 37, 42, 43, 45, 46, 48, 49, 55–58, 60, 63, 64, 67–71, 91, 93–95, 97–100, 102, 104, 105, 108, 111–114, 118, 120–123, 127–131, 133, 136, 138, 139, 146–148, 151–155
Temperate, 1, 18, 23, 57
Temperature, 20, 49, 66, 67, 112
Tethys, 2, 29, 31, 33–35, 37, 55, 56, 58, 59
Tibetan-Tethys zone, 2, 29, 31, 34, 35
Topography, 16, 112
Trans-Himalayan, 19, 29, 36, 59
Tropical, 1, 17, 23, 57
Tsangpo Suture Zone, 29, 34
Tundra, 19, 57, 128

U
Udochrepts, 60
Udolls, 60
Udorthents, 60
Ustalfs, 60
Ustifluvents, 68–71
Ustochrepts, 60, 68–70, 72, 114, 118
Ustorthents, 60, 70, 75

V
Valleys, 2, 6, 7, 29, 34–37, 45, 53, 54, 56–59, 64–66, 93–95, 108, 112–114, 122, 129, 136
Vedic, 1, 5, 6, 12
Vertisols, 1, 53, 59, 90
Vesicular Arbuscular mycorrhiza, 107
Volumetric water content, 92

W
Water-holding capacity, 95
Water infiltration, 95
Water quality, 148
Weathering, 1–3, 23, 24, 35, 53, 54, 56, 59, 93, 113, 151
Wetlands, 146
Wind erosion, 135
Winter, 20

Y
Yellow soils, 63
Young Nepalese soils, 1